SOLID FREEFORM FABRICATION:
A New Direction in Manufacturing

Solid Freeform Fabrication: A New Direction in Manufacturing

with

Research and Applications in Thermal Laser Processing

by

JOSEPH J. BEAMAN
JOEL W. BARLOW
DAVID L. BOURELL
RICHARD H. CRAWFORD

University of Texas
Austin, Texas

HARRIS L. MARCUS

University of Connecticut
Storrs, Connecticut

KEVIN P. McALEA

DTM Corporation
Austin, Texas

Springer Science+Business Media, LLC

ISBN 978-1-4613-7905-8 ISBN 978-1-4615-6327-3 (eBook)
DOI 10.1007/978-1-4615-6327-3

Library of Congress Cataloging-in-Publication Data

A C.I.P. Catalogue record for this book is available
from the Library of Congress.

Originally published by Kluwer Academic Publishers in 1997
Softcover reprint of the hardcover 1st edition 1997

Printed on acid-free paper.

CONTENTS

PREFACE

Solid Freeform Fabrication is a set of manufacturing processes that are capable of producing complex freeform solid objects directly from a computer model of an object without part-specific tooling or knowledge. In essence, these methods are miniature manufacturing plants which come complete with material handling, information processing and materials processing. As such, these methods require technical knowledge from many disciplines; therefore, researchers, engineers, and students in Mechanical, Chemical, Electrical, and Manufacturing Engineering and Materials and Computer Science will all find some interest in this subject. Particular subareas of concern include manufacturing methods, polymer chemistry, computational geometry, control, heat transfer, metallurgy, ceramics, optics, and fluid mechanics. History of technology specialists may also find Chapter 1 of interest.

Although this book covers the spectrum of different processes, the emphasis is clearly on the area in which the authors have the most experience, *thermal laser processing*. In particular, the authors have all been developers and inventors of techniques for the Selective Laser Sintering process and laser gas phase techniques (Selective Area Laser Deposition).

This is a research book on the subject of Solid Freeform Fabrication. Although there are now several commercial entities producing and marketing Solid Freeform Fabrication systems, there is no attempt in this book to compare or contrast these systems. I believe any such comparison is premature in this rapidly changing field. But, rather, the authors have described in detail the basic freeform fabrication techniques and the engineering fundamentals on which they depend. I wish to recommend future *Solid Freeform Fabrication Symposia Proceedings* for updates to the technical information presented in this work.

Joseph J. Beaman
Austin, Texas

ACKNOWLEDGMENTS

Many of the ideas presented in this work were made possible by the hard work and guidance of others. I would like to single out Carl Deckard at Clemson University, who got our program started at the University of Texas, Ralph Wachter and Steve Fishman at the Office of Naval Research for their early belief and continued support for this area, the Defense Advanced Research Projects Agency for their support, and the National Science Foundation for their backing, especially for providing the Freeform Fabrication Laboratory with early seed money almost 10 years ago. The authors would also like to thank the many students, both past and present, who worked in our laboratory. You will find much of their work described in this book. I would like to personally thank Ken Wei Chen, Suman Das, Nicole Harlan, Larry Jepson, Martin Wohlert, and Britt Birmingham for giving invaluable assistance to produce this work. Finally, I would like to give special thanks to my wonderful wife Lisa for putting up with me during this editorial task.

Joseph J. Beaman

1
INTRODUCTION

1.1 BACKGROUND AND DEFINITIONS

Today, several new technologies address the rapid creation of models, prototypes, patterns and limited run manufacturing. These technologies are capable of producing complex freeform solid objects directly from a computer model of an object without part-specific tooling or knowledge. They are, for the most part, additive processes and have been termed rapid prototyping, desk top manufacturing, and *Solid Freeform Fabrication*. In this book, a general technical review of freeform fabrication technologies is given with an in-depth discussion of those technologies which have been actively studied at the University of Texas Laboratory for Freeform Fabrication.

As a prelude to this discussion, consider first conventional manufacturing. Traditionally, there are three general methods to create a complex solid object. They are:

> (1) Subtractive - Take simple stock material (bar, block, etc.) and remove unwanted material as in machining,
> (2) Net-Shape - Take simple stock material (ingot, powder, melt, etc.) and reshape the material in a complex die or mold as in forging, molding or casting,
> (3) Additive - Take relatively complex subunits and add material(s) selectively as in a joining processes like welding to create a more complex object.

If done manually, the manufacturing cost and the time to produce a few artifacts with these methods often cannot be justified for commercialization. Even the time and cost of construction for design prototypes can be prohibitive. (A typical design prototype might be $50,000 with a 6 month lead time.) If these methods are automated, the manufacturing cost of a particular mechanical component is a direct function of geometric complexity and production quantity. For removal and reshaping processes, complex tooling and tool path planning must be designed and constructed before a complex object can be made. For conventional additive processes, complex subunits must be pre-fabricated and special fixtures must be constructed in order to hold and position the workpiece. When automated, each conventional manufacturing operation is well-suited for incremental addition of complexity and large production numbers.

Conversely, Solid Freeform Fabrication technologies are well-suited for integral construction of highly complex parts at lower production numbers. If successful, Solid Freeform Fabrication will provide a basis for entirely new manufacturing methods in which low production numbers could become economic. Other product benefits are greatly reduced prototyping cost and design time, and the capability to achieve desired part geometry in one operation. Solid Freeform Fabrication has the capacity to dramatically shorten development cycle times for engineering systems by breaking the modelmaking/prototyping bottleneck and also to improve configuration management by simplifying the link between a CAD database, materials processing, and the fabricated part.

To achieve this, freeform fabrication technologies use a computer graphic representation and simple stock material (powder, liquid, gas, sheets, etc.) to fabricate complex parts in an additive fashion. The features of a Solid Freeform Fabrication process can be examined in Figure 1.1.

In this figure, a mechanical component is placed in a space divided into an N^3 array of small cubes. These smaller cubes are termed voxels. They are the three-dimensional counterpart to the two-dimensional pixel. Assuming a binary state for each voxel with state one corresponding to a structural solid state and state zero corresponding to a nonstructural support state, we can describe any three-dimensional shape within the large cube in terms of these states, only being restricted by the resolution of the array.

Constraints occur if one aspect of the shape of a component dictates how other aspects are to be manufactured. One type of manufacturing constraint is tool clearance considerations in conventional machining which, for example, would result in machining the two holes on the boss depicted in Figure 1.1. This problem can be avoided by building the component in thin layers by material addition rather than by material removal. Each layer is one voxel thick and corresponds to a particular cross-section of the component. Tool access constraints are eliminated because the operation on the part always occurs on the surface. Depending on the particular Solid Freeform Fabrication process, a component is constructed at the level of individual voxels, horizontal lines of voxels, or horizontal areas of voxels rather than the entire object as a whole.

Additional manufacturing constraints are requirements for molds and dies in conventional forging or molding and fixtures in conventional joining processes. For instance, besides constraints in actual machining of a casting mold for a part like that in Figure 1.1, this mold would have requirements for part-specific draft and release angles. Also, consider joining the boss to the main part by welding. In order to achieve this, both parts would have to be accurately placed and held by a fixture during the welding process. The positioning of this fixture and likely even the fixture itself would be part-specific. These problems can also be avoided by additive

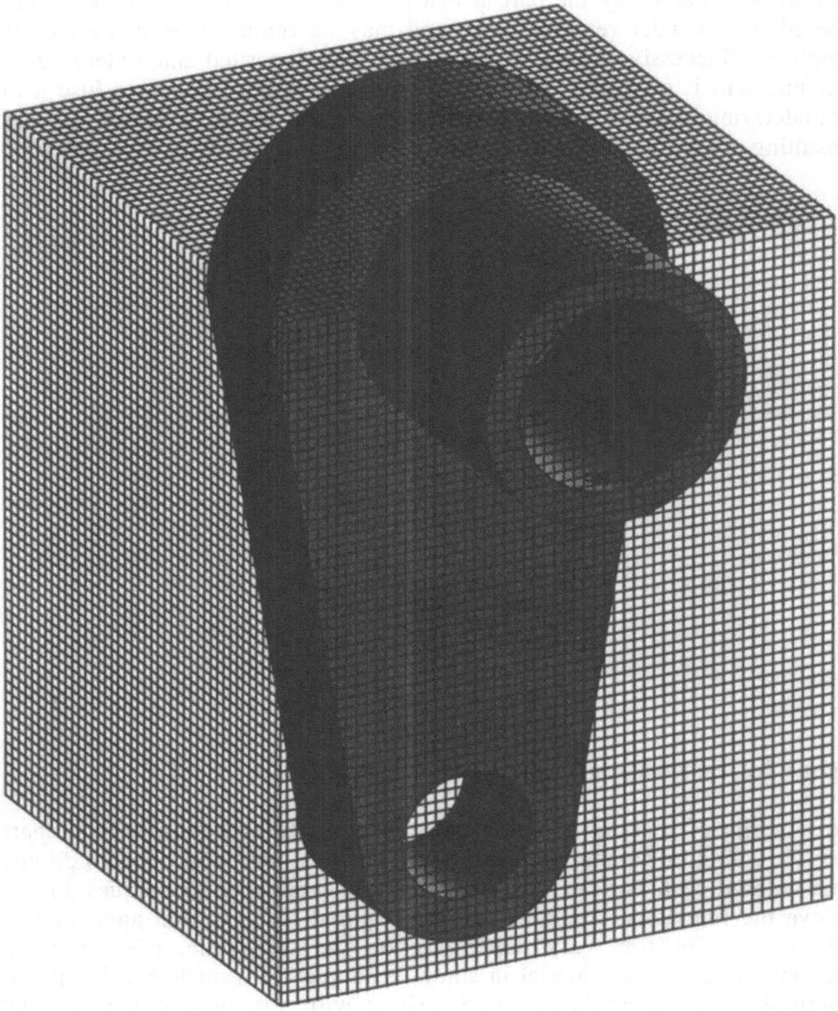

Figure 1.1 Mechanical component embedded in a voxel cube.

manufacturing. While the part is being constructed, sacrificial support structures can be simultaneously built or, even better, the nonstructural support state acts explicitly as a mold/fixture to automatically constrain the part.

A Solid Freeform Fabrication process which is being studied at the University of Texas at Austin is Selective Laser Sintering (SLS). This process begins by first depositing a thin layer of powder into a container. The powder surface is raster-scanned with a high power energy beam such as a laser or electron beam. Beam intensity is modulated to fuse the powder in areas to be occupied by the part at that particular cross-section. In areas not sintered, the powder remains loose and may be removed once the part is completed. Successive layers of powder are then deposited and sintered until the entire part is produced. Each layer is sintered deeply enough to fuse it to the underlying layer. A schematic of this process is shown in Figure 1.2 and a resulting complex part built in glass filled nylon is shown in Figure 1.3.

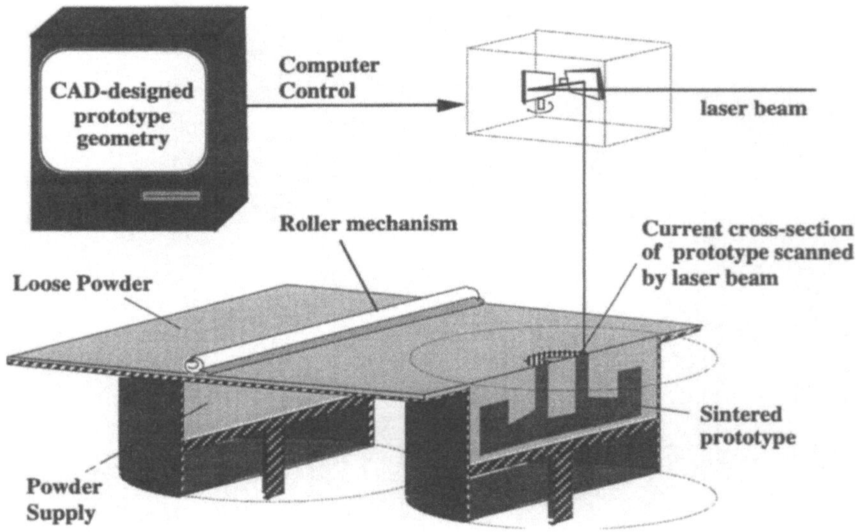

Figure 1.2. Schematic of SLS process (Nelson, 1993).

The benefits of making complex parts in a single operation without part specific tooling or human intervention are obvious, but these benefits do not come without cost. For the most part, freeform fabrication techniques do not achieve the functional strengths that are required for structural applications. The reasons for this vary depending on the technique, ranging from inherently weak stock material in some techniques to insufficient density of structural stock materials in others. Along with structural properties, the geometric accuracy and surface finish of these techniques may be somewhat limiting for certain applications. But these problems are not necessarily inherent in freeform fabrication techniques. Researchers are just now beginning to address these issues from a basic engineering science viewpoint. The structural and geometric properties one can achieve with conventional manufacturing processes are well known, with many of these techniques in existence for hundreds and sometimes thousands of years. This is not true for freeform fabrication techniques. In many ways, the tenets of standard manufacturing processes are changed in freeform fabrication. Rather than starting with structurally sound material and then removing material in order to obtain geometric complexity as in standard machining, freeform fabrication incorporates geometric complexity in concert with structural properties by material addition. The manufacturing, material, and information science required to optimally produce parts by additive methods is in its infancy since these techniques have only been in existence for a little more than a decade. As this science base is being developed one may only wonder at the capabilities that will be eventually realized. For as any other type of manufacturing process, freeform fabrication will have its own unique sensitivities to materials and how they are processed. The part shown in Figure 1.3 consists of a set of interconnected gears and would be impossible to manufacture with machining techniques.

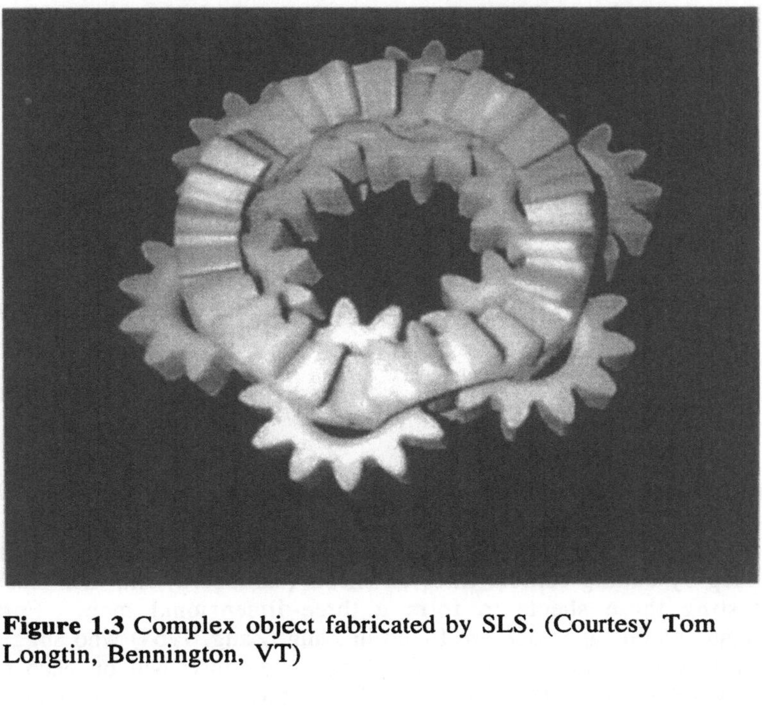

Figure 1.3 Complex object fabricated by SLS. (Courtesy Tom Longtin, Bennington, VT)

1.2 HISTORICAL PERSPECTIVE

1.2.1 Technology

The early roots of Solid Freeform Fabrication technology can be traced to at least two technical areas: topography and photosculpture.

Topography

As early as 1890, Blanther (Blanther, 1892) suggested a layered method for making a mold for topographical relief maps. The method consisted of impressing topographical contour lines on a series of wax plates and cutting these wax plates on these lines. After stacking and smoothing these wax sections, one obtains both a positive and negative three dimensional surfaces which correspond to the terrain indicated by the contour lines. After suitable backing of these surfaces, a printed paper map is then pressed between the positive and negative forms to create a raised relief map. This is shown in Figure 1.4.

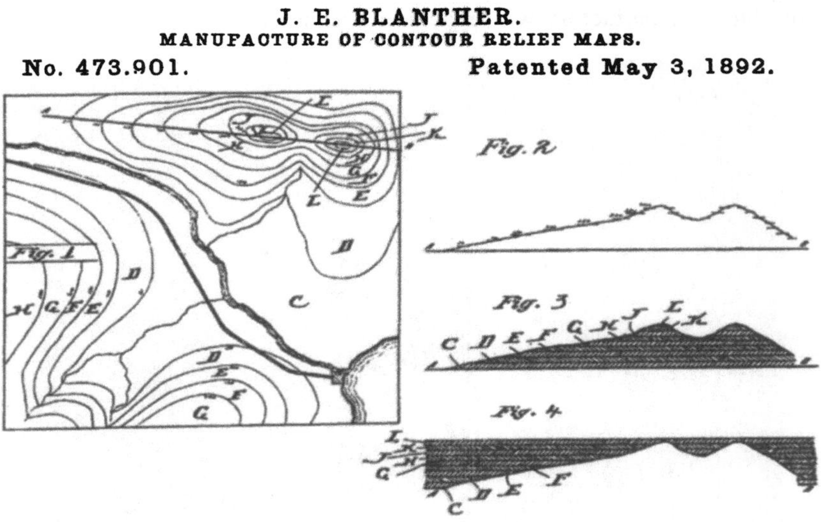

Figure 1.4. Layered mold for relief map proposed in 1892 (Blanther, 1892).

In a similar fashion, Perera (Perera, 1940) proposed a method for making a relief map by cutting contour lines on sheets (cardboard) and then stacking and pasting these sheets to form a three-dimensional map. Further refinements of this approach are found in Zang (Zang, 1964) who suggested using transparent plates with topographical detail inscribed on each plate and Gaskin (Gaskin, 1973) who described a three dimensional geological teaching device. In 1972, Matsubara of Mitsubishi Motors (Matsubara, 1974) proposed a topographical process that uses photo-hardening materials. In this process, a photopolymer resin is coated onto refractory particles (e.g., graphite powder or sand). These coated particles are then spread into a layer and heated to form a coherent sheet. Light (e.g., Mercury vapor lamp) is then selectively projected or scanned onto this sheet to harden a defined portion of it. The unscanned, unhardened portion is dissolved away by a solvent. The thin layers formed in this way are subsequently stacked together to form a casting mold. In 1974, DiMatteo (DiMatteo, 1976) recognized that these same stacking techniques could be used to produce surfaces which are particularly difficult to fabricate by standard machining operations. Examples he mentions include propellers, air foils, three-dimensional cams, and forming of dies for punch presses. In one embodiment, contoured metallic sheets are formed by a milling cutter, these sheets are then joined in layered fashion by adhesion, bolts, or tapered rods as depicted in figure 1.5. This process has obvious similarity to the earlier 19th century work.

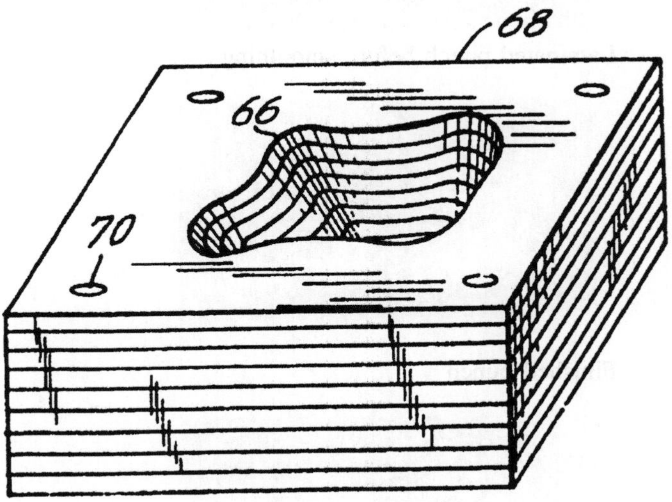

Figure 1.5 Layered mold of stacked plates proposed in 1974 (DiMatteo, 1976)

In 1979, Professor Nakagawa of Tokyo University began to use lamination techniques to produce actual tools such as blanking tools (Nakagawa, *et al* 1979), press forming tools (Kunieda and Nakagawa 1984), and injection molding tools (Nakagawa, *et al* 1985). A laminated punch tool and the resultant part is shown in Figure 1.6.

Photosculpture

Photosculpture arose in the 19th century as an attempt to create exact three-dimensional replicas of any object - including human forms (Bogart,

1979). One somewhat successful realization of this technology was designed by Frenchman François Willème in 1860. As shown in Figure 1.8, a subject or object was placed in a circular room and simultaneously photographed by 24 cameras placed equally about the circumference of the room. The silhouette of each photograph was then used by an artisan to carve out 1/24th of a cylindrical portion of the figure in order to create a solid image (Figure 1.7).

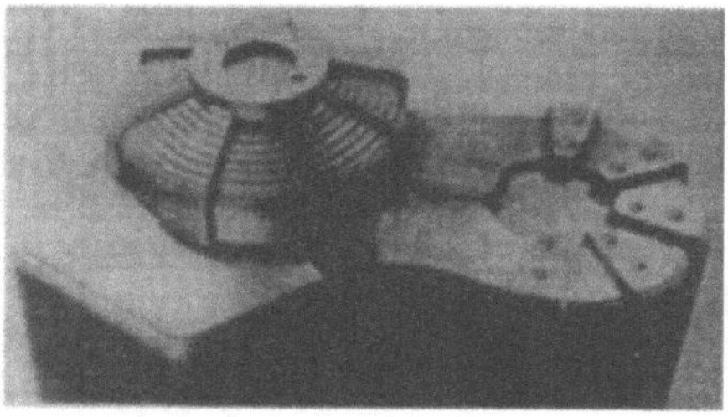

Laminated punch before smoothing

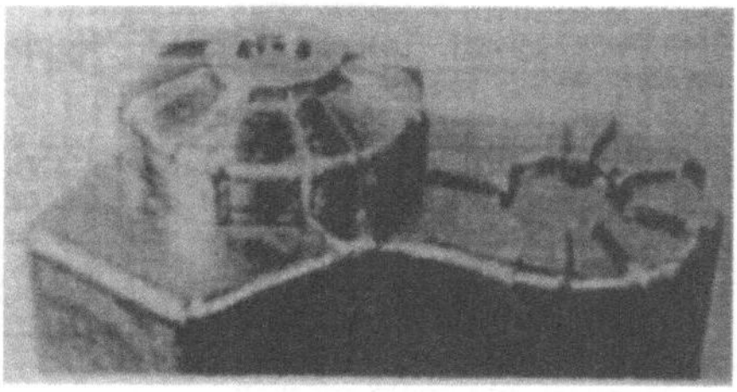

Finished punch

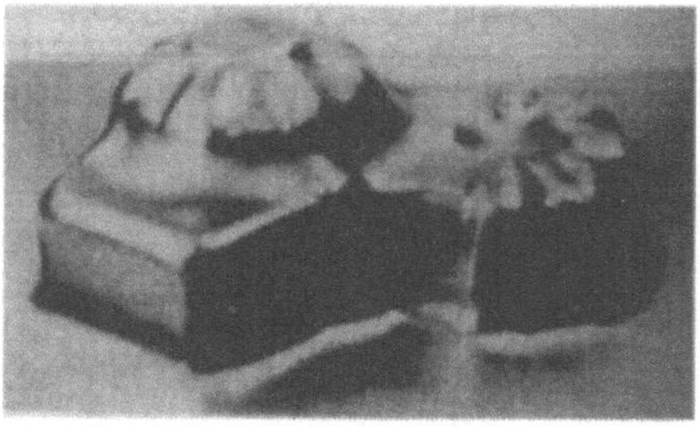

Sample part

Figure 1.6 Laminated tool created by Nakagawa (Kunieda and Nakagawa, 1984)

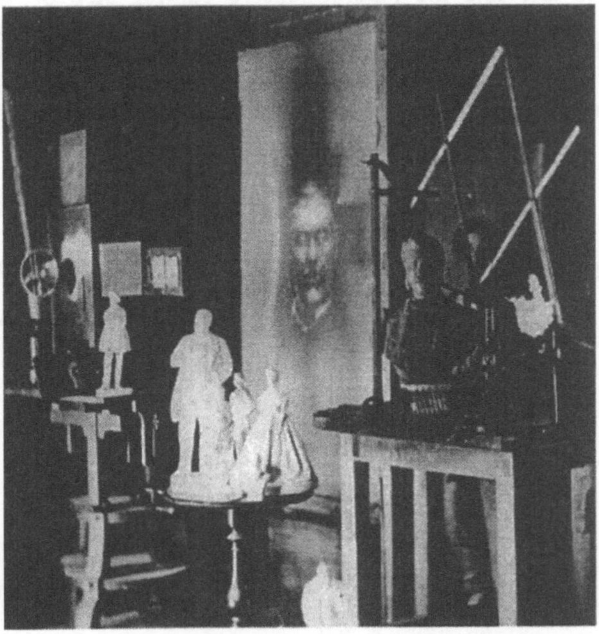

Figure 1.7 François Willème's photosculpturing studio in Paris about 1870. (Courtesy of George Eastman House)

Figure 1.8 Admiral Farragut sits, late 1860's, for photosculpture. (Courtesy of George Eastman House)

In an attempt to alleviate the labor intensive carving step of Willème's photosculpture, Baese (Baese, 1904) described a technique using graduated light to expose photo sensitive gelatin which expands in proportion to exposure when treated with water. Annular rings of this treated gelatin could then be fixed on a support to make a replica of an object as shown in Figure 1.9.

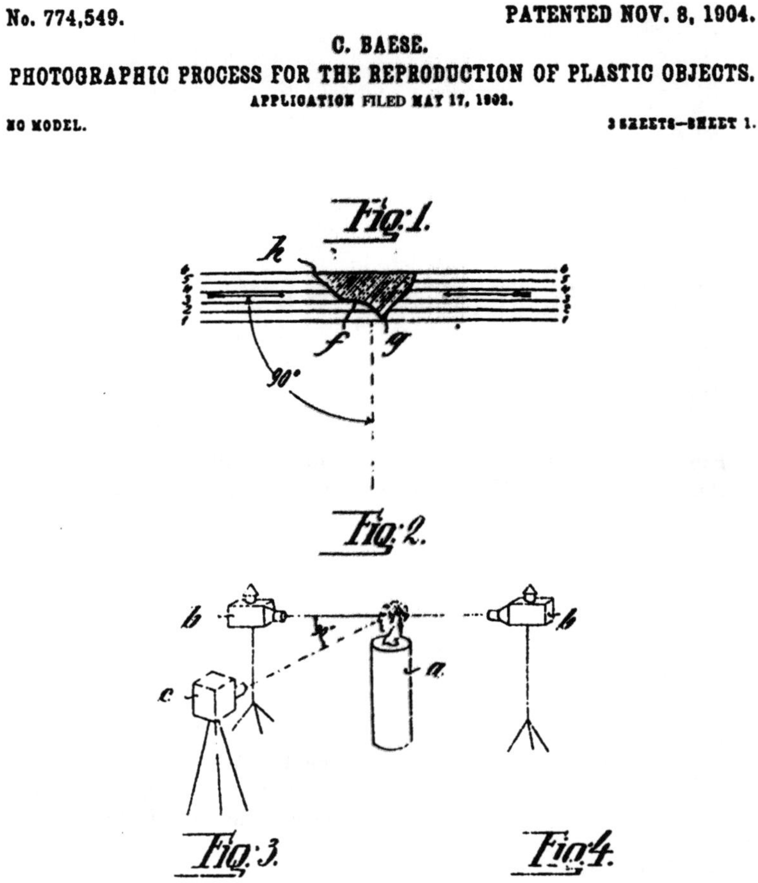

Figure 1.9. Photographic process for the reproduction of plastic objects (Baese, 1904)

Similar techniques and improvements were developed by Monteah (Monteah, 1924). In some of the earliest work in Japan, Morioka (Morioka, 1935,1944) developed a hybrid process between photosculpture and topography. This method uses structured light (black and white bands of light) to photographically create contour lines of an object. These lines could then be developed into sheets and then cut and stacked or projected onto stock material for carving.

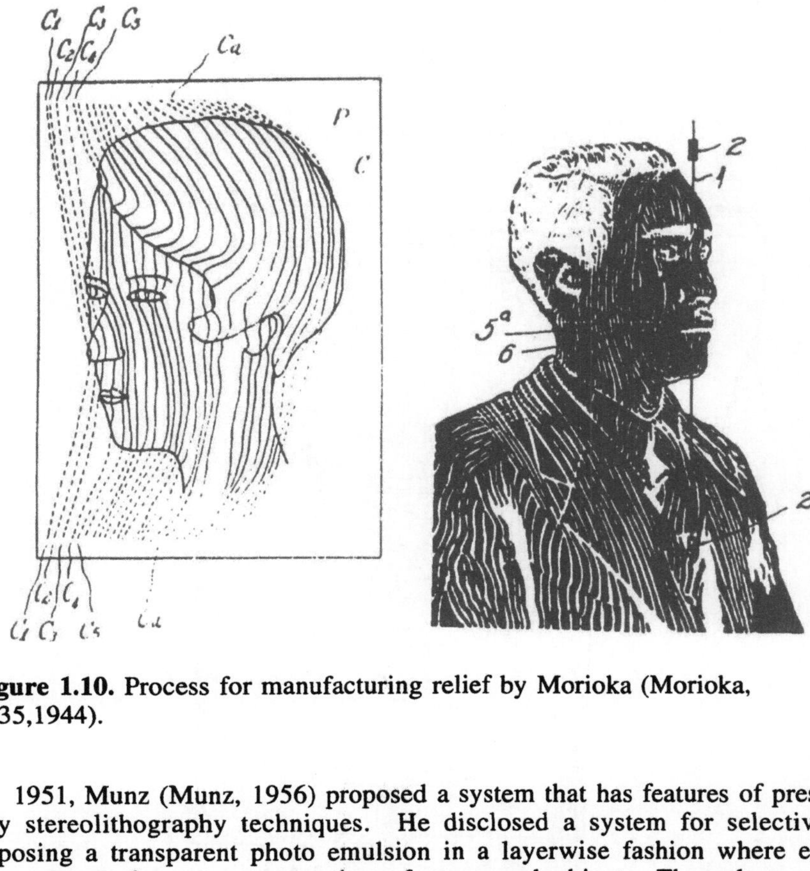

Figure 1.10. Process for manufacturing relief by Morioka (Morioka, 1935,1944).

In 1951, Munz (Munz, 1956) proposed a system that has features of present day stereolithography techniques. He disclosed a system for selectively exposing a transparent photo emulsion in a layerwise fashion where each layer comes from a cross section of a scanned object. These layers are created by lowering a piston in a cylinder and adding appropriate amounts of photo emulsion and fixing agent. After exposing and fixing, the resulting solid transparent cylinder contains an image of the object. Subsequently this object can be manually carved or photochemically etched out to created a three-dimensional object.

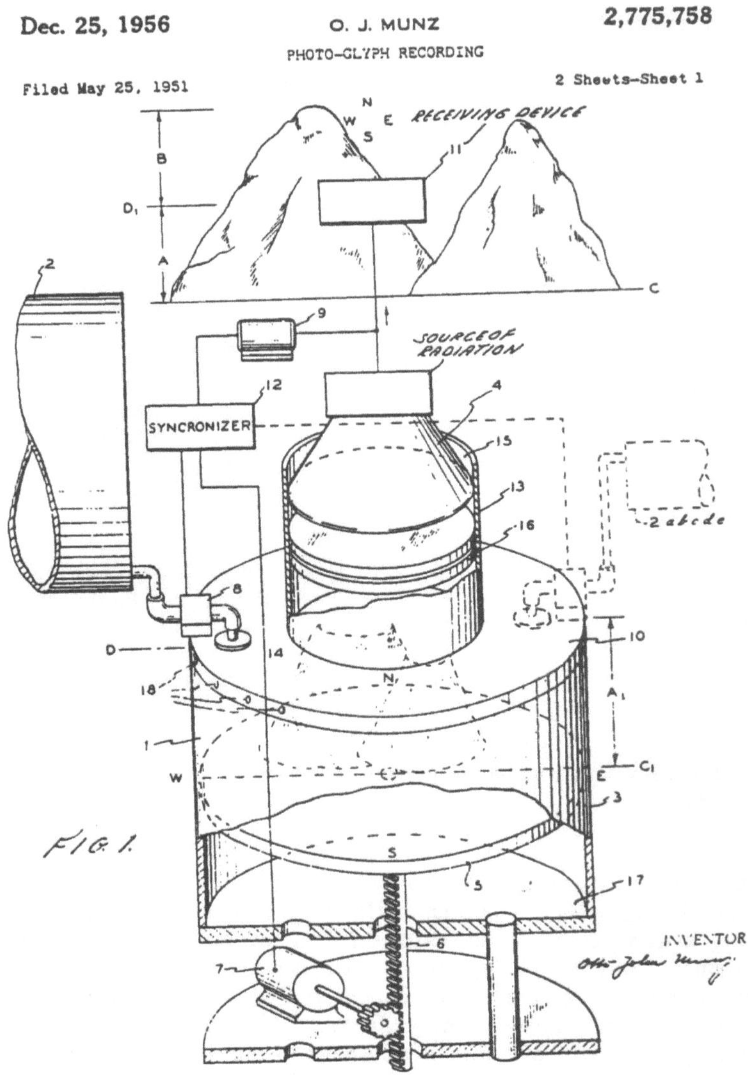

Figure 1.11. Process by Munz (Munz, 1956) to reproduce a three-dimensional image of an object

Early Solid Freeform Fabrication

In 1968, Swainson (Swainson, 1977) proposed a process to directly fabricate a plastic pattern by selective, three dimensional polymerization of a photosensitive polymer at the intersection of two laser beams. Parallel work was conducted at Battelle Laboratories (Schwerzel, 1984). The essential features of this process, termed Photochemical Machining, are depicted in Figure 1.12. The object is formed by either photochemically crosslinking or degrading a polymer by simultaneous exposure to intersecting laser beams. Although laboratory hardware was constructed, it is not believed that a commercially viable process was achieved.

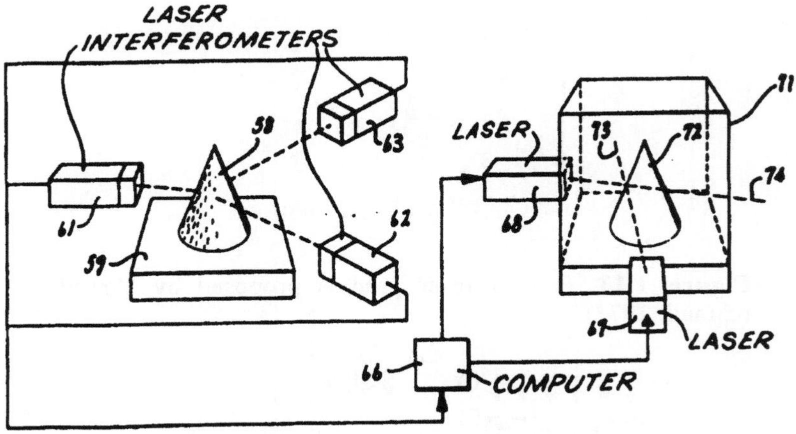

Figure 1.12. Schematic of Photosculpture process using intersecting laser beam (Swainson, 1977)

A powder process that has more in common with laser surface cladding techniques than photosculpture was proposed by Ciraud in 1971 (Ciraud, 1972). This disclosure describes a process for the manufacture of objects from a variety of materials which are at least partially meltable. In order to produce an object, small particles are applied to a matrix by gravity, magnetostatics, electrostatics, or positioned by a nozzle located near the matrix. The particles are then heated locally by a laser, electron beam, or plasma beam. As a consequence of this heating, the particles adhere to each other to form a continuous layer. As shown in Figure 1.13, more than one laser beam can be used to increase the strength of the union between the particles.

Hideo Kodama of Nagoya Municipal Industrial Research Institute was the first to publish an account of a functional photopolymer rapid prototyping system (Kodama, 1981). In his method, a solid model is fabricated by building up a part in layers where exposed areas correspond to a cross-section in the model. He studied three different methods for achieving this:

(a) Using a mask to control exposure of UV source and immersing the model downward into a liquid photopolymer vat to create new layers.
(b) Using a mask, as in (1), but the mask and exposure is positioned on the bottom of the vat and the model is drawn upward to create a new layer.
(c) Immersing the model as in (1) but using an x-y plotter and an optical fiber to expose the new layer. (See Figure 1.14).

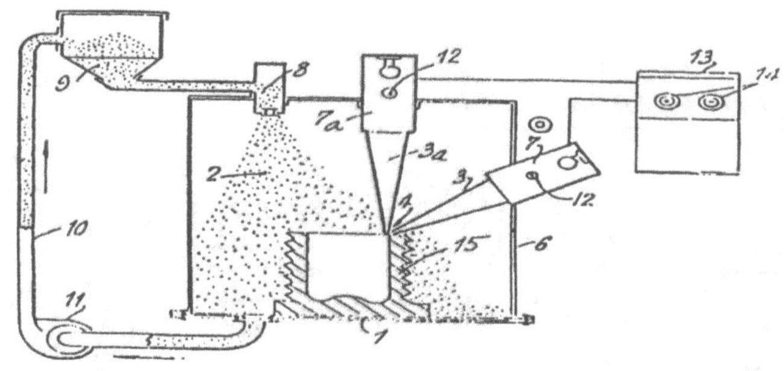

Figure 1.13. Powder laser process proposed by Ciraud (Ciraud, 1972).

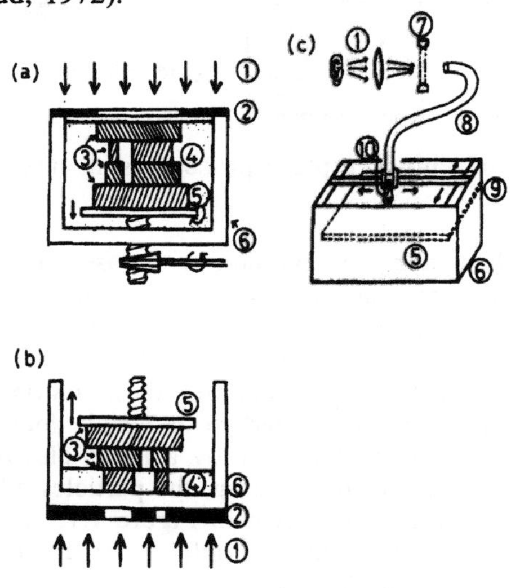

Figure 1.14 Schematics of three systems studied by Kodoma (Kodama, 1981)

A second, parallel but independent, effort was conducted by Herbert at 3M (Herbert,1982). Herbert describes a system that directs a UV laser beam to a photopolymer layer by means of a mirror system on a x-y plotter. (See Figure 1.15) In Herbert's experimental technique, a computer was used to command a laser beam across a layer, the photopolymer vessel was then lowered (≈1mm), and additional liquid photopolymer was then added to create a new layer.

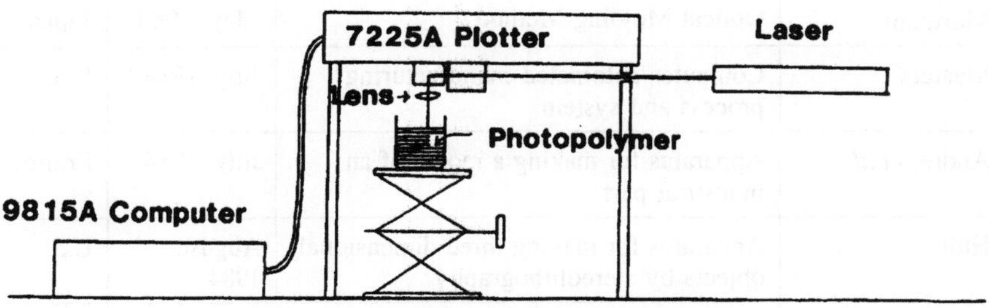

Figure 1.15 Herbert's photopolymer process (Herbert, 1982).

Active Patents

Besides the discussion above, there are numerous active patents that cover existing commercial processes. Some prominent among these are shown in Table 1.1.

Table 1.1 Some prominent active patents in SFF technology

Name	Title	Filed	Country
Housholder	Molding Process	December, 1979	U.S.
Murutani	Optical Molding Method	May, 1984	Japan
Masters	Computer automated manufacturing process and system	July, 1984	U.S.
Andre, *et al*	Apparatus for making a model of an industrial part	July, 1984	France
Hull	Apparatus for making three-dimensional objects by stereolithography	August, 1984	U.S.
Pomerantz, *et al*	Three dimensional mapping and modelling apparatus	June, 1986	Israel
Feygin	Apparatus and method for forming an integral object from laminations	June, 1986	U.S.
Deckard	Method and apparatus for producing parts by selective sintering	October, 1986	U.S.
Fudim	Method and apparatus for production of three-dimensional objects by photosolidification; radiating an uncured photopolymer	February, 1987	U.S.
Arcella, *et al*	Casting shapes	March, 1987	U.S.
Crump	Apparatus and method for creating three-dimensional objects	October, 1989	U.S.
Helinski	Method and means for constructing three-dimensional articles by particle deposition	November, 1989	U.S.
Marcus	Gas phase selective beam deposition: three-dimensional, computer controlled	December 1989	U.S.
Sachs, *et al*	Three-dimensional printing	December, 1989	U.S.
Levent, *et al*	Method and apparatus for fabrication of three-dimensional articles by thermal spray deposition	December, 1990	U.S.
Penn	System, method, and process for making three-dimensional objects	June, 1992	U.S.

Figure 1.16 Three early Solid Freeform Fabrication parts form top to bottom by Kodama (1981), Herbert (1982), Housholder.

Early Parts

Although we are now accustomed to very intricate parts produced by freeform fabrication equipment, the first parts out of these types of systems required a good deal of faith that improvements would occur. Shown in Figure 1.16 are three early parts from different systems. The Housholder part was made from an embodiment that includes a grid for separating mold material (concrete and water) from casting material (dry concrete). The Herbert part was created in August, 1979. The authors do not know exactly when the Kodama and Housholder parts were created.

1.2.2 Commercial Development

Early Development

In the earliest commercial development, Willème's photosculpture studio was commercially successful for 1861 to 1868 but eventually closed, probably due to the labor involved in hand sculpting with a pantographic (tracing) instrument. The next known commercial attempt was the formation of Formagraphic Engine Co. in 1977 by Swainson. Formagraphic later formed an alliance with Battelle Laboratories and changed its name to Omtec Replication. It appears that this effort was abandoned before a commercial process was developed. Also in 1977, DiMatteo formed an company called Solid Photography which was spun out of Dynell Electronics Corporation when Dynell merged with United Technologies. As a result, an affiliated retail outlet called Sculpture by Solid Photography was opened in New York City. In this commercial endeavor, a subject was first photographed with structured lighting. From this photograph, a computer was used to create layered cutting paths in order to carve out a rough rendition of the subject which was subsequently smoothed to create a three dimensional likeness. In 1981, Solid Photography changed its name to Robotic Vision. Solid Photography and another company, Solid Copier, operated as subsidiaries of Robotic Vision until 1989 (Lightman, 1996). The following photographs show the process steps.

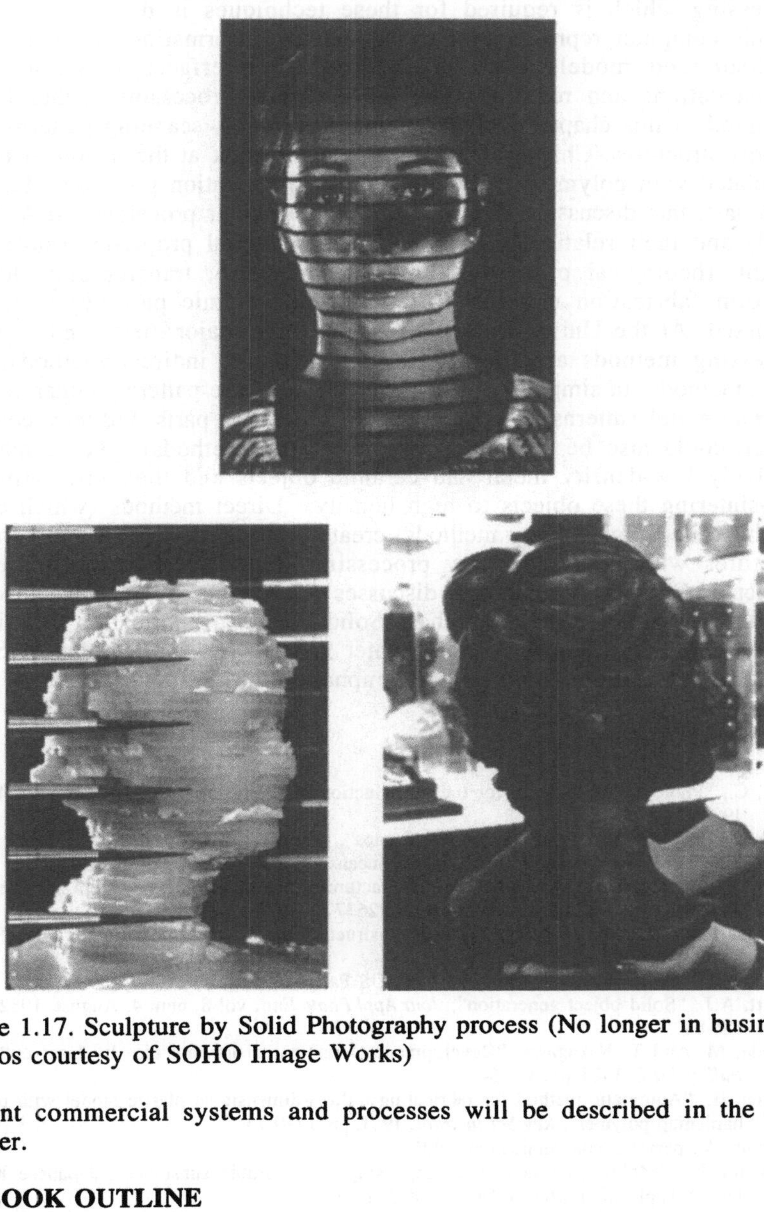

Figure 1.17. Sculpture by Solid Photography process (No longer in business) (Photos courtesy of SOHO Image Works)

Present commercial systems and processes will be described in the next chapter.

1.3 BOOK OUTLINE

In the next chapter, a description of the variety of processes that make up Solid Freeform Fabrication techniques are described. These techniques are roughly categorized into those using photopolymers, powders, sheet feedstock, and material deposition. In Chapter 3, the necessary information processing which is required for these techniques is described. Topics include computer representations of geometric information in the form of solid/surfaced models and the interprocess interfaces between these representations and required files for materials processing. Other topics discussed in this chapter include slicing algorithms, scanning patterns, and support structures. Chapter 4 takes an in-depth look at the technical issues associated with polymers in Solid Freeform Fabrication processes. For the most part, this discussion focuses on three particular processes (SLA, SLS, FDM) and their relationships to polymeric material properties (molecular weight, rheological properties, etc). In Chapter 5, transfer and indirect freeform fabrication methods for metal and ceramic part fabrication is discussed. At the University of Texas, the three major metal and ceramic processing methods are termed transfer methods, indirect methods, and direct methods. In simple terms, transfer methods use patterns, either casting or spray metal patterns, to create metal and ceramic parts. Indirect methods (which could also be called low-density matrix methods) involve making relatively low-density metal and ceramic objects and then infiltrating or post-sintering these objects to high density. Direct methods (which could also be called high-density methods) create high density metal and ceramic structures without a secondary processing step. Direct methods are the subject of Chapter 7. Chapter 6 discusses process modeling and control of Selective Laser Sintering. Gas phase Solid Freeform Fabrication techniques is the subject of Chapter 8 and Chapter 9 discusses applications of Solid Freeform Fabrication with a primary emphasis on SLS applications.

REFERENCES

Baese, C., "Photographic process for the reproduction of plastic objects", US Patent 774,549, 1904.

Blanther, J.E., "Manufacture of contour relief maps", US Patent 473,901, 1892.

Bogart, M., "In art the end don't always justify means", *Smithsonian*, 1979 , pp. 104-110.

Ciraud, P.A., "Process and device for the manufacture of any objects desired from any meltable material", FRG Disclosure Publication 2263777, 1972.

DiMatteo, P.L., "Method of generating and constructing three-dimensional bodies", US Patent 3,932,923, 1976.

Gaskin, T.A., "Earth science teaching device", US Patent 3,751,827, 1973.

Herbert, A.J., "Solid object generation", *Jour Appl Photo Eng*, vol 8, num 4, August, 1982 , pp. 185-88.

Kunieda, M. And T. Nakagawa, "Development of laminated drawing dies by laser cutting", *Bull of JSPE*, 1984, pp. 353-54.

Kodama, H., "Automatic method for fabricating a three-dimensional plastic model with photo-hardening polymer", *Rev Sci Instrum*, 1981, pp. 1770-73.

Lightman, A., personal communication, 1996.

Matsubara, K., "Molding method of casting using photocurable substance", Japanese Kokai Patent Application, Sho 51[1976]-10813, 1974.

Monteah, F.H., "Photochemical process for producing BAS reliefs", US Patent 1,516,199, 1924.

Morioka, I., "Process for manufacturing a relief by the aid of photography", US Patent 2,015,457, 1935.

Morioka, I., "Process for plastically reproducing objects", US Patent 2,350,796, 1944.

Munz, O.J., "Photo-glyph recording", US Patent 2,775,758, 1956.

Nakagawa, T. , Et al, "Blanking tool by stacked bainite steel plates", *Press Technique*, 93-101,

1979.
Nakagawa, T., *et al*, "Laser cut sheet laminated forming dies by diffusion bonding", *Proc 25th MTDR Conf*, 1985 , pp. 505-510.
Nelson, J.C., "Selective laser sintering: a definition of the process and an empirical sintering model", *PhD Dissertation*, University of Texas, May, 1993.
Perera, B.V., "Process of making relief maps", US Patent 2,189,592, 1940.
Swainson, W.K., "Method, medium and apparatus for producing three-dimensional figure product", US Patent 4,041,476, 1977.
Schwerzel, R.E., *et al*, "Three-dimensional photochemical machining with lasers", *Appl of Lasers to Ind Chem*, SPIE,1984 , pp. 90-97.
Zang, E.E., "Vitavue relief model technique", US Patent 3,137,080, 1940.

2

PROCESS METHODS[1]

2.1 BACKGROUND

Many Solid Freeform Fabrication (SFF) methods are based upon a layered manufacturing approach. In this layered approach (Figure 2.1), a computer representation of an object's geometry is decomposed into slices which represent 2-1/2 dimensional information. A slice represents a planar cross-section of an object with an associated small, but finite, thickness. During processing, each slice is physically deposited and the slice interior is fused to the previous layer to create an object sequentially. While the interior region of a slice represents the geometry of the object, the exterior region of a slice also serves an important function in SFF technologies. In most processes, it becomes a supporting structure with two primary roles. First, it serves as an implicit fixture for the part. Second, it serves as a supporting

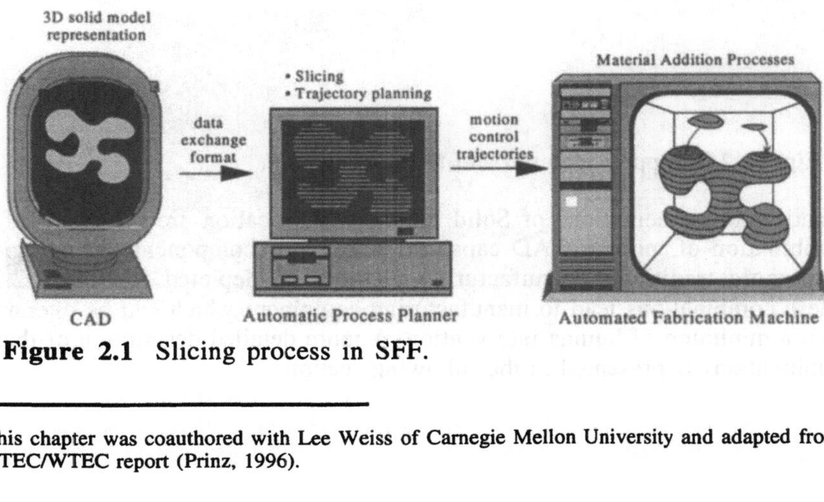

Figure 2.1 Slicing process in SFF.

[1]This chapter was coauthored with Lee Weiss of Carnegie Mellon University and adapted from a JTEC/WTEC report (Prinz, 1996).

matrix upon which initially unconnected regions which occur in overhangs can be constructed. The unconnected regions require this support since they are not joined with the main body until subsequent layers are deposited.

The support areas of a layer are handled in various ways and provide a clear distinction among many of the process methods. In some of the methods, explicit support structures are deposited with the same material as the object being formed (left side of Figure 2.2), but are built in such a way as to be easily removed, usually by manual means. There can be substantial process planning to create these explicit support structures (see Chapter 3); post processing to remove them after the part is built may also be substantial.

In other process methods, the entire exterior of the slice becomes a sacrificial portion of the layer which is later removed and discarded or recycled (right side of Figure 2.2). The support material is often different from the part material and has the feature that it can be selectively removed (by dissolution, melting, etching, etc).

Still other process methods use base materials which are the same as the part but these materials are in powder form which intrinsically supports the part but is still removable.

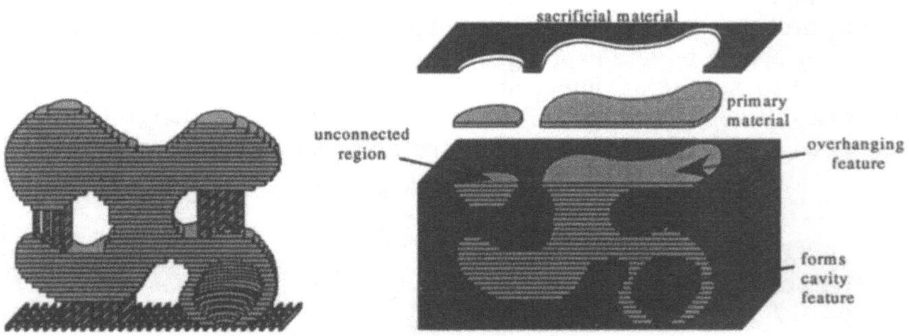

Figure 2.2 Support material and fixturing in SFF.

Practical implementations of Solid Freeform Fabrication are the result of a combination of modern CAD capabilities and new component technologies with more traditional manufacturing methods as depicted in Figure 2.3. These combinations lead to manufacturing equipment which can be operated with a minimum of human intervention. A more detailed description of these combinations is presented in the following section.

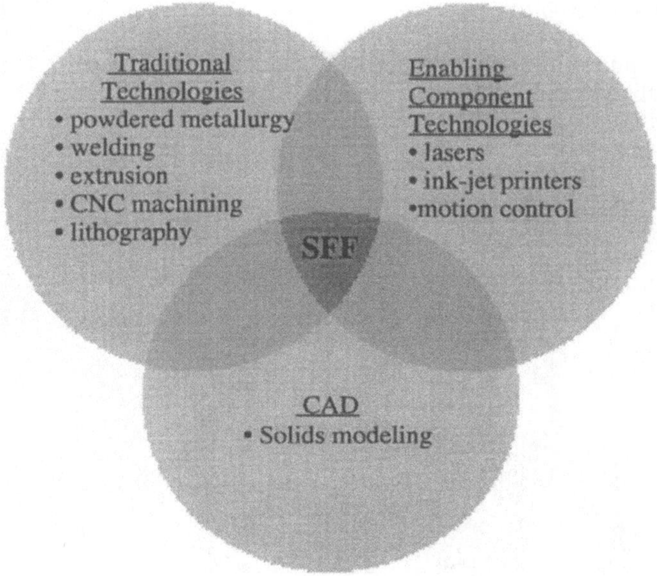

Figure 2.3 SFF enabling technologies.

A layered approach with explicit or complementary supports allows for part fabrication without part specific-knowledge or human intervention for several reasons:

(1) the shape decomposition operation maps complex 3D geometry into relatively simple 2-1/2D representations,
(2) custom fixturing is not required, and
(3) the machinery required to implement this approach is relatively easy to operate.

2.2 PROCESS TECHNIQUES

In this section, a sampling of SFF building strategies and deposition/fusion processes will be described. The figures which are included here only schematically represent these systems.

2.2.1 Photopolymers

Photopolymer systems build shapes using light to selectively solidify liquid photocurable resins. There are two basic approaches: laser photolithography (voxel lines) and photomasking (voxel areas)[2].

Laser Photolithography

The laser photolithography approach depicted in Figure 2.4 was first commercialized by 3D Systems.

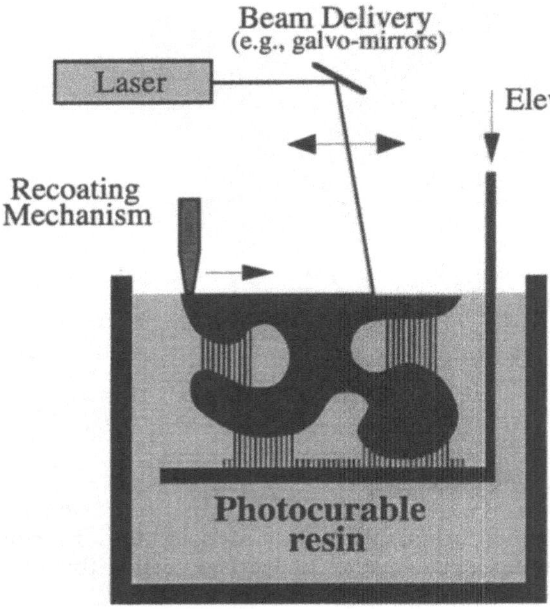

Figure 2.4 Stereolithography

Laser photolithography systems have since been developed and manufactured in both Europe and Japan (see Table 2.1). While most laser photolithography systems use the building strategy represented in Figure 2.4, there are significant differences in the lasers and in the recoating and beam delivery mechanisms employed. Laser photolithography creates acrylic or epoxy parts directly from a vat of liquid photocurable polymer by selectively solidifying the polymer with a scanning laser beam. Parts are built up on an elevator platform which incrementally lowers the part into the vat by the distance of the layer thickness. To build each layer, a laser beam is guided across the surface (by servo-controlled galvanometer mirrors, for example), drawing a cross-sectional pattern in the x-y plane to form a solid section. The platform is then lowered into the vat and the next layer is drawn and adhered to the previous layer. These steps are repeated, layer-by-layer, until the complete part is built up.

[2]For a discussion of voxels see Chapter 1.

Table 2.1 Commercial Laser Photolithography Systems.

Manufacturer	Process Name	Start	Ship	Notes
USA				
Aaroflex	Solid Imager	95	n/a	License from Dupont
Dupont	Solid Object Modeling System (SOMOS)	87	n/a	License to Teijin Seiki and Aaroflex
3D Systems	Stereolithgraphy Apparatus (SLA)	86	88	First commercial shipment
Japan				
CMET (Mitsubishi)	Solid Object Ultra-violet Laser Plotter (SOUP)	88	90	
Denken Engineering	Solid Laser Plotter (SLP)	85	93	Scan from bottom
D-MEC (JSR/Sony)	Solid Creation System (SCS)	88	90	
Meiko Corp.	Meiko	91	94	Jewelry appl.
Mitsui Zosen	Computer Operated Laser Active Modeling Machine (COLAMM)	91	91	Scan from bottom
Teijin Seiki	Solid Forming System (Soliform)	91	92	License from DuPont
Europe				
EOS (Germany)	STEREOS	89	90	
Fockele & Schwarze (Germany	Laser Modeling System (LMS)	91	94	
Laser 3D (France)	StereoPhotoLitho-graphy (SPL)	92	n/a	Operates as a service bureau

Since the photopolymers are relatively viscous, simply lowering the elevator the small distance of the layer thickness (i.e., ~50μm to ~500μm) into the vat does not permit the liquid to uniformly recoat the upper surface of the part in a timely fashion. A recoating mechanism is therefore required to facilitate this process. There are legally important distinctions in recoating mechanisms among the various laser photolithography vendors. For example, 3D System's stereolithography has used a "deep dip" recoating whereby the elevator is first lowered several millimeters so that the liquid entirely flows over the current upper surface of the part. The elevator is then raised to the desired height and a "doctor-blade" (wiper arm) traverses the surface to quickly level the excess viscous material. Other methods of recoating include:

1. A metered pump which draws up the resin and then disperses it uniformly through a nozzle.
2. An electrostatic recoating blade which attracts photopolymer above the vat surface and then spreads the material over the surface.
3. Double recoating blades which attract resin up between the blades due to surface tension in a fashion similar to 2.

With laser photolithography, features with gradually changing overhangs can be built up without support structures. Large overhanging features, however, require supports since the initial thin layers which form them can warp or break off as the part moves down into the liquid. The supports are typically built up as thin wall sections which can easily be broken away from the part upon completion.

There are a few laser photolithography systems which build using slightly different approaches than those depicted in Figure 2.4. In the machine manufactured by Denken Engineering, (see Figure 2.5a), the part is built inverted, attached to a platform which rises up as each successive layer is drawn and attached to the bottom-most face.[3] The liquid resin layer is deposited on a specially prepared window that is transparent to the laser and to which cured polymer adheres poorly. The platform and previously built structure are lowered into the resin leaving a liquid film, having the correct thickness for the next layer, between the part and the plate. The new layer is drawn from beneath the plate. After the layer is drawn, the new structure is raised, separating the layer from the plate, and the process is repeated until all layers are fabricated. Mitsui Corporation also manufactures a process, "COLAMM," which scans from below.

The process represented in Figure 2.5b is being developed and optimized by Professor Koji Ikuta of Nagoya University for producing micro-scale parts,

[3]See also the discussion of Kodama's research in Chapter 1.

i.e., laser spot size is less than 5µm (Inada, 1995). In this system a transparent plate is lowered into the vat to form a thin layer of liquid film over the part being built up. The growing part remains stationary in the vat, and the vat is moved relative to a fixed laser beam which passes through the plate, drawing the cross-section. The laser moves up with the plate to maintain precise focusing on the film layer.

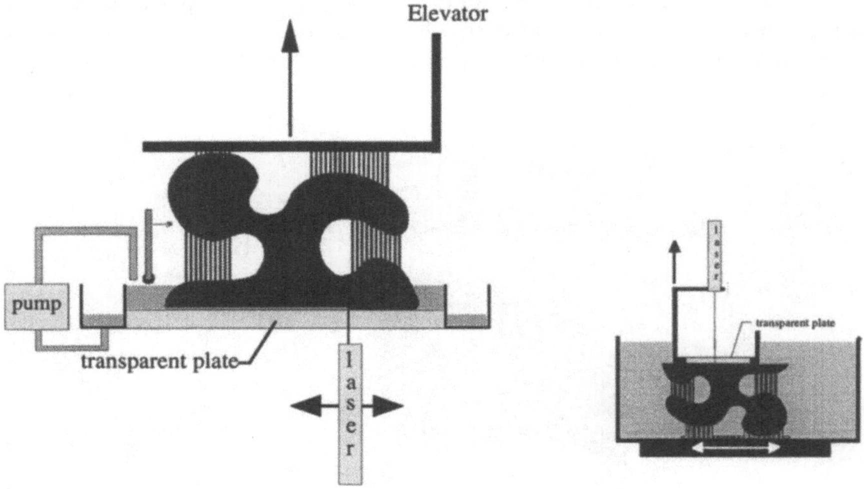

Figure 2.5 Other laser photolithography techniques.

Photomasking

In contrast to "drawing" each cross-section with laser photolithography, it is possible to image an entire cross-section in a single operation by using photomasks. This approach was originally developed and commercialized by Cubital (Israel/Germany). Their system, Solid Ground Curing (SGC), is depicted in Figure 2.6. The various processes used to implement SCG are performed at different stations. In SGC, each cross-section is imaged onto an erasable mask produced by charging a plate via an ionographic process and then developing the image with an electrostatic toner (e.g., similar to the Xeroxgraphy process). The mask is then positioned over a uniform layer of liquid photopolymer and an intense pulse of UV light passed through it to selectively cure the material. Uncured photopolymer is removed from the layer with a vacuum system and replaced with a low melting point, water-soluble wax which serves as the sacrificial support. After the wax has cooled, the layer is milled to produce a flat surface. The pattern on the exposed mask is erased by wiping off the toner, and the entire process is repeated until the part has been completed. Then the wax is removed by melting.

A unique feature of the photomasking approach is the capability to build multiple parts in a timely fashion in a single batch. Since the building time to form each layer is independent of the part geometry or size, multiple parts (e.g., the four replicates next to each other in Figure 2.6) can be fabricated in the same time that it normally takes to build a single part. Furthermore, the SGC system builds parts in a solid sacrificial wax material which permits multiple parts to be packed into a single batch (e.g., the two rows of parts shown in Figure 2.6). This approach also produces superior downward facing surfaces. There is a trade-off, however, with the complexity of the system required to implement this approach.

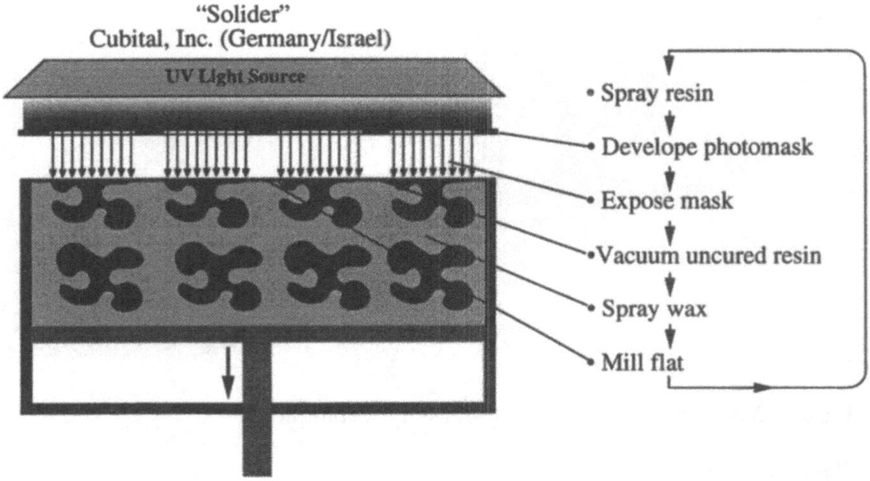

Figure 2.6 Photomasking.

Another photomasking system from Light Sculpting (Milwaukee, Wisconsin) was also developed by Efrim Fudim (Fudim, 1988). In this system, masks are created by a photoplotter. In at least one rendition of this process, parts are built inverted in a similar fashion to the laser photolithography process of Denken and Mitsui. Light Sculpting was the first commercial photopolymer entity to start business, as a service bureau, in 1986. To date, this company has not sold any machines.

On a research level, Stanford Research Institute is developing a method similar to the commercial process of Cubital but utilizing ceramic filled photopolymers as feedstock. The resulting parts are to be used as green ceramic parts in traditional ceramic processing (Ventura, 1996).

Table 2.2 Photomasking systems.

Manufacturer	Process Name	Start	Ship	Notes
Cubital	Solid Ground Curing (SGC)	87	90	
Light Sculpting	Design Controlled Automated Fabrication (DESCAF)	86	n/a	First photopolymer service bureau
Stanford Research Institute		95	n/a	Ceramics research project

2.2.2 Powders

Several systems use powders as their stock material. There are three current approaches: laser sintering (voxel lines), three-dimensional printing (individual voxels), and 3D cladding (voxel lines, no implicit support).[4]

Selective Laser Sintering

Several systems use lasers to build up shapes by selectively fusing powdered material. The Selective Laser Sintering approach, which was described in Chapter 1 and is depicted in Figure 2.7, was originally developed at The University of Texas at Austin and then commercialized by DTM Corporation. EOS Corporation has also developed and marketed their own laser sintering machines. In these systems, a layer of powdered material is spread out and leveled over the top surface of the growing structure. A laser then selectively scans the layer to fuse those areas defined by the geometry of the cross-section; the laser energy also fuses subsequent layers together. The powders are joined by a variety of fusion mechanisms including melting, surface bonding, sintering aids, and polymer coatings. The unfused material remains in place as the support structure. After each layer is deposited, an elevator platform lowers the part by the thickness of the layer and the next layer of powder is deposited. When the shape is completely built up, the part is separated from the loose supporting powder. Several types of materials are commercially available including plastics, waxes, and low melting metal alloys, as well as polymer coated metals and ceramics for making green preforms. Direct fusing of metals and ceramics (i.e., uncoated powders) is also being investigated.

---

[4]Masking techniques (voxel areas) have also been proposed (Housholder, 1981 and Jog, 1994).

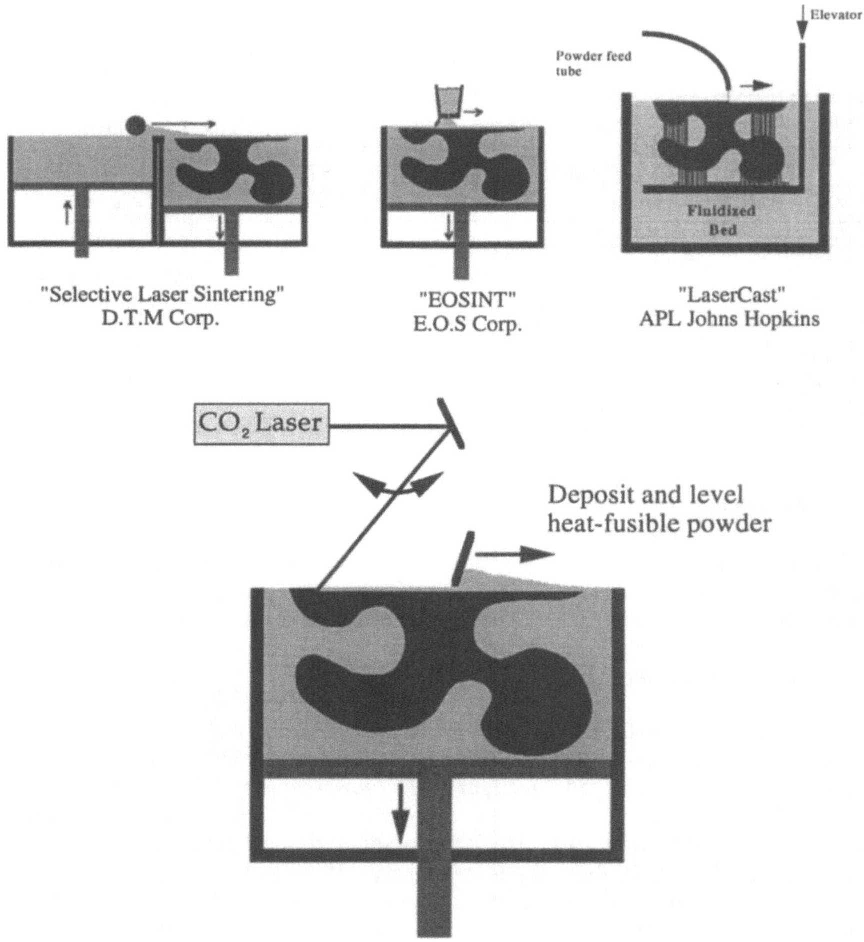

Figure 2.7 Selective Laser Sintering.

While both the DTM and EOS machines are based on the same underlying methodology, there are differences in machine implementations, including their respective material delivery approaches. The DTM machine delivers powder from a cylinder adjacent to a second cylinder in which the part is grown; a roller is used to spread and level the powder. In the EOS system, powder spreading is accomplished by a slot nozzle whose leading and trailing edges are contoured to supply force normal to the surface as the head is vibrated from side to side.

Another process based on Selective Laser Sintering is being pursued by the Applied Physics Laboratory at Johns Hopkins University (Arcella, 1989,

House, 1996). In this process, LaserCast, powdered material is delivered by fluidizing the powder bed. A new layer of powder can be delivered to the top surface in several ways (Figure 2.7):

(1) Lowering the elevator in the fluidized powder.
(2) Increasing the fluidizing gas flow to increase the surface height.
(3) Adding additional powder through a feed tube.

The initial application for this process is planned to be rough net shape titanium parts suitable for further processing either through forging or machining.

Three Dimensional Printing

The Three dimensional Printing (3DP) process, depicted in Figure 2.8, was developed at MIT as a method to form green preforms for powdered ceramic and metal applications. While different powdered materials can be used, 3DP is currently commercialized by Soligen Corporation under the name Direct Shell Production Casting (DSPC) for creating ceramic shells/cores for casting applications. In 3DP, the part is built-up in a bin which is fitted with a piston to incrementally lower the part into the bin. Powder (e.g., alumina) is dispensed from a hopper above the bin, and a roller is used to spread and level the powder. An ink-jet printing head scans the powder surface and selectively injects a binder (e.g., colloidal silica) into the powder. Each drop from the printhead binds the underlying powder to create one voxel cell. Individual cells are knit together by overlapping. The bound cells join the powder together into those areas defined by the geometry of the cross-section. The unbound powder becomes the support material. When the shape is completely built up, the green structure is fired, and the part is then removed from the unbound powder. 3DP of metal powders, such as stainless steel bound with a polymeric binder, is also being explored; subsequent infiltration of the matrix is then required for densification.

3D Cladding

3D cladding[5] is a generalization of laser cladding (Breinan, 1983).[6] As shown in Figure 2.9a, the process is initiated on a metal substrate by a focused laser beam. The laser beam creates a weld pool into which powder is injected and melted. The substrate is scanned by the laser/powder system in order to trace a cross-section. Upon solidification, the trace forms a cross-section of a part. Consecutive layers are then additively deposited, thereby producing a three-dimensional component. The process is under development

---

[5]These processes could also be categorized as deposition processes and are very similar to 3D welding which is described in Section 2.2.4.

[6]See also the discussion of Ciraud in Chapter 1.

at Sandia National Laboratories under the process name Laser Engineered Net Shaping (LENS) (Griffith, 1996) and at Los Alamos National Laboratory under the process name Directed Light Fabrication (DLF) (Lewis, 1995). The present DLF process is more flexible in that it has 5 axis positioning capability. This allows for the manufacture of more complex parts (i.e., overhangs) at the expense of increased equipment cost and process planning complexity.

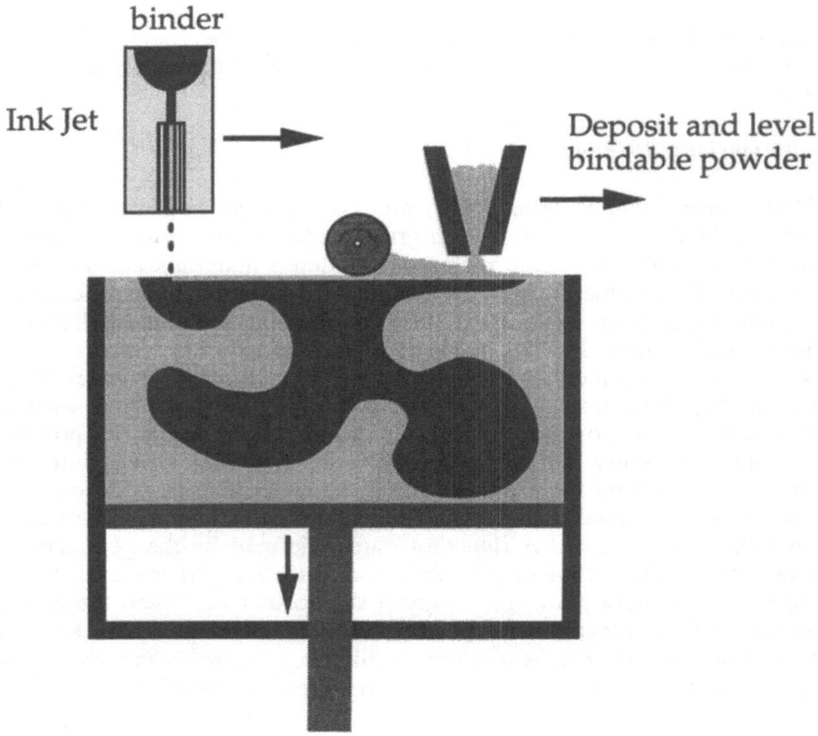

Figure 2.8 Three Dimensional Printing

A variant of this approach is under development at the Fraunhofer Institute for Production Technology (IPT) under the process name Controlled Metal Build Up (CMB) (Klocke, 1996) (Figure 2.9b). This process works in a similar fashion to LENS or DLF, but there is a 2-1/2 D milling cutter within the work chamber to finish the walls of the metal part and improve tolerances.

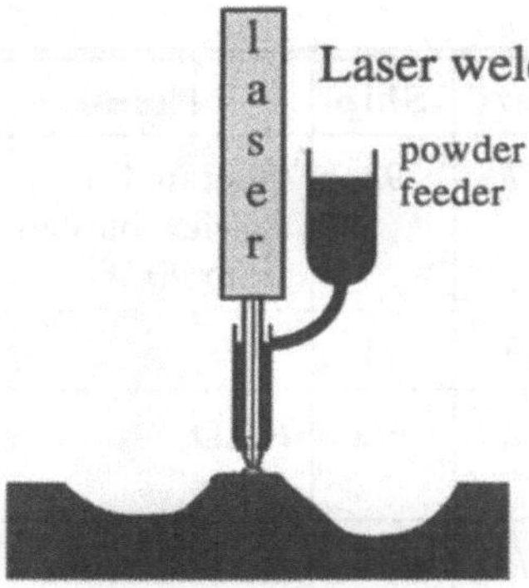

a. LENS (Sandia National Lab)
DLF (Los Alamos National Lab)

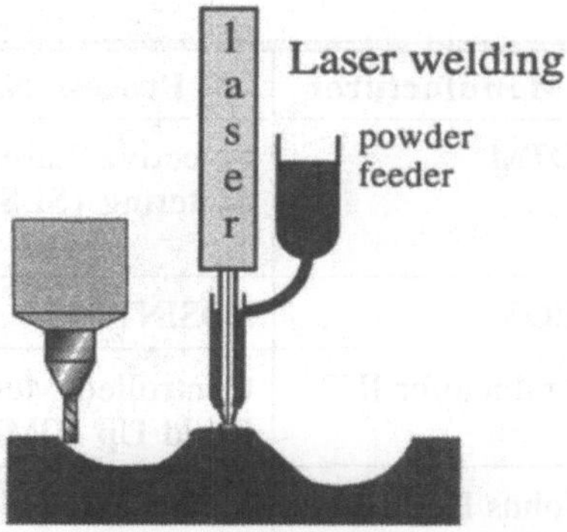

b. Controlled Metal Build Up (CMB)
(Fraunhofer IPT)

Figure 2.9 3D Cladding.

Table 2.3 contains a list of the powder SFF processes.

2.2.3 Laminations

Lamination systems, which trace their roots to the topography described in Chapter 1, operate with a variety of feedstocks from paper sheets to metal plates. Laminated object manufacturing (LOM) is a method which was developed and commercialized by Helisys Corp. LOM currently builds shapes with layers of paper or plastic (see Figure 2.10a). The laminates, which have a thermally activated adhesive, are glued to the previous layer with a heated roller. A laser cuts the outline of the part cross-section for each layer. The laser then scribes the remaining material in each layer into a crosshatch pattern of small squares (see insert in Figure 2.10).

As the process repeats, the crosshatches build up into tiles of support structure. The cross-hatching facilitates removal of this tiled structure when the part is completed. LOM builds up large parts relatively rapidly because only contours are scanned. LOM is also being investigated by Helisys, the University of Dayton (Klosterman, 1996), and Lone Peak Corp. (Griffin, 1994) for building up ceramic and reinforced composite shapes using layers of green tape castings (i.e., sheets of bound powder); the final part must subsequently be sintered. Lone Peak has also demonstrated the use of a flexible metal tape in the LOM process. The resulting parts are sintered and then infiltrated (Griffin, 1996).

Table 2.3 Powder SFF Processes.

Manufacturer	Process Name	Start	Ship	Notes
DTM	Selective Laser Sintering (SLS)	87	92	Operated a service bureau from 90-93
EOS	EOSINT	89	94	
Fraunhofer IPT	Controlled Metal Build Up (CMB)	92	n/a	R&D
Johns Hopkins	LaserCast	n/a	n/a	R&D
Los Alamos National Lab	Directed Light Fabrication (DLF)	n/a	n/a	R&D
Soligen	Three Dimensional Printing (3DP)	91	93	Operating a service bureau
Sandia National Lab	Laser Engineered Net Shaping (LENS)	93	n/a	R&D

A lamination system being produced in Japan is manufactured by Kira Corp. While their basic building approach is the same as that used for the LOM machine, the Kira Solid Center (SC) machine (Figure 2.10b) is implemented in a significantly different manner. The SC machine uses standard printing paper which is fed into the machine using a conventional laser printer. The printer uses an adhesive-based toner to print the outline of the cross-section as well as a cross-hatched bonding pattern on each piece of paper. A hot-plate then laminates the paper to the previous layers. The cross-sectional outline is then cut with a carbide knife which is mounted on a swivel base. Additional segments of 'parting-plane' sections are also cut to facilitate removal of the excess remaining support material. An inexpensive, manual stacking version of this process was developed at the University of Utah and is being marketed by Schroff Development Corp. Another inexpensive LOM process was introduced by Sparx AB of Molndal, Sweden. This system, Hot Plot, consists of a flat-bed plotter equipped with a heated cutting electrode and a manual mounting fixture. Feedstock material is 1 mm thick polystyrene sheet.

The lamination systems described above are relatively quick since only the cross-section outlines need to be cut. However, internal cavities are hard to form since it is difficult to remove the sacrificial material from internal regions. To address this issue, Case Western Reserve University (Wei, 1996) and CAM-LEM, Inc. are developing a lamination system using green tape castings with a separate supporting fugitive tape material. Each section is individually cut with a laser and then stacked in place. The fugitive tape is then burned out during the final firing process.

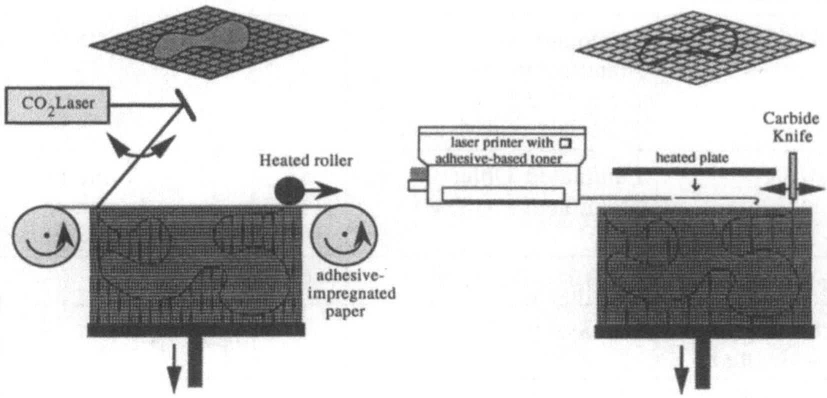

a. "Laminated Object Manufacturing" Helisys, Corp. (USA)

b. "Solid Center", Kira Corp. (Japan)

Figure 2.10 Lamination systems.

As discussed in Chapter 1 and shown in Figure 2.11, Professor Nakagawa (Nakagawa, 1985) is developing lamination processes to build up large scale tooling such as forming dies for automobile bodies. Individual sections of material are shaped with CNC milling or laser cutting and then stacked to form the tool .

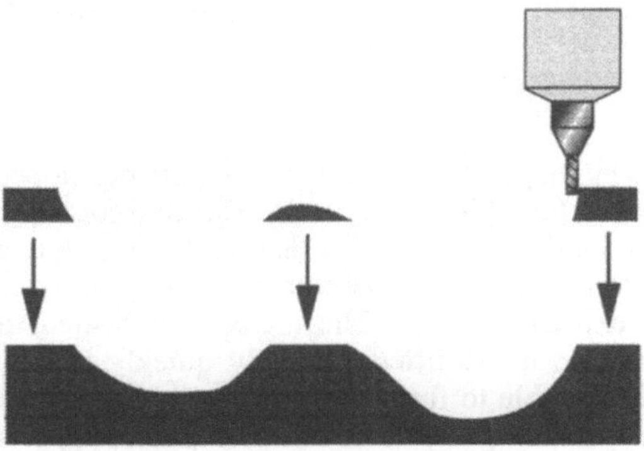

Figure 2.11 Large scale lamination.

Table 2.4 contains a summary listing of lamination processes.

Table 2.4 Lamination SFF Processes.

Manufacturer	Process Name	Start	Ship	Notes
CAM-LEM Inc.	Computer Aided Manufacture of Laminated Engineering Materials (CAM-LEM)	?	n/a	R&D
Helisys	Laminated Object Manufacture (LOM)	85	91	Formally known as Hydronetics
Kira	Solid Center (SC)	92	94	
Schroff Development	JP5 System	?	95	
Sparx AB	Hot Plot	?	94	
University of Tokyo	Laser cut sheet lamination	n/a	n/a	R&D

2.2.4 Depositions

SFF deposition techniques include: extrusion (voxel lines), ink-jets (voxels), 3D welding (voxel lines and areas), and gas phase (voxels/voxel lines).

Extrusion

The process of extruding freeform shapes (Figure 2.12) was first developed and commercialized by Stratasys, Inc. Their approach, Fused Deposition Modeling (FDM), deposits a continuous filament of a thermoplastic polymer or wax through a resistively heated nozzle. In the current system, material is delivered as a wire into an extrusion head. The material is heated to slightly above its flow point so that it solidifies relatively quickly after it exits the nozzle. Therefore, it is possible to form short overhanging features without the need for explicit support. In general, however, explicit supports are needed. In one method, supports, which are built from the same material as the part, are formed by creating perforations where supports adjoin the model, making it easy to snap them off. In a newer method, two modeling materials are dispensed through a dual tip mechanism. A primary modeling material is used to produce the model geometry. A secondary material is used to produce support structures. When force is applied to these supports, they break away easily due to the weaker nature of the bond between the model and the support structure. Various US investigators, including Rutgers University (Agarwala1996), Allied-Signal, Lone Peak Engineering (Griffin, 1995) and AdvancedCeramics Research (Hilmas, 1996), are also exploring the use of FDM with thermoplastic wires and rods loaded with ceramic powders to build green preforms.

Multi-Phase Jet Solidification (MPJS) is another extrusion-based process. MPJS is being jointly developed by the Fraunhofer-Institutes for Applied Materials Research (IFAM, Bremen) and Manufacturing Engineering and Automation (IPA, Stuttgart) (Geiger, 1994). IPA is working on software development and IFAM on material aspects. MPJS extrudes metal or ceramic slurries using metal injection molding technology. The slurry, which is a ~50/50 mixture of wax and metal or ceramic powder, is contained in a heated vessel and pumped through an attached nozzle by means of a screw-activated plunger.

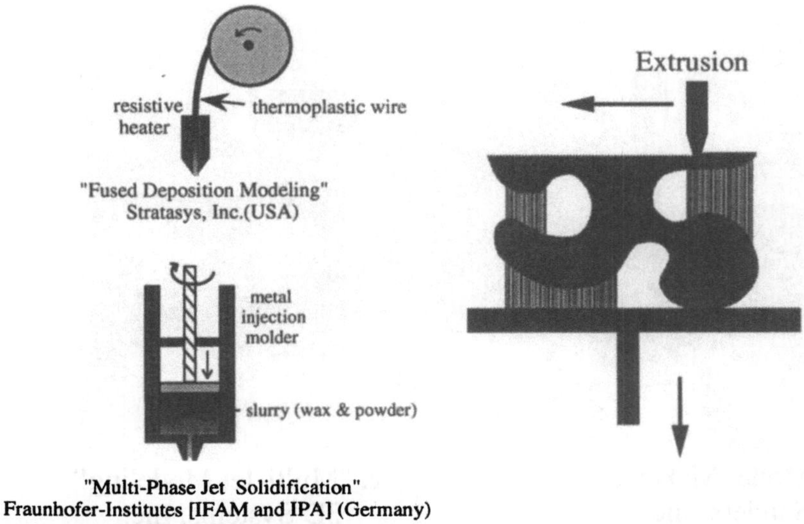

Figure 2.12 Extruding freeform shapes.

Ink Jets

Several SFF processes use ink jets to directly deposit, low-melting target materials. Ballistic Particle Manufacturing (BPM) (Masters, 1987), which was developed and commercialized by BPM Technology, Inc., uses a piezoelectric jetting system to deposit microscopic particles of molten

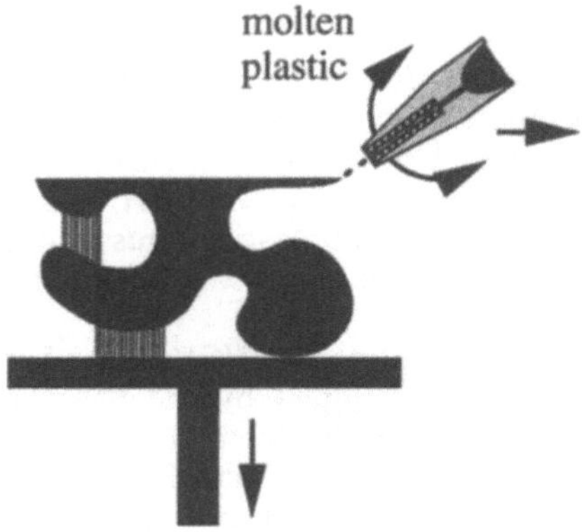

a. "Ballistic Particle Manufacturing",
BPM, Inc.

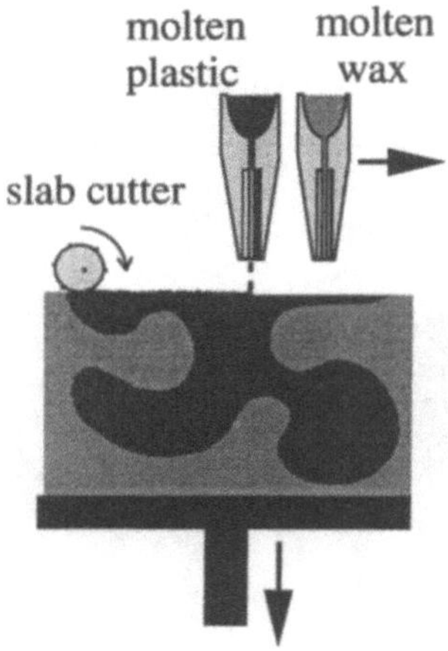

b. "Model Maker",
Sanders, Inc.

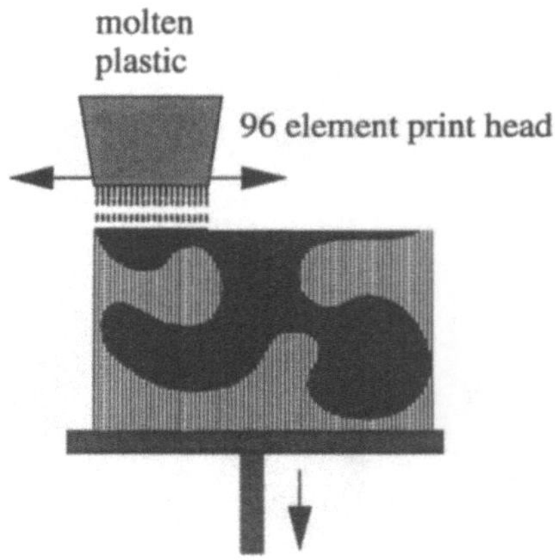

c. "Multi-Jet Modeling"
3D Systems, Inc.

Figure 2.13 Ink Jet Systems.

thermoplastic (Figure 2.13a). Like FDM and SLA, support structures are required for "unconnected" features. In this case, the supports are deposited in a perforated pattern to facilitate removal. The BPM jet head, however, is mounted on a 5-axis positioning mechanism so that overhanging features can be deposited without support as is shown in the figure. The Sanders, Inc. Model Maker system (in Figure 2.13b), dispenses both a low-melting thermoplastic, as well as a separate wax support material. In addition, it incorporates a slab cutter to plane each layer to a precise thickness. Similar systems have been investigated by Texas Instruments (Penn, 1993) and Helinski (Helinski, 1992). In other commercial developments, 3D Systems, Inc. has just introduced a new ink-jet prototyping system, "Multi-Jet Modeling," which uses a printing head with 96 individual jets that deposit a low melting thermoplastic. The support structure, which is made from the same material, is deposited as thin needle-like structures.

In other research efforts, The Technical University of Munich is developing a modified Three Dimensional Printing process which injects streams of UV-curable binder resins under a UV light. And, investigators at The University of California Irvine (Orme, 1995), Incre Corp. (Burns, 1993), and MIT are separately developing jetting systems to deposit metal alloys.

3D Welding

Several groups are investigating the use of conventional welding processes to build up three dimensional shapes. The 3D welding process that is being developed at the University of Nottingham (Ribeiro, 1996) uses an arc welding robot to deposit successive layers of metal in order to create a 3D solid component. A version of this process which uses an e-beam welder, electron beam Solid Freeform Fabrication (EBSFF), is under development at M.I.T. (Davé, et al, 1995). Previous work in this area was performed at Babcock and Wilcox (Doyle, 1991) under the process name Shape Melting and in Germany in the 1960's (Kassmaul, 1983) under the name Shape Welding. In general, machining operations must be used to finish parts built up with welding processes in those cases where the parts require accurate dimensions or smooth surfaces.

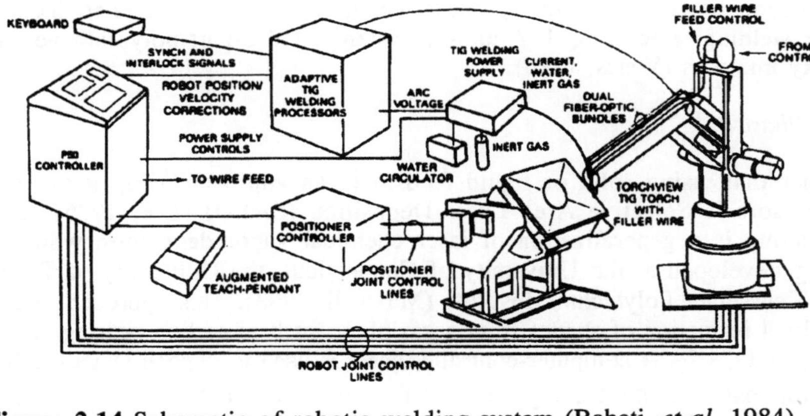

Figure 2.14 Schematic of robotic welding system (Baheti, *et al*, 1984)

Shape Welding grew out of a project sponsored by the German Federal Ministry for Research and Technology to cope with increased competition in the forging industry (Schmidt, et al, 1990, Gnriss, 1987). This project resulted in a production machine capable of producing large items, up to 10.5 m in length and 5.8 m in diameter and weighing up to 500 tons. Generally speaking, this technique requires considerable part specific tooling or human intervention either in terms of fixtures or multi-degree-of-freedom actuators. As an example, see the schematic of a robotic welding system shown in Figure 2.14. These types of systems require extensive process planning for parts with any geometric complexity. Due to this limitation, Shape Welding was not considered in the historical development of Chapter 1.

One reason for the complexity in the above process is the need for special purpose supports. Neither the laser cladding nor the 3D welding processes described above incorporate an automatic support structure. Carnegie Mellon and Stanford universities are developing an addition/removal process, Shape Deposition Manufacturing (SDM) (Fessler, 1996), which incorporates support structures. In SDM, a CAD model is first sliced into 3D layered structures (i.e., the outer surface of each layer maintains the 3D geometry of the original model). Layer segments are then deposited as near-net shapes and then machined to net-shape before additional material is deposited (Figure 2.15). The sequence for depositing and shaping the primary and support materials is dependent upon the local geometry; the idea is to decompose shapes into layer segments such that undercut features need not be machined, but are formed by previously shaped segments. SDM can use alternative deposition sources. One process, microcasting (Merz, 1994), is a novel non-transferred welding process which deposits discrete, super-heated molten metal droplets in order to build up fully dense, metallurgically bonded structures. For example, stainless steel is deposited as the primary material and copper as the sacrificial material. Other types of deposition processes for this process which are being investigated include laser welding, gelcasting, UV curable systems, and 2-part polyurethane and epoxy mixtures (Weiss, 1996).

Gas Phase

Rather than using solid or liquid feedstock, an approach using gas phase precursors is Selective Area Laser Deposition (SALD) (Lee, 1995). This technique is a generalization of laser chemical vapor deposition which is being developed at the University of Connecticut, the University of Texas, and Rensselaer Polytechnic Institute (Maxwell, 1993). This approach results in direct formation of shapes from a gas phase such as carbon bearing gases, CH_4 and C_2H_2. A computer-controlled laser is used to deposit shapes either by thermal or photo decomposition of the gas. A schematic of the process is shown Figure 2.16.

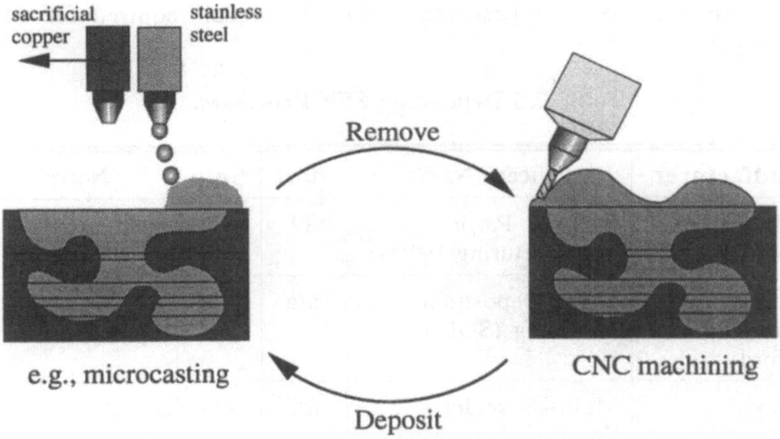

Figure 2.15 Shape deposition manufacturing.

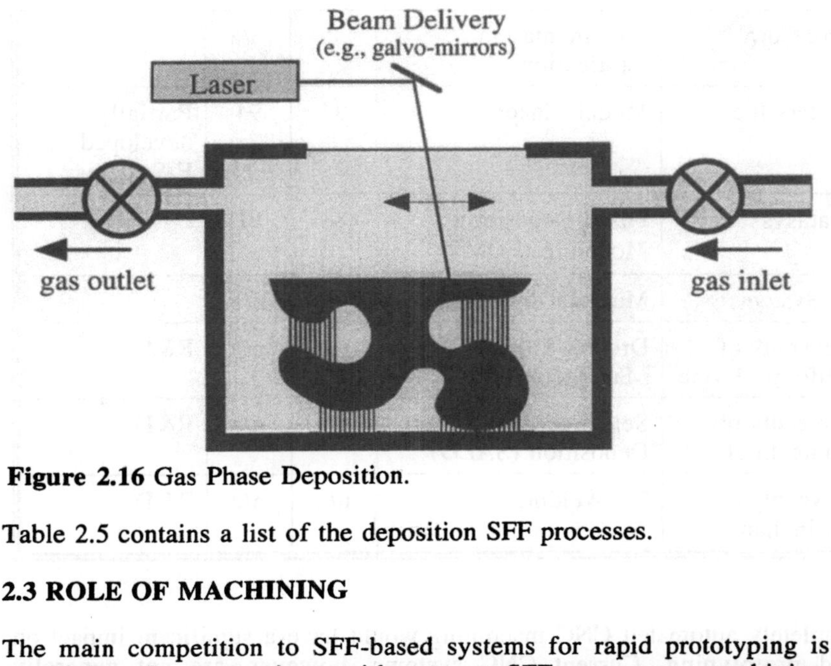

Figure 2.16 Gas Phase Deposition.

Table 2.5 contains a list of the deposition SFF processes.

2.3 ROLE OF MACHINING

The main competition to SFF-based systems for rapid prototyping is NC machining. In comparison with current SFF processes, machining can produce parts with superior accuracy and surface finish; in addition, machining processes can operate on a much broader range of materials, especially tooling steels. And, if only 2D drawings are available, then machining by a skilled model maker can often be executed more quickly than the time it takes to first create the 3D model required for SFF processing.

Table 2.5 Deposition SFF Processes.

Manufacturer	Process Name	Start	Ship	Notes
BPM Technology	Ballistic Particle Manufacturing (BPM)	89	95	
Carnegie Mellon and Stanford	Shape Deposition Modeling (SDM)	n/a	n/a	R&D
Fraunhofer IFAM	Multi-Phase Jet Solidification (MPJS)	n/a	n/a	R&D
MIT	Electron Beam Solid Freeform Fabrication (EBSFF)	n/a	n/a	R&D
Incre Corp	Incremental Fabrication	?	n/a	
Sanders Inc.	Model Maker	94	94	Partially developed at E-Systems
Stratasys	Fused Deposition Modeling (FDM)	88	91	
3D Systems	Multi-Jet Modeling	?	n/a	
University of California Irvine	Droplet Stream Manufacturing	n/a	n/a	R&D
University of Connecticut	Selective Area Laser Deposition (SALD)	n/a	n/a	R&D
University of Nottingham	3D Welding	n/a	n/a	R&D

Completely automated CNC machining would have a significant impact on rapid prototyping. Current CNC systems, however, are not generally considered to be SFF technologies because: they still require skillful human intervention to help plan the operations and to operate the equipment; custom fixturing and special tooling is often required; and machining has inherent geometric limitations (see Figure 2.17). However, just as SFF process performance capabilities are expected to improve, automated CNC planning capabilities are also expected to continue to improve with the proliferation of 3D CAD modeling systems. In the future, both CNC and SFF will remain important technologies for prototyping needs.

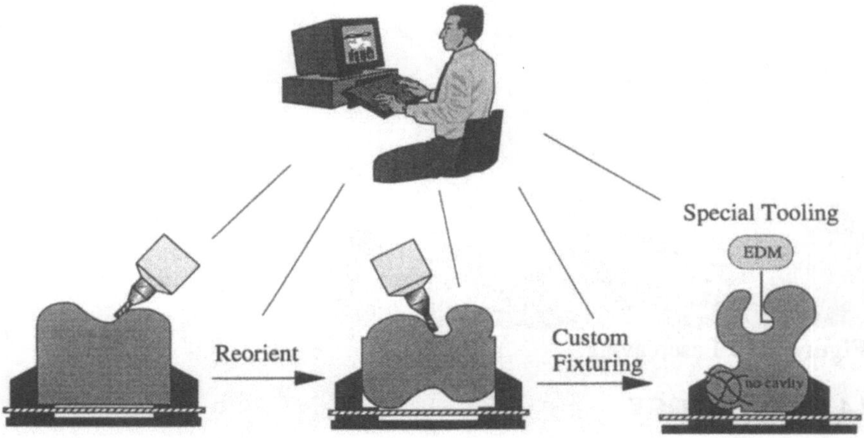

Figure 2.17 Machining as an SFF process?

In addition to improving the capabilities of automated CNC planning systems, there is also a need to improve the machining processes themselves. There is interesting work in both Japan and Germany on machining and other material removal processes. The main application area for these processes is in rapid tool manufacturing. Tooling fabrication relaxes some geometry constraints, e.g., undercut features are not normally required. Lasercaving (Eberl, 1992), which is manufactured by LCTec, Inc. (Germany) and being refined by the Bavarian Laser Center (BLZ), is a material removal process which can be used to cut cavities into metal or ceramic stock in a downward, layer-by-layer fashion. A high-power laser and oxygen source are simultaneously directed at the surface to be cut, and swept across the cutting path. The heated metal oxidizes, and oxidized chips break away due to differential thermal expansion between the underlying unoxidized material and the oxide fragment (Figure 2.18). The system includes a 5-axis CNC vertical milling machine and a 750W CO_2 laser. Parts are mounted on the mill and moved relative to the fixed beam. The claimed accuracy is 0.05 mm, and the material removal rate is 5 mm^3/min. The overall machining times can be reduced by first roughing out material at a higher rate using the laser in a melting mode which removes material at a rate of 1000 mm^3/min. Surface roughness is about 5 microns. The primary applications which BLZ is exploring are cutting cavities in tools and texturing tooling surfaces.

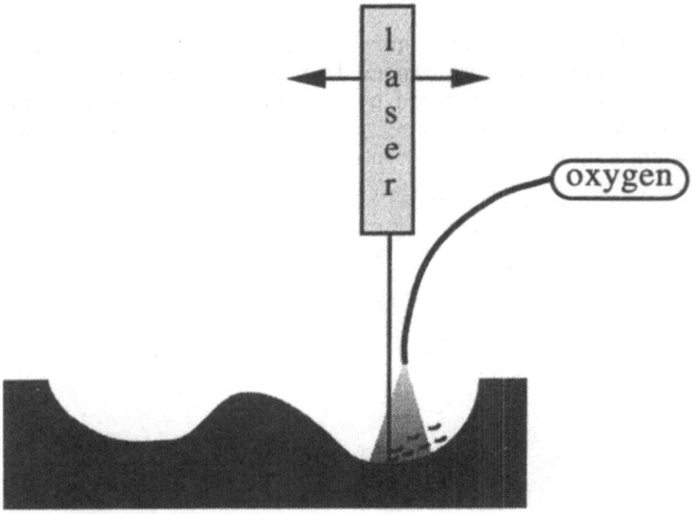

Figure 2.18 Lasercaving.

2.4 CHRONOLOGY

Figure 2.19 displays the overall chronology of Solid Freeform Fabrication. This chronology reflects the discussion found in Chapters 1 and 2, but it should not be considered complete; it indicates some (but not all) of the major time events in this rapidly changing field.

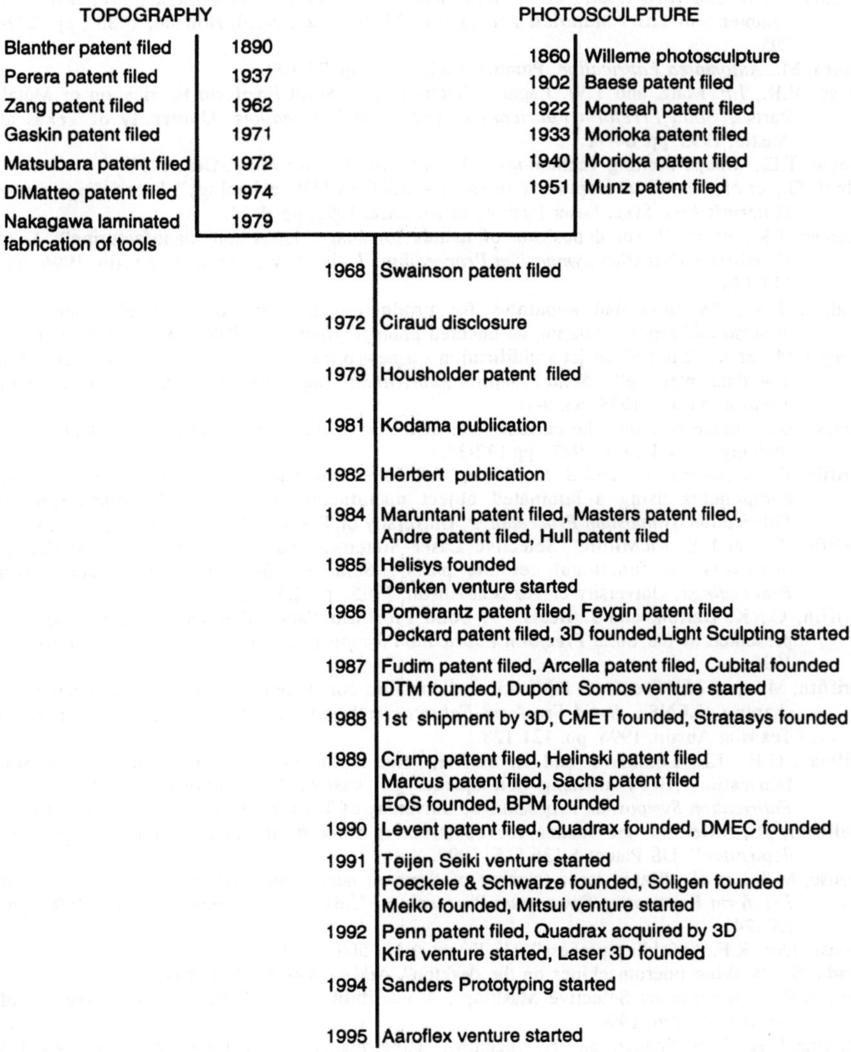

Figure 2.19 Solid Freeform Fabrication Chronology.

REFERENCES

Agarwala, M.K., et al, "Fused deposition of ceramics (FDC) for structural silicon nitride components", *Solid Freeform Fabrication Symposium Proceedings,* University of Texas at Austin, 1996, pp. 326-334.

Arcella, F., "Casting shapes", US Patent 4,818,562, 1989.

Baheti, R.S., K.B. Haefner, and L.M. Sweet, "Operational performance of vision-based arc welding robot control system", in *Sensors and Control for Automated Manufacturing and*

Robotics, ed. K.A. Stelson and L.M Sweet, ASME,1984, pp. 93-106.

Breinan, E.M. and B.H. Kear, "Rapid solidification laser processing at high power density", Chapter 5 in *Laser Materials Processing* , M. Bass ed., North-Holland, 1983 , pp. 237-295.

Burns, M., *Automated Fabrication*, Prentice-Hall, 1993, pp. 97-100.

Davé, V.R., J.F. Matz, and T.W. Eagar, "Electron beam Solid Freeform Fabrication of Metal Parts", *Solid Freeform Fabrication Symposium Proceedings*, University of Texas at Austin, 1995, pp. 64-71.

Doyle, T.E., "Shape Melting Technology", *3rd Intl Conf Desktop Manf*, Oct., 1991.

Eberl, G., *et al*, "New developments in the LASERCAV**R technology",*Proceedings Laser Materials Proc Sym.*, Laser Institute of America, 1992, pp. 1-14.

Fessler, J.R., *et al*, "Laser deposition of metals for shape deposition manufacturing", *Solid Freeform Fabrication Symposium Proceedings*, University of Texas at Austin, 1996, pp. 112-120.

Fudim, E.V., "Method nad apparatus for production of three-dimensional objects by photosolidification; radiating an uncured photopolymer", US Patent 4,752,498, 1988.

Geiger, M., *et al*, "Multiphase jet solidification - a new process toward metal prototypes and a new data interface", *Solid Freeform Fabrication Symposium Proceedings*, University of Texas at Austin, 1994, pp. 9-16.

Gnirss, G., "Shape welding: the current state of the art in the Federal Republic of Germany", *Welding in the World*, 1987 , pp.132-38.

Griffin, C., J. Dautenbach, and S. McMillin, "Solid Freeform Fabrication of functional ceramic components using a laminated object manufacturing technique", *Solid Freeform Fabrication Symposium Proceedings*, University of Texas at Austin, 1994 , pp. 17-24.

Griffin, E. and S. McMillin, "Selective Laser Sintering and Fused Deposition Modeling processes for functional ceramic parts", *Solid Freeform Fabrication Symposium Proceedings*, University of Texas at Austin, 1995 , pp. 25-30.

Griffin, C., K. Barton, and J. Beals, "A Solid Freeform Fabrication technique for metals", presented at the *Solid Freeform Fabrication Symposium*, University of Texas at Austin, 1996.

Griffith, M.L., *et al*, "Freeform fabrication of metallic components using laser engineered net shaping (LENS), *Solid Freeform Fabrication Symposium Proceedings*, University of Texas at Austin, 1996, pp. 121-128.

Hilmas, G.E., J.L. Lombardi, and K. Stuffle, "Recent developments in extrusion freeform fabrication (EFF) utilizing non-aqueous gel casting formulations", *Solid Freeform Fabrication Symposium Proceedings*, University of Texas at Austin, 1996. Pp. 432-439.

Helinski, R., "Method and means for constructing three-dimensional articles by particle deposition", US Patent 5,136,515, 1992.

House, M.A., *et al*, "Rapid laser forming of titanium near shape articles: LaserCast", *Solid Freeform Fabrication Symposium Proceedings*, University of Texas at Austin, 1996, pp. 232-241.

Housholder, R.F., "Molding process", US Patent 4,247,508, 1981.

Inada, S., "Making micromachines on the desktop", *Nikkea Weekly*, July, 1995.

Jog, A.V., "Sintering by Selective Masking : A Feasibility Study," *MS Thesis*, University of Texas at Austin, 1994.

Kassmaul, K., F.W. Schoch, and H. Lucknow, "High quality large components: Shape Welded by a SAW process", *Welding Jour*, Sept, 1983.

Klocke, F., H. Wirtz, and W. Meiners, "Direct manufacturing of metal prototypes and prototype tools", *Solid Freeform Fabrication Symposium Proceedings*, University of Texas at Austin, 1996, pp. 137-144.

Klosterman, D., *et al*, "Structural composites via laminated object manufacturing", *Solid Freeform Fabrication Symposium Proceedings*, University of Texas at Austin, 1996, pp. 100-111.

Lee, Y.L., *et al*, "Deposition rates of silicon carbide by Selective Area Laser Deposition", *Solid Freeform Fabrication Symposium Proceedings*, University of Texas at Austin, 1995, pp. 433-436.

Lewis, G.K., "Direct laser metal deposition process fabricates near-net-shape components rapidly", *Materials Tech,* March, 1995.

Masters, W.E., "Computer automated manufacturing process and sytem", US Patent 4,665,492, 1987.

Merz, R., Prinz, F.B., Ramaswami, K., Terk, M. and Weiss, L.E., "Shape Deposition Manufacturing," *Solid Freeform Fabrication Symposium Proceedings*, The University of Texas At Austin, 1994, pp. 1-8.

Maxwell, J.L., J. Pegna, and A.G. Ostrogorsky, "Thermal analysis and modeling of steady-state rod growth during gas-phase Solid Freeform Fabrication", *Solid Freeform Fabrication Symposium Proceedings*, University of Texas at Austin, 1993, pp. 253-270.

Nakagawa, T., *et al*, "Laser cut sheet laminated forming dies by diffusion bonding", *Symposium Proceedings 25th IMTDR,* Univ Birmingham, 1985, pp. 505-510.

Orme, M.E. and C. Huang, "Thermal design parameters critical to the development of Solid Freeform Fabrication of structural materials with controlled nano-liter droplets", *Solid Freeform Fabrication Symposium Proceedings*, University of Texas at Austin, 1995, pp. 88-95.

Penn, S.M., "System, method, and process for making three-dimensional objects", US Patent 5,260,009, 1993.

Prinz, F. ed., *JTEC Panel Report on Rapid Prototyping in Japan and Europe,* Loyola College, 1996.

Sshmidt, J., *et al*, "Manufacture of complex parts by shape welding", *Jour Nuclear Materials*, 1990 , pp. 120-27.

Ventura, S.C., *et al*, "A new SFF process for functional ceramic components", *Solid Freeform Fabrication Symposium Proceedings*, University of Texas at Austin, 1996, pp. 318-325.

Wei, P., *et al,* "Metal and ceramic components made via CAM-LEM technology", *Solid Freeform Fabrication Symposium Proceedings*, University of Texas at Austin, 1996, pp. 366-373.

Weiss, L.E., Neplotnik, G., Prinz, F.B., Schultz, L., Padmanabhan, P, Krishnan, R., and Merz, R.., "Shape Deposition Manufacturing of Wearable Computers," *Solid Freeform Fabrication Symposium Proceedings,* University of Texas At Austin, 1996, pp. 29-36.

3

INFORMATION PROCESSING

Producing parts by layer-based manufacturing requires the transformation of the geometric description of a part into a form suitable for processing by the particular SFF process. The process proceeds by first slicing the geometric description of the part into layers. The slicing operation generates the contours of the part for each layer. The contours are then processed in a manner dependent upon the particular SFF technology. For instance, for Selective Laser Sintering (SLS), the contours are discretized into "toggle points" at which the laser beam must be modulated to produce the desired solid. In this chapter, we discuss the representations of parts that are required to support geometric processing for SFF, and the interfaces between geometry sources and SFF software. We also describe slicing algorithms and processing within layers. The chapter ends with a short discussion of process planning for SFF based on geometric considerations.

3.1 GEOMETRIC MODELS FOR SFF

Producing a part by SFF requires a complete, unambiguous mathematical description of the part's geometry. This means that users of SFF must generate valid surface or solid models of parts to fabricate them successfully. Two types of modelers are currently used by designers to generate the necessary geometric description: surface modelers and solid modelers. Each of these is discussed briefly, along with other sources of geometric data which are inherently layer-based.

3.1.1 Solid Modelers

Solid modeling technology was developed in the early 1980's to overcome limitations of early wireframe CAD systems. Solid models are distinguished by their ability to determine unambiguously whether a point lies inside, outside, or on the surface of the part. This is accomplished by providing explicit topological information about the part, or by constructing the part from the combination of simple solids with rigorously defined set operations that ensure the validity of the result. While many schemes for representing solids have been developed over the past decade, two schemes have emerged as dominant representations in commercial CAD systems: boundary representations (Brep) and constructive solid geometry (CSG).

In a boundary representation, the boundaries of the part are explicitly represented in terms of both topology and geometry. The representation is based on a graph structure, in which the faces, edges, and vertices of the part are nodes. Arcs between the nodes represent adjacency relationships. For instance, an arc between two face nodes indicates the two faces share an edge, and the adjacency arc contains a reference to the edge shared by the faces. Arcs between edges indicate connectivity between them. Geometric information is attached to each topological node; face nodes contain surface equations, edge nodes contain curve equations, and vertex nodes contain coordinates of points. A Brep is also built around an orientation convention, usually the outward pointing normal of each face. This information allows the Brep to answer inside/outside/on queries.

Constructive solid geometry is often used as a construction technique, but CSG is also a valid representation scheme for solids. With CSG the geometry of a part is modeled as a binary tree whose leaf nodes are scaled and oriented instances of primitive shapes, and whose non-terminal nodes are the regularized Boolean set operations union, intersection, and difference. The primitive set consists of surfaces for which intersections can be computed, in closed form or with robust numerical algorithms, and usually includes the parallelepiped, quadric surfaces (*e.g.*, sphere, cylinder, cone, etc.), and the torus. Other surfaces provided include those generated by revolving or linearly sweeping 2D profiles. The set operations are termed regularized because they are modified to remove certain non-solid artifacts, such as dangling edges and repeated faces. The CSG scheme is a much more compact representation than the Brep, but it is an implicit, unevaluated representation. For many applications, such as finite element mesh generation, which requires the part's boundaries, the CSG tree must be evaluated to produce a boundary representation of the solid.

3.1.2 Surface Modelers

For many applications, such as airfoil design or design of automobile body surfaces, solid models are not needed. These applications are marked by the need to design flexible, sculptured surfaces. Modelers for these applications typically provide the ability to interactively define a complex surface in terms of a series parametric surface patches "stitched" together such that continuity conditions are maintained at the patch boundaries. Typical parametric surface forms include Bézier surfaces and B-spline surfaces, including non-uniform rational B-spline (NURBS) surfaces. The main difference between surface models and solid models is the lack of explicit topological information. This lack makes certain validity checks difficult with surface models.

3.1.3 Features

Another geometric representation that is gaining acceptance among engineering designs is based on the notion of features. Although a general formal definition of a feature is elusive, a *form feature* may be informally defined as a collection of low-level geometric and topological entities (typically edges and faces) with which higher level semantic information may be associated. The notion of features grew out of the desire to perform automated manufacturing process planning from a solid model of a part (Kyprianou, 1980, Henderson, 1984). In light of this application, features have historically been viewed as the results of unit manufacturing processes, most typically machining processes. Thus, typical features in feature-based modeling systems include holes, slots, pockets, steps, etc.

Initial efforts to utilize features in process planning focused on feature recognition or feature extraction, in which various algorithms are applied to the solid model (usually a boundary representation) to locate features in the part. To date, no general algorithm for feature recognition has been demonstrated. Approaches that have been investigated include artificial intelligence methods (Henderson, 1984, Vandenbrande and Requicha, 1990), graph-matching algorithms (DeFloriani, 1989, Falcidieno and Giannini, 1989, Gadh and Prinz, 1992, Gahd, 1993, Joshi and Chang, 1988, Rosen *et al*, 1992), and surface extension/volume decomposition methods (Kim, 1992, Kim, 1994, Sakurai, 1994, Sakurai and Chin, 1994, Sakurai and Gossard, 1990, Shah *et al*, 1994, Shen and Shah, 1994, Tseng and Joshi, 1994, Woo, 1992). Problems with general feature recognition are related to the domain-specific nature of features (i.e., the features of interest depend upon the application for which they are used) and the combinatorial nature of the recognition and matching process. With these limitations in mind, many researchers have investigated feature-based design, in which the part definition is created with features (Cunningham and Dixon, 1988, Cutkosky *et al*, 1988, Shah *et al*, 1990, Turner and Anderson, 1988). Provided that the features used in creating the geometry are the features of interest for the application, this approach obviates the need for computationally expensive recognition algorithms. Additionally, users generally agree that features provide intuitive construction interfaces for creating complex geometry. Commercial feature-based design systems are beginning to appear, most notably the Pro/ENGINEER system from Parametric Technology, Inc., Waltham, MA.

The impact of features on Solid Freeform Fabrication is subtle, yet distinct. Pro/ENGINEER is a popular geometric modeling system that is used by many major manufacturers who are users of SFF technology. Most SFF service bureaus will accept Pro/ENGINEER files as part descriptions. Additionally, Pro/ENGINEER has one of the more robust tessellation implementations among CAD systems, and reliably produces accurate faceted models for those SFF technologies that accept only such input (as described in Section 3.2).

Suh and Wozny (1995) have investigated the use of features for SFF. The authors specify a system based on a *fundamental features* that are used as building blocks to define features of interest. Additionally, *fundamental spatial relations* are used to represent relationships among features. Defining features based on the set of fundamental features simplifies the feature recognition problem, and facilitates feature translation from one domain of interest to another (*e.g.*, from design features to SFF features to injection molding features). The authors illustrate the usefulness of features for SFF with an application that computes the optimum orientation for building a part. They demonstrate reasoning about orientation with respect to such criteria as the volume of trapped liquid (for SFF processes with liquid precursors), support structure minimization, part strength, surface finish, dimensional accuracy, and build time.

3.1.4 Layered Geometry

Many potential applications of SFF naturally provide data in layer-based formats. For example, Rogers *et al* (1991) report the use of a laser digitizer to provide data for the fabrication of prosthetic devices using Selective Laser Sintering. Levy *et al* (1992) used Computed Tomography (CT) data to produce models of human temporal bones using SLS. Bartels *et al* (1993) used imaging data from a confocal microscope as input to an SLS machine to create models of pollen grains. These examples illustrate the potential of SFF for applications in areas other than product engineering. In each of these cases the data is presented to the SFF machine as layers.

Slice data can be used in SFF processes by first reconstructing the surface from the data, a problem that has been researched well (Hoppe *et al*, 1992, Pavlakos *et al*, 1989). The resulting surface is then processed in the same manner as data from a CAD system. However, the first operation on the reconstructed surface is reslicing. Intuitively at least, it seems that a more efficient approach is using the original slice data to fabricate the part. In most cases, however, the thickness of the data slices does not coincide with the thickness required for the SFF process (typically 0.005 to 0.01 in). One strategy to compensate for this is for the SFF control software either to skip data layers (for thinner layers) or to replicate layers (for thicker layers). To achieve higher accuracy, however, more sophisticated strategies must be developed and adopted by the SFF software. The laser digitizer software described in (Rogers *et al*, 1991) actually interpolates between slices to realize the required resolution for the SLS process. Similar schemes must be developed for processing CT and confocal microscope imaging data.

Both CT data (Levy *et al*, 1992) and confocal microscope data (Bartels *et al*, 1993) consist essentially of raster images of the physical objects being imaged. In each case, the data provides a measure of the relative density of the material at each pixel in the imaging plane. However, SFF processes require boundary information rather than interior density data. In the studies cited above, the raster images were first converted to contour images of the sample slice in each plane by specifying a minimum threshold density below which the data are ignored. For these studies, the threshold value was selected manually and varied from layer to layer. However, to realize the full potential of fabricating such models with SFF, threshold values must be determined automatically. Algorithms for determining optimal threshold values are needed for preprocessing layer-based SFF input.

3.2 INTERPROCESS INTERFACES

3.2.1 Faceted Geometry

The form of the geometric description of mechanical parts to be produced by SFF significantly affects the accuracy of the final part. The current state-of-the-art for most SFF technologies consists of tessellating the surfaces of the geometric model into a mesh of non-overlapping triangular facets. The resulting geometry is transmitted in a standard file format, the so-called "STL" file format, established by 3D Systems, Inc. (3D Systems, 1988) This format has been adopted by many CAD vendors, is readily available, and is considered adequate for most visualization applications. However, for producing accurate patterns and functional parts, the adequacy of the STL format is unclear (Darrah and Wielgus, 1990). There is a trade-off between the accuracy and the size of the geometric description. Highly non-linear surfaces, such as those that comprise turbine blades, manifolds, etc., must be tessellated into a large number of small facets, resulting in very large data files, and the accuracy of such descriptions is still suspect.

In many cases, the tessellation operation introduces errors in the model. Tessellation of surfaces with large curvature can result in errors at the intersections between such surfaces (Wozny, 1992), leaving gaps or "holes" along edges of the part model (see Figure 3.1). Tessellation of fine features is susceptible to round-off error, which leads to non-manifold models of parts, where more than two facets are adjacent to a single edge, or facets with opposing outer normals meet at a single vertex (see Figure 3.2). These problems are difficult for slicing algorithms to handle and cause fabrication problems for SFF processes, which require valid solids as input. Bøhn and Wozny (1992) have developed a method to repair faceted geometric models. Their technique uses the adjacency information to locate gaps between facets. They then use a "shell-closure" algorithm to add missing facets to the close gaps.

Mäkelä and Dolenc (1993) likewise use adjacency information to detect and repair errors in faceted models. The authors use a spatial subdivision approach to develop a representation of the object that explicitly models vertex and edge adjacencies. Using this data structure, their algorithm can identify gaps between facets, detect interior facet intersections, and compute correct facet normals. Simple cases of gaps are repaired by appropriately connecting sequences of vertices into triangles, using normal directions to handle concave vertices. The authors state that an implementation of their technique is available as a commercial product.

Besides CAD vendors, several software companies offer products based on the STL format. Imageware (Ann Arbor, MI) supports three-dimensional point data manipulation and surface reconstruction with the Surfacer product. The RPM module provides tools to generate, visualize, modify, repair, and verify STL data files. Materialise (Belgium and Ann Arbor, MI) provides software tools, as well as prototyping services. The Magics product provides visualization and manipulation of STL data, as well as interactive support generation. Contour Tools supports repair of STL files at the contour level. Solid Concepts, Inc. (Valencia, CA) offers an optimized STL slice program.

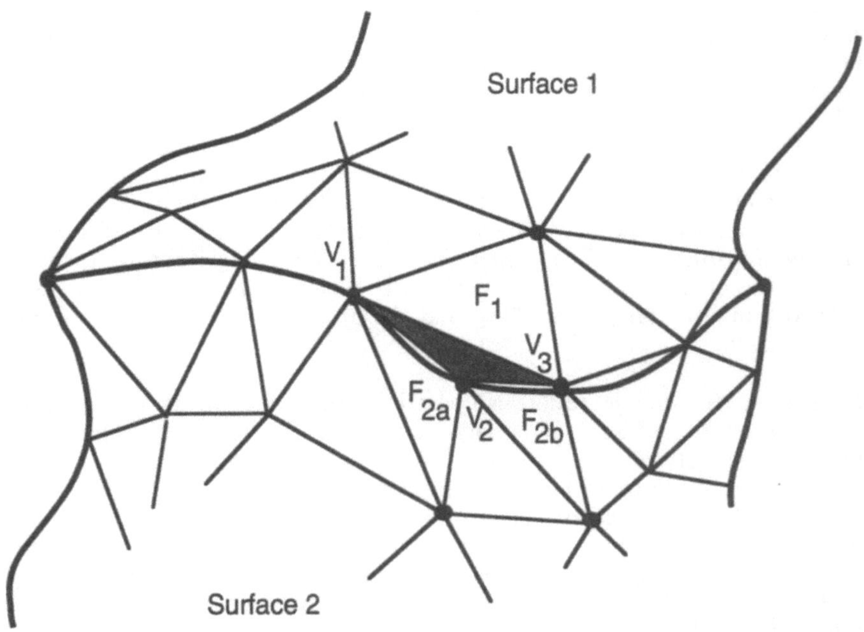

Figure 3.1 A gap in the seam at the intersection of two surfaces (Bøhn and Wozny, 1992).

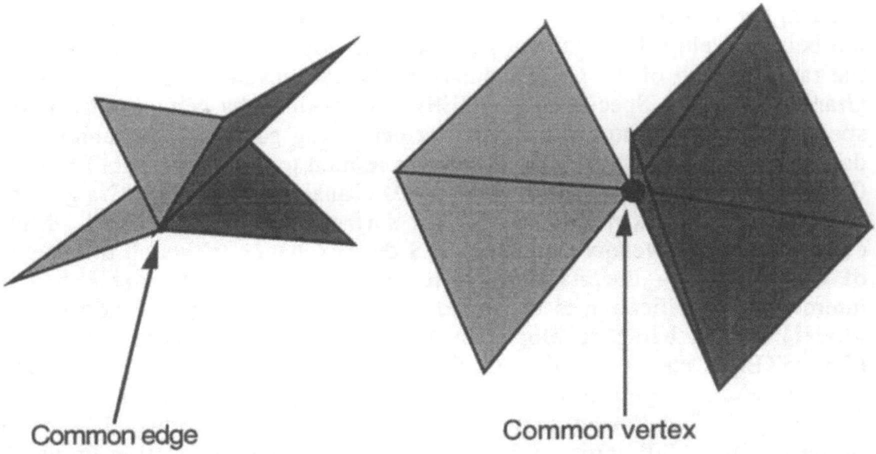

Figure 3.2 Other examples of non-manifold topology in a faceted SFF model.

3.2.2 Higher Order Geometric Descriptions

Processing higher order geometry for SFF processes offers several advantages over exchange of faceted geometric descriptions. Generally speaking, files containing higher order information will be smaller and more accurate than comparable faceted geometry files. Also, many of the problems that result in non-manifold geometric information in faceted descriptions can be avoided. Instead, it is incumbent upon the SFF geometry processor to ensure that the results are realizable for the particular SFF technology that is used to fabricate the part. Potential problems in the slicing operation can be solved because more information is available about the intended geometry of the part; thus, higher order descriptions are easier to troubleshoot when necessary. Finally, when approximations are necessary for the given input geometry, the approximation process is driven by the particular SFF technology rather than by generic criteria meant to satisfy the requirements of many SFF technologies. This provides a rational basis for approximating the geometry when necessary.

There are disadvantages to higher order geometric data exchange as well. First, there is no single geometry form that is satisfactory for all applications. There are many different geometric descriptions that are used in product design, each with different requirements for a slicing algorithm. Designers of commercial SFF processing software will have to make compromise decisions about which geometric forms to support or risk losing potential customers from lack of geometric coverage. Also, because the geometric input is more complex, algorithms for processing the geometry are more complex as well.

Finally, the SFF industry and the CAD industry must agree on standards for exchanging higher order geometric data. Fortunately, interchange standards are being developed for general geometric descriptions. In the early 1980's, the rapid growth of the CAD industry led to the development of the Initial Graphic Exchange Specification (IGES). The philosophy behind IGES is the specification of a neutral file format for describing geometric and annotation data in a CAD model. Initially IGES was limited to wireframe data from 2D CAD drafting systems. IGES Version 4.0 (Panzica *et al*, 1988, Craig *et al*, 1992) now specifies representations for surfaces and solids as well. Most commercial CAD vendors support IGES data exchange, although the degree of compliance to the standard varies widely. Another emerging data interchange specification is the Standard for the Exchange of Product Data Model (STEP), being developed by the International Standards Organization. STEP is meant to supersede the IGES format by providing the means for representing much more information about a product than its geometry, including manufacturing technology, materials, and product life cycle support. The STEP approach is to establish so-called Application Protocols (AP) to allow partial implementation of the standard as it develops. Each AP contains data requirements, testing scenarios, and a test suite for the particular application. Currently, APs have been established for both boundary representation and surface modeling, with another AP being developed for CSG modeling (Ryan, 1992). Both IGES and STEP represent viable alternatives to the STL format. Bjørke (1991) has demonstrated the use of both these standards, as well as others, for stereolithography. Both DTM's and Stratyays's software accept IGES models as input. Gilman and Rock (1995) describe an architecture that uses STEP to provide CAD data to a SFF system. Carleberg (1994) proposes an extension to the existing boundary representation application protocol for STEP that includes process parameters, scanning patterns, and material data.

3.2.3 Layered Data Interfaces.

Before the use of layered data (CT, digitized, etc.) can expand, data interchange standards must be established. Currently there are no standards to specify the form of layer-based geometry for exchange among programs, a necessity for accommodating such data in SFF processing. In the examples cited in Section 3.1, Rogers presented the data to the SLS machine in the form of contours (Crawford *et al*, 1991). The other two sources, as described above, present their data as raster bit maps. These examples suggest that two exchange standards should be developed: one for contour data and one for raster data. In any case, the details for processing the input geometry, *i.e.*, interpolation or thresholding as described above, should be performed with respect to the particular SFF technology to be used to ensure that maximum part quality can be realized. Reconstruction of surfaces represented by layered data avoids these problems, allowing the use of faceted or higher order geometric interchange standards already established.

The burgeoning interest in SFF technologies for medical, reverse engineering, and forensic applications has produce some initial efforts at establishing standards for representing layered geometric data. Recently, 3D Systems has announced the "SLC" data format (Jacobs, 1996). The SLC format represents a part in terms of 2-1/2D layers that describe the boundaries of the part as polygonal contours. The format does supports up to 256 different layer thicknesses per part. New SLA processing software accepts both the SLC and STL data formats and processes each appropriately. The Common Layer Interface (CLI) is a layered geometry data format for SFF developed by researchers in Europe. Like the SLC format, CLI describes parts as 2-1/2D layers in terms of polygonal contours. The format allows layers to have different thicknesses. Additionally, CLI supports optional layer hatch pattern descriptions, where a hatch consists of a set of independent straight lines in the layer, each described by the coordinates of its start and end points.

3.3 SLICING ALGORITHMS

3.3 Slicing Algorithms

Geometric processing for SFF proceeds by first slicing the geometric description of the part into layers representing the build planes, typically at 0.005-0.010 in intervals (see Figure 3.3). The slicing operation generates the curves, or contours, that represent the boundaries of the part for each layer (see Figure 3.4). Slicing consists of computing the intersection of the geometry of the part with a series of planes. The intersection calculation method depends upon the form of the input geometry; specific examples are discussed below.

3.3.1 Slicing Faceted Representations

A faceted representation of a part, such as the STL format, is by definition an approximation of the surface geometry. The main advantage of this representation is the simplicity of the intersection calculations for slicing such models. Conceptually, the slicing operation consists of computing the intersection of the slicing plane with each of the planar facets, resulting in a set of line segments. The triangular facets used in the STL format simplify the calculations further, since each facet is necessarily convex. Mathematically, determining the intersection of a triangular facet and a slicing plane reduces to computing the intersection of the slicing plane with each of the three line segments that define the triangular facet. If the coordinates of the facet vertices are denoted (x_i, y_i, z_i, $i = 1\ldots3$), and the slicing plane is given by $z = z_0$, where z_0 is the height of the slicing plane, then the coordinates of the intersection point are determined by solving the following equations:

$$\frac{z_0 - z_i}{z_{i+1} - z_i} = \frac{x_0 - x_i}{x_{i+1} - x_i} = \frac{y_0 - y_i}{y_{i+1} - y_i} \qquad (3.3.1)$$

where (x_0, y_0, z_0) are the coordinates of the intersection point. Note that the indices of the coordinates in equation (3.3.1) are computed using modulo 3 arithmetic. For non-degenerate triangular facets, there will be two such intersection points for each slicing plane. These intersection points are then assembled in the proper order to produce the approximate planar contour of the part. Assembly of the final contour is the most complex part of the process, as care must be taken to properly account for multiply-connected contours that result from slicing non-convex objects, such as the example in Figure 3.3. Also, a robust slicing algorithm must handle the special cases where a vertex, edge, or entire facet lie in the slicing plane.

Crawford *et al* (1991) have implemented another approach to slicing faceted geometries based on building a rich topological data structure that removes redundant data present in the input file and explicitly represents facet adjacency through a face-edge-vertex structure. The algorithm begins by determining the intersection of the current slicing plane with a given triangular facet. The most common case will result in an intersection with two of the facet's edges and generate one segment of the contour. The next facet to intersect is then chosen based on the explicit adjacency information in the geometric data structure. The contour edge generated by this facet is known to connect to the previous contour edge by virtue of the adjacency of their respective facets. This approach greatly simplifies the process of building the final contours, since the contours are built incrementally as the intersections are computed. The degenerate cases of facet-in-plane, edge-in-plane, and vertex-in-plane are handled separately, and require slightly more complicated decisions. The final step for each layer is determining the correct orientation for each contour. This operation is based on in-plane ray-tracing to determine which contours are contained by others. Orientation is then alternated between CCW and CW, with the outermost contours being oriented CCW. Rock and Wozny (1991b, 1992) have also developed a similar slicing algorithm to take advantage of the explicit adjacency information when constructing slice contours. Rock and Wozny (1991a) have also proposed an alternative to the STL file format that captures the adjacency information in the exchange file.

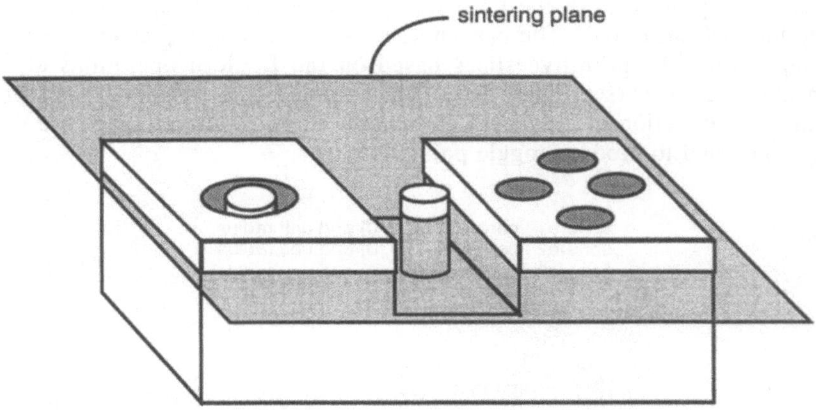

Figure 3.3 Slicing operation for a typical sintering plane.

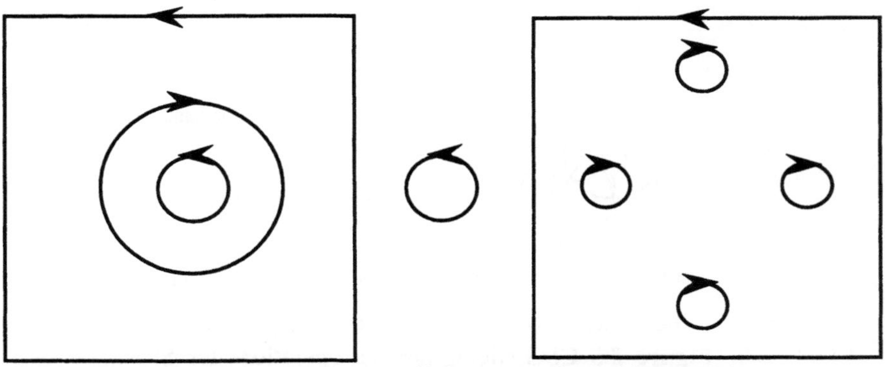

Figure 3.4 Oriented contours from slicing operation of figure 3.1.

3.3.2 Slicing Higher Order Geometry

The problems with the STL geometry exchange format arise because tessellation is a first-order approximation of more complex geometric entities. An obvious solution to these problems is to exchange higher order geometric entities, preferably the source geometry with which a part is designed. This approach has been adopted by Guduri *et al* (1992, 1993), where an interface based on constructive solid geometry (CSG) has been developed to provide input to SFF processes.

Normal evaluation of a CSG tree involves computing the intersection between pairs of primitive or intermediate surfaces and performing set classification in three dimensions. The process implemented by Guduri is based on evaluation of *sliced* primitives rather than 3-D primitives. In this process the primitives in the CSG tree are sliced individually, generating a slice for each primitive. The part contour in the slice plane is then produced by combining the primitive slices based on the Boolean operations at the non-terminal nodes (see Figure 3.5). The contour of the part in a given slice plane is a collection of piecewise continuous curves. These curves are then scan-converted to produce toggle points.

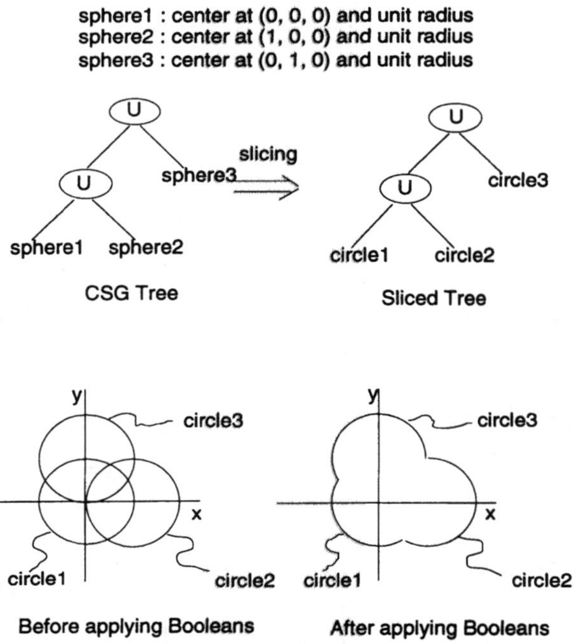

Figure 3.5 CSG slicing and set operations for SFF (Guduri *et al*, 1992).

The net result of this approach is a boundary evaluator developed specifically for SFF applications. The approach provides a more accurate, compact part description. Also, all surface-surface intersections are performed in the 2D slice plane, resulting in considerably simpler and more robust algorithms. For the common quadric surfaces (spheres, cones, cylinders, and ellipsoids), the surface-surface intersection calculations are exact (subject to round-off errors). For higher order surfaces, such as the torus, the method requires an approximation of the slice contour of the primitive. Such an approximation is still more accurate and efficient that the linear approximations obtained from a faceted model. In fact, the method provides a rational basis for approximating the geometry, since it is based on error measures in the slice plane.

Other researchers have also demonstrated the advantages of exchanging higher order geometry between CAD systems and SFF technologies. Gursoz *et al* (1990) describe the use of the NOODLES non-manifold geometric modeler, developed at Carnegie Mellon University, to slice solid models, and to generate hatch patterns as well. Drake (1990) describes research to develop a slicer in the Alpha_1 modeling system, which uses NURBS surfaces as the underlying geometric representation, to drive SLA machines directly. Chari and Hall (1993) have developed a slicer for CAD data in the form of NURBS surfaces that produces layer contours represented as NURBS curves in IGES format. They use a Newton-Raphson algorithm for computing surface and curve intersections, and ray casting to generate laser toggle points. Rajagopalan *et al* (1995) describe the use of the CAD system itself to slice the geometric model of a part for fabrication by stereolithography. The authors implemented their slicer within the I-DEAS modeler of SDRC Inc., Cincinnati, OH. NURBS surfaces are used as the underlying representation of objects. The slicing operation thus results in layer contours described as NURBS curves. These contours are processed numerically to generated hatch patterns for the SLA

3.3.3 Ray Casting

Another approach to processing geometric data that has intrigued researchers in the SFF community is based on a technique for high quality computer graphics image generation, namely, ray casting. In this technique, mathematical rays are "cast" from the image plane (one ray for each pixel in the image) and intersected with all objects in the scene (see Figure 3.6). Hidden surface removal is accomplished by sorting the intersections in increasing distance from the image plane and choosing the pixel color based on the properties of the surface of closest intersection. A variety of special effects are naturally incorporated in the ray casting algorithm, including reflection, refraction, and shadow generation.

For SFF processing, e.g., for SLS, rays are cast along the "fast axis" of laser travel in each slicing plane (see Figure 3.7). The intersections (entry and exit points) with the object data are the laser toggle points. With this approach, the slicing operation is in effect bypassed, and ray casting produces the desired laser scanning pattern for each layer of the part. In fact, any desired scanning pattern composed of straight lines can be generated with ray casting.

The appeal of ray casting is based on several advantages. For instance, algorithms are available for intersecting many different types of geometry with rays. For CSG models, rays can be cast at primitives and Boolean operations can be performed on the rays. Ray casting algorithms are also amenable to hardware implementation, potentially providing dramatic speed enhancement in SFF geometry processing. However, the speed of ray casting is dependent on the number of entities that must be intersected. For instance, STL file processing is relatively slow due to the large number of facets necessary for reasonable accuracy of the part description. Also, for SFF systems that require layer contour information, ray casting requires post-processing to determine contours for each slice.

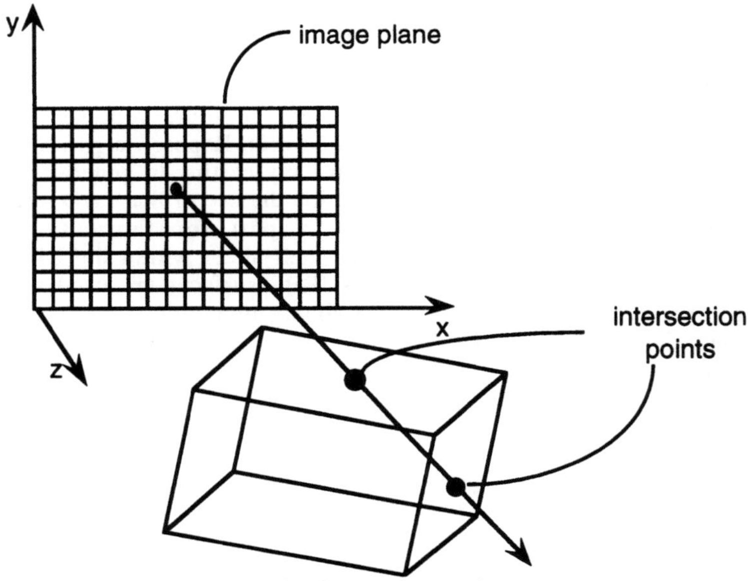

Figure 3.6 Ray casting for image generation.

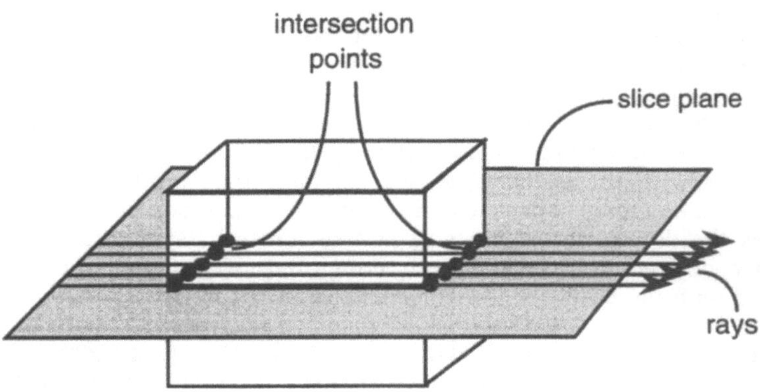

Figure 3.7 Slicing based on ray casting.

3.3.4 Other Considerations For Slicing

Research continues to improve the efficiency and accuracy of slicing algorithms. Kirschman and Jara-Almonte (1992) describe a parallel slicing algorithm that gives superlinear speed improvements over a serial processor implementation of the same algorithm. Dolenc and Mäkelä (1994) describe heuristics to correctly handle planar areas in the build plane of the part, as well as approaches to improving the aliasing due to discretization of the part surface during slicing. The authors also describe a method for offsetting the part surface for creating hollow tooling patterns. Sachs *et al* (1994) use procedural CAD models to represent surface textures that are manufactured by 3D Printing. The applications include textures with sub-millimeter sized features.

The slicing operation required for layered SFF technologies introduces errors in final geometry and surface finish. This is the so-called "stair-step" or "aliasing" effect caused by approximating angled surfaces with stacked layers of material. This source of error is exacerbated by the constant slice thickness assumed by most SFF control software. Aliasing can be reduced by computing variable, or adaptive, slice thicknesses based on, for instance, the curvature of the part's surface. This method is illustrated in Figure 3.8. Suh and Wozny (1994) propose such an approach. Their implementation operates on the source geometry of the part, which is transferred in the STEP format. In their approach, the thickness of the current layer is based on the curvature of the surface at a set of sample points on the previous layer. Successive sample points are computed by approximating the contour of the previous layer as an arc of radius $\rho = 1/\kappa$, where κ is the curvature of the contour at the current sample point. Likewise, the surface curvature at each sample point is computed based on approximating the surface locally as a sphere. Their algorithm accounts for small features and horizontal planar faces. The authors report that, in many cases, adaptive slicing results in improved accuracy with fewer total slices, resulting in increased build speed.

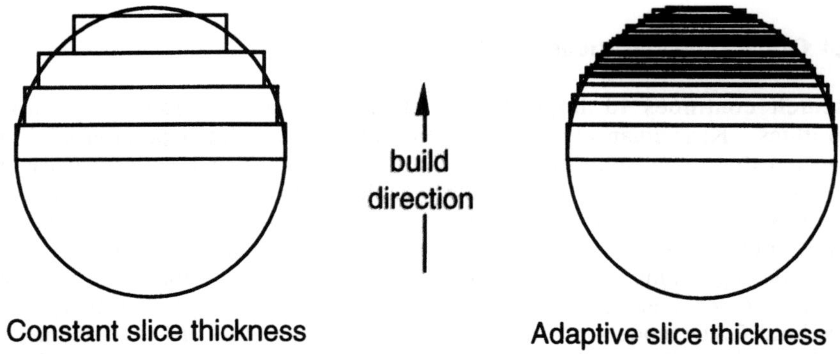

Figure 3.8 Effect of adaptive slice thickness on part accuracy

Thompson and Crawford (1995) propose a metric for part surface quality with respect to aliasing for parts approximated by triangular facets. The metric is meant to characterize the influence of part orientation on surface quality by computing the difference in volume between the intended part and the layered approximation. This volume difference can be approximated on a face-by-face basis for a given orientation, and then summed to give a total "alias volume", $V_{a,t}$. There are two cases to consider in computing $V_{a,t}$: when the height, h, of a face projected onto the build direction, **b**, is less than one layer thickness, l_t, and when h is larger that l_t.

When the projected height is larger than one layer, the alias volume can be written as:

$$V_a = \left\{\frac{A_S}{L}\right\}\left\{\left(\frac{l_t}{2}\right)^2 \tan\theta\right\}\left\{\frac{L\sin\theta}{2l_t}\right\}, \qquad (3.3.2)$$
$$= \frac{A_S l_t \tan\theta \sin\theta}{8}$$

where A_S is the surface area of the face, L is the length of the line formed by projecting the face onto a plane normal to the face and the build direction, and q is the angle between the face and the build direction. These dimensions are illustrated in Figure 3.9. The first term in equation (3.3.2) is the effective width of the face. The second term is the area of two triangular alias volumes enclosed in one layer thickness. The third term is the number of triangular alias volumes on the surface, divided by 2 to account for the second term. Equation (3.3.2) may be taken as a minimum volume since the conditions at the boundaries of the face may add volume. Clearly, the larger a face is, the more accurate the metric becomes.

When a face is smaller than a layer thickness for a given build direction, the alias volume is minimally:

$$V_a = \left\{\frac{A_S}{L}\right\}\left\{\frac{(Lcos\theta)(Lsin\theta)}{2}\right\},\qquad (3.3.3)$$
$$= \frac{A_S L sin2\theta}{2}$$

with the dimensions shown in Figure 3.10. Computation of the surface area, A_S, for a triangular facet is straightforward. The "height" of the surface is determined by projecting the facet edges onto the build direction. The distance that must be compared to to layer thickness, l_t, is then:

$$Lcos\theta = max\{\mathbf{V_i}\cdot\mathbf{b} - \mathbf{V_j}\cdot\mathbf{b},\ \forall i,j \in [0,1,2]\}\ ,\qquad (3.3.4)$$

where $\mathbf{V_i}$ is the *i*th vertex of the facet, and **b** is the build direction.

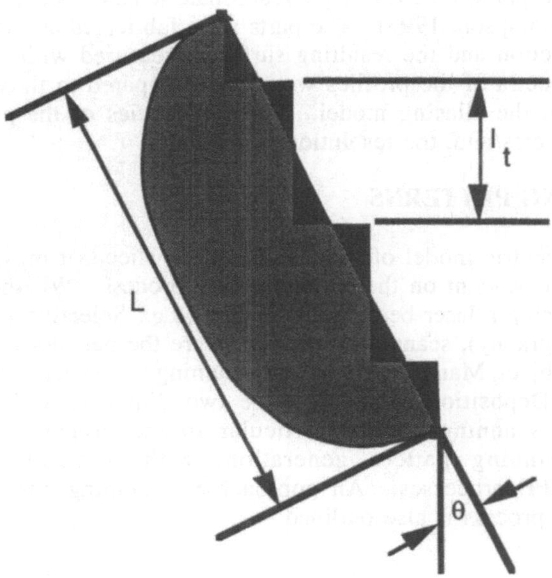

Figure 3.9 Alias volume when the face height *h* is larger than l_t.

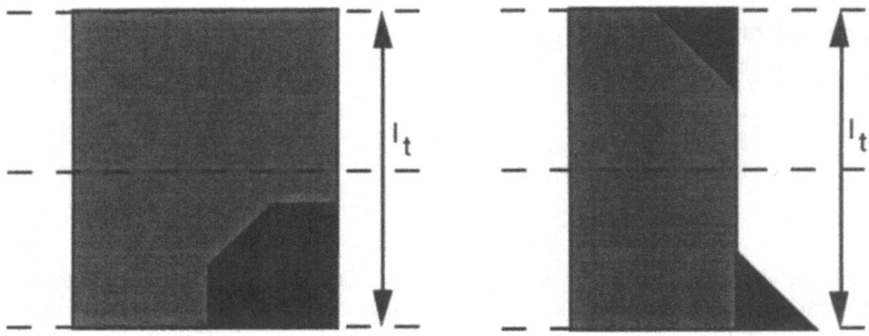

Figure 3.10 Alias volume when the face height h is smaller than l_t

To ensure that the alias volume modeled is a significant phenomenon on the surface of SFF parts, a series of polycarbonate tensile specimens were made using SLS (Thompson, 1995). The parts were fabricated at varying angles to the build direction and the resulting surfaces measured with a profilometer. The power spectra of the profiles were then compared to those of a surface produced with the aliasing model. The frequencies of the peaks match to within 0.10 cycles/mm, the resolution of the data.

3.4 SCANNING PATTERNS

After the geometric model of the part has been sliced, it must be processed in a manner dependent on the particular SFF process. Whether the process involves scanning a laser beam over the part (*e.g.*, Selective Laser Sintering or Stereolithography), scanning the layer where the part does not exist (*e.g.*, Laminated Object Manufacturing), or scanning a material extrusion head (*e.g.*, Fused Deposition Modeling), the two dimensional layers must be converted to scanning pattern particular to the process. This section describes scanning pattern generation, with particular attention to laser-based SFF processes. An approach to scanning based on physical models of the process is also outlined

3.4.1 Toggle Point Generation

The simplest scanning pattern for laser-based SFF processes is raster scanning, in which the laser beam is energized along one axis only. Scanning pattern creation for raster scans consists of locating points on the layer boundary at which the laser must be turned on or off, or *toggled.* The process of toggle point generation for each layer is akin to rasterization in computer graphics image generation (Crawford *et al*, 1991), but more correctly might be called boundary discretization, since the goal for SLS is to generate discrete points on the boundary contours at which the energy beam must be toggled on or off, rather than shades for each pixel between such points. The process consists essentially of planar ray casting, in which mathematical rays directed along each scan line within each layer are intersected with each segment of the boundary contours in the layer. These intersections are then ordered according to increasing distance from a given datum (the x-z plane for SLS). Each successive intersection thus represents a point where the energy beam is alternately turned on or off. The algorithm described by Crawford *et al* (1991) accounts for the special cases of vertices and boundary segments that lie entirely on a given scan line. Rock and Wozny (1991b) provide an elegant solution to this problem by offsetting the vertices in the layer contour to ensure that they do not lie on a scan line.

3.4.2 Stereolithography Patterns

Processing for stereolithography consists of scanning the layer boundaries, followed by hatching the internal areas of the part contour to solidify a portion of the inside of the part. A fully solidified part is then produced by post-processing to cure the entrained resin. 3D Systems (Jacobs, 1992) has developed a number of different hatching patterns for different resins and applications. The Tri-Hatch pattern scans the layer along lines oriented at 0°, 60°, and 120° from the x axis. The resulting scanning pattern consists of equilateral triangles. Research by 3D Systems determined that the large amount of uncured resin resulting from this pattern leads to an unacceptable amount of distortion. To alleviate this problem, the WEAVE pattern was introduced. WEAVE consists of scanning orthogonally, first along the y axis, followed by the x axis, at much closer scan spacing than Tri-Hatch. Additionally, the laser exposure is altered such that the cure depth for y axis scan lines is less than the layer thickness. Thus, the y axis scan lines do not adhere to the previous layer.

Further enhancements in part quality were made possible with the development of the STAR-WEAVE pattern. This pattern is different from the WEAVE pattern in three ways. First, the scan lines in alternate layers are staggered such that each vector on a given layer is scanned over a gap between scan lines in the previous layer. Second, the scanning sequence for the x and y axes is altered from layer to layer such that, for instance, the x vectors are scanned first on even layers and the y vectors are scanned first on odd layers. Third, the end of each scan line is "retracted" slightly such that each scan line is attached to only one contour border, allowing shrinkage of the cured resin to occur with minimum distortion.

Applications in investment casting pattern fabrication, along with the desired for increased accuracy and the development of epoxy resins, drove the creation of the QuickCast scanning pattern (Pang and Jacobs, 1993,

Jacobs, 1996). For investment casting, the uncured resin trapped in the interior of the part must be removed, rather than cured in post processing as in normal stereolithography part construction. Likewise, for these applications, there is no need to build a completely solid part; only the surface of the part is needed to create the investment casting mold. The QuickCast build pattern was developed to take advantage of these characteristics of casting applications. In QuickCast, the hatch spacing of the scanning pattern is much larger, resulting in a final part consisting of a thin shell supported by a minimal internal skeleton. Additionally, the hatch pattern for alternate layers of the part is offset by up to half the hatch spacing. This results in an internal structure consisting of relatively large interconnected cells. This structure, combined with the creation of drain and vent holes on the part surface, allows the internal trapped resin to drain from the part once the part build is complete. The QuickCast style can be applied to various hatch patterns, including equilateral triangles and square patterns.

Further improvements in accuracy of parts fabricated with epoxy resins are realized by the ACES (accurate, clear, epoxy, solid parts) scanning pattern. Development of this pattern was based on research that indicated that much part distortion is due to the internal stresses created by shrinkage of the top layer with respect the previous layer to which it is bonded. The ACES pattern alleviates this problem by scanning each layer twice to different curing depths. During the first scan, the curing depth does not reach the previous layer. This allows the current layer to float in uncured resin and shrink freely as it is created. The second scan then cures the resin deep enough to complete the bond to the previous layer. This sequence of improvements in scanning patterns, from Tri-Hatch to ACES, have resulted in almost an order of magnitude improvement in accuracy for stereolithography parts in a period of approximately 5 years (Jacobs, 1996).

3.4.3 Contour scanning

Recent research at the University of Texas has focused on developing an optimal laser control system for scanning along curves (Wu and Beaman, 1990, Wu and Beaman, 1991, Wu and Beaman, 1992). The scanning algorithm and associated hardware maintain constant laser power density by simultaneously controlling laser speed and laser power. The goal is to produce parts with better surface resolution than can be obtained from vector scanning the part contour (see Figure 3.11).

Wu's tracking control algorithm is based on the assumption that the tracking path (*i.e.*, contour) is represented parametrically. Wu did not explicitly specify the requirements of the parametric form, but these can be obtained from the derivation of his control algorithm. In general, the parametric form must satisfy the following three requirements: (1) $n-1$ differentiability, except at corners; (2) parametrization by arc length, so that the first and second derivatives represent tangential velocity and acceleration; and (3) uniform distribution of points. Wu derived the dynamics of scanner galvanometer as a second order system. Therefore, the tracking path must exhibit C^1 continuity. Note that corners have specific meanings. Wu defined a corner as "a point in the prescribed path where tracking velocity must be zero due to the available bounded control torque". To handle corners accurately, the tracking path is divided into several segments such that corners only exist at the start and end points of the segments. The third requirement allows "near constant tracking speed to gain uniform exposure of laser power".

This research emphasizes the need for higher level geometric descriptions for SFF processes. The algorithm uses information about the curvature of the contour to determine appropriate laser parameters to achieve the desired power density. A polygonal approximation of the contour, such as that obtained from slicing a faceted part model, is not accurate enough to support this control scheme.

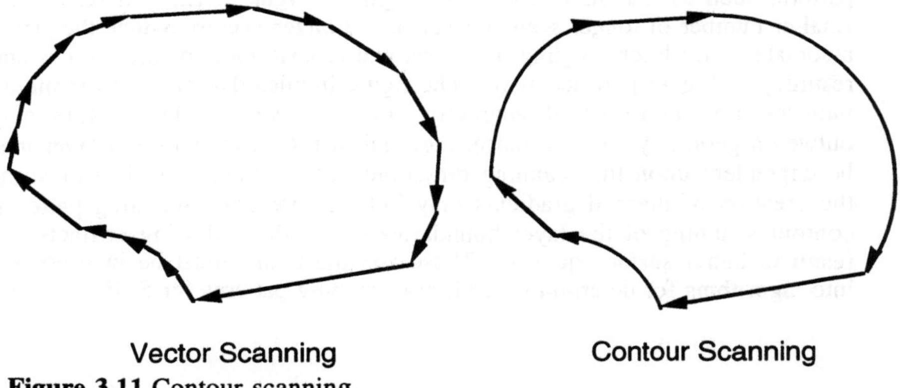

Figure 3.11 Contour scanning.

3.4.4 Off-Line Control and Process Planning

Solid Freeform Fabrication has the potential for producing accurate, structurally sound three-dimensional solid versions of objects. To develop accurate patterns and functional parts with adequate strength, however, requires in-depth basic research to understand the correlation between the mechanical properties and geometric accuracy of the final part with respect to the SFF process parameters (Beaman, 1992). These relationships must be captured in computer models that can be used to control SFF processing. To enable real time control, these models must represent a compromise between accurately modeling the physics of the SFF process and intelligent use of geometric information to approximate optimal processing patterns.

While process planning for SFF is considerably reduced compared to conventional fabrication technologies, there are several considerations that will require reasoning about the geometry of the final part. Scaling and orientation of the part within the work space of the SFF machine have a significant impact on the efficiency of the process. Aside from other factors, the part should be oriented in a manner that minimizes the number of layers. Other factors, however, may override this consideration. For instance, tolerances tend to be directionally dependent. Likewise, the mechanical properties of the final part will depend upon its orientation during the process. These issues require geometric reasoning on both global and local scales. Global reasoning will indicate the best part orientation within the workspace of the SFF machine. Local reasoning refers to considerations of the geometry of each layer to determine scanning and build patterns that maximize geometric accuracy and mechanical properties of the part.

As discussed above, some SFF processes build each layer of a part by raster scanning the powder bed along a single axis, as indicated in Figure 3.12. For geometries such as this, a unidirectional scanning pattern will result in a large number of very short scanning vectors. A multidirectional scanning pattern, such as that indicated on the right in Figure 3.12, will result in a smaller number of longer scanning vectors. Longer vectors reduce the errors associated with laser toggling transients and repositioning of the laser beam, resulting in higher part accuracy. The figure implies that preferred scanning patterns are the result of geometry alone; however, other factors may outweigh geometry. For instance, mechanical properties within a layer may be dependent upon the scanning direction; heat transfer considerations and the creation of thermal gradients may indicate preferred scanning patterns; contour scanning of the layer boundaries may reduce aliasing artifacts and result in better surface quality. These considerations must be incorporated into algorithms for determining arbitrary scanning patterns for SFF processes.

Figure 3.12 Directional scanning versus raster scanning.

3.4.5 Model-Based Scanning

Advances in control of SFF processes will depend on developing a better understanding of the physics of each process. For example, the process control program for SLS administers three areas: laser control, control of powder delivery and leveling, and control of the environment within the machine. Input to the module consists of a suitable geometric description of the part layers (at this time, scan lines and laser toggle points) and settings for process parameters, including material properties, bed temperature, gas flow rate, scanning speed, laser power, beam diameter, and scan spacing. The current SLS process control software is built under the assumption that these parameters will be constant for an entire part. However, initial investigations indicate that fabrication of metal parts will require local control of laser beam parameters, allowing these parameters to change from layer to layer or even within different areas in a given layer. Such model-based scanning is depicted conceptually in Figure 3.13, where a part layer has been divided into several regions based on part quality predictions from a physical model of the process. A scanning pattern is then generated for each of the simple regions, again based on predictions from a process model. Such a scanning scheme could be precomputed off-line before the part is fabricated, or it may updated on a layer-by-layer basis, with input from sensors about the status of previous layers. The key to solving this problem is development of physical models of the process. While initial models of the SLS process have been developed (Sun *et al*, 1990, Sun and Beaman, 1991, Weissman and Hsu, 1991, Nelson *et al*, 1993), they are not structured to support real-time process control of this kind. The solution to this problem will likely require a combination of hardware and software development.

Figure 3.13 Model-based scanning.

3.5 SUPPORT STRUCTURES

While the goal of SFF is to reduce the overhead of process planning in prototyping and manufacture, such considerations are not completely eliminated. In particular, SFF technologies such as stereolithography, selective area laser deposition, and fused deposition modeling, require the design of auxiliary supports for overhanging features and "floating" components. Third-party software vendors provide tools to facilitate this auxiliary design work. Bridgeworks, from Solid Concepts, Inc. (Valencia, CA) provides automatic support design and is reportedly in use at over 300 sites. Magics, from Materialise (Ann Arbor and Belgium) provides interactive support generation for STL files. C-SUP, also from Materialise, provides contour level support generation.

Chalasani *et al* (1995) describe support generation methods that have been incorporated by Stratasys, Inc. in their QuickSlice software for FDM. The FDM process has been modified to allow an alternate material to be used for supports. Thus, the supports can be fabricated in a softer material, which facilitates their removal and impact on the surface quality of the part. The methods operate with layer contours and embody three support generation strategies. The containment strategy essentially determines the shadow of the part with respect to the build direction. This outline is computed by merging the contours of all layers such that a minimal containing outline is produced. This outline is expanded slightly to contain the entire part. Containment produces a sub-optimal support system with respect to fabrication time, since a portion of the containment volume is above the part, where support is not necessary. The regional support strategy allows the user to interactively specify regions requiring support. This strategy eliminates some of the unnecessary supports generated by containment, but potentially requires significant user input. The third strategy employed in QuickSlice generates custom supports based on the part geometry. This strategy is based on considerations of overall part stability and rules of thumb with respect to feature dimensions. This direct strategy generates supports by computing the "shadow" of the current layer on the previous layer.

3.5.1 Part Orientation

Allen and Dutta (1994) describe a technique for determining the best orientation for creating a part based on minimal support requirements and part stability. Their approach consists of three steps: (1) a list of candidate orientations is generated; (2) for each candidate orientation, a support structure is created; and (3) the best orientation is selected from the candidate list. Candidate orientations are determined by considering large faces of the convex hull[7] of the part, where "large" is defined by a user-determined threshold value for the surface area of a face. The convex hull is computed from a faceted approximation of the surface of the part. The inward-pointing normals of the faces whose surface areas are larger than the threshold form the list of candidate orientations.

To generate the support structures for a particular candidate orientation, the object is first transformed into that orientation. Next, the sample points (the vertices of the tessellation) are classified by simulating the growth of the object. As layers are added in the simulation, new floating components are detected and their sample points are identified as needing supports. Likewise, when the center of mass of a component is no longer directly above its based, the sample points are identified. After the simulation is complete, supports are grown along a half-line from the x-y plane through each identified sample point. Support material is added only for those portions of the half-line that lie outside the object.

Finally, the orientation that has minimal contact area with the object is chosen as the best orientation. If two orientations have the same contact area, the one with the lowest center of mass (*i.e.*, the most stable) is chosen.

[7]The convex hull of a set **A** is the minimal convex set that contains all points in **A**. For a solid model, the convex hull can be visualized as the surface formed by a rubber sheet stretched to enclose the solid.

Thompson (1995) describes another method for determining the orientation of a part that minimizes the volume of the supports necessary to fabricate the part using SALD in particular, or any other SFF technique requiring supports. The approach is based on identifying a set of candidate orientations using visibility maps (Chen and Woo, 1992, Gan *et al*, 1994, Tang *et al*, 1992, Woo, 1994). The visibility map is a tool developed to orient parts for NC machining. Most manufacturing processes cannot construct a surface of any arbitrary geometry without repositioning the workpiece or forming supports of some kind. For this reason, it is useful to think of the machine in terms of the set of surface normals that it can produce without repositioning or supports. For SALD, this set is a hemisphere whose pole is the unit normal of the substrate, **b**. This means that any face whose normal, when projected onto **b**, is codirectional, does not need support. The only exception to this is the base of the part, which does not require supports even though the faces point downwards.

The goal of the approach, then, is to find a hemisphere that contains all of the normals of the part, save the base. If all the part normals are contained in a hemisphere, then the part can be fabricated without any supports, as illustrated in Figure 3.14. Rather than perform a search for one hemisphere that will meet the criterion, we can transform the collection of normals into a visibility map, which contains the solution to the problem.

The visibility map is derived from the Gaussian map, which is simply the collection of all normals from the part on a unit sphere. To transform the Gaussian map into the visibility map, consider a set of hemispheres, one for each point in the Gaussian map and each with its pole centered at that point, as shown for the simple case of two faces in Figure 3.15. By intersecting each of these hemispheres, we obtain the visibility map. Each hemisphere represents the possible set of directions from which the manufacturing process can form the face being considered. By intersecting two of these, we are left with the possible directions from which both faces can be formed simultaneously. Thus, any point in the visibility map indicates an orientation for the part where the machining process can "see" every face. This means that if **b** is contained in the region on the far right of Figure 3.15, both faces are visible.

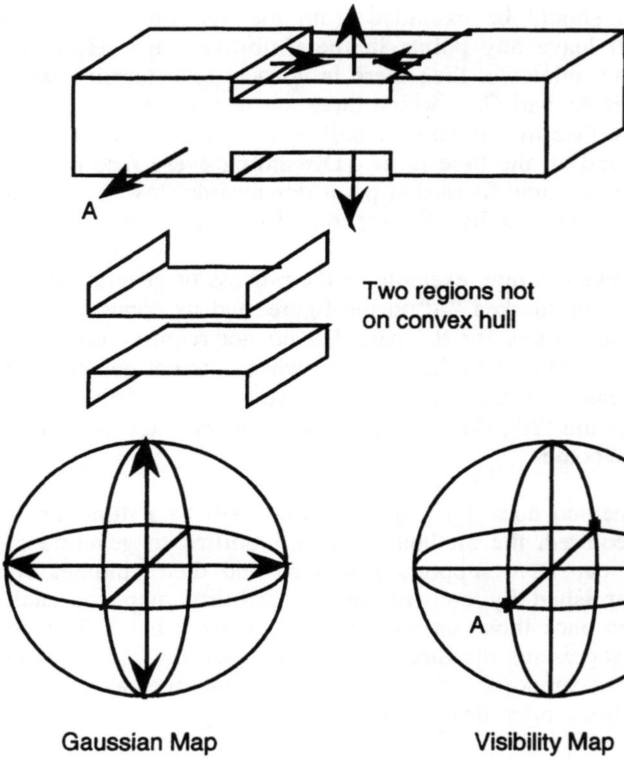

Figure 3.14 Visibility map of a part.

Note that in figure 3.14 the Gaussian map includes the boundaries of the hemisphere, so that two antipodal hemispheres intersect along a great circle[8].

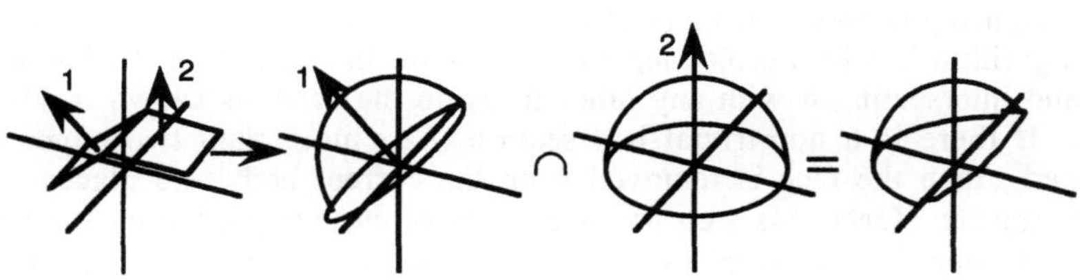

Figure 3.15 Construction of a visibility map.

[8]A great circle is a circle inscribed on a sphere whose center is coincident with that of the sphere.

When attempting to orient a part for SFF so that no supports are required, the base faces should be excluded from the visibility map, or else no orientation will leave any points in the visibility map. However, if the orientation is an unknown, then there is no way to determine which faces form the base of the part (*i.e.*, which faces are in contact with the substrate). By treating any face on the convex hull of the part as a base face, we are guaranteed to find all the base faces. Discarding every face on the convex hull may eliminate some faces that point downwards (toward the substrate), but will leave a candidate list of directions that may be evaluated quickly.

Figure 3.14 shows a simple example of the process of generating a visibility map. The direction marked "A" in the figure, and its opposite, are the only possible build directions for the part that do not require support volumes. The convex hull is shown in dashed lines when it is not coincident with the part. For this case, because there is no curvature on the faces that form the base along direction "A", the removal of all faces on the convex hull has no effect on the outcome.

Because this method does not guarantee there will be a direction where no supports are required, the method must be modified to generate candidate directions that can have supports generated and their volumes compared. This is accomplished by constructing a visibility map for each set of connected faces once those on the convex hull have been discarded. The visibility map represents the directions from which that group of faces can be made without supports. These directions can serve as trial directions even if the visibility map for the entire part is empty.

To generate supports, each face whose normal, $\mathbf{n}$, satisfies $\mathbf{n.b} < 0$ must have support material below it. As a starting point, consider sweeping each downward-pointing face to the bottom of the bounding box of the part. Each swept volume can then be unioned with all the others and, finally, the part subtracted from the union. While this approach generates a support for the part, it involves N–1 boolean union operations, where N is the number of faces requiring support. A more efficient method is shown in Algorithm 3.1. The algorithm involves projecting each face onto the bottom of the bounding box and intersecting it with any other faces on the base, as shown in Figure 3.16. If there is a non-trivial intersection (*i.e.*, more than two edges are returned), then the face is removed from the current body and placed on a list of rejects. Once this step has been completed, every edge on the body with only one face attached (a lamina edge) is projected to the base plane and used to create a "side panel" connecting the original face to the base, shown in the far right of Figure 3.16. Finally, the base is stitched onto the body and the entire process is repeated for any rejected faces. All of the resulting bodies are united and then the original part is subtracted from the union. The stitch function attaches two bodies to each other at common vertices and edges without checking to see if there are any intersections, and is used to efficiently collect faces into a body when it is known there is no intersecting geometry.

The VS-PROJECT-LUMP function creates the base and "side panels' mentioned earlier. The base is created by making copies of a face's vertices and setting the z coordinate to the lower bound on the part. The vertex list is then converted to a face. The side panels are created by defining a set of vertices that form the corners of a polygonal face and stitching the face to the body. The vertices are placed in counterclockwise order by reversing the order of the vertices on the existing co-edge to which the panel will be attached.

Algorithm 3.1 An algorithm to generate volumetric supports.

```
VOLUME-SUPPORT(body,b,thick,DotAngTol}
    z_b ← GET-BOUNDING-BOX(body).low.z – thick
    F ← Get-Faces(body)
    for i ← 1 to length[F]
    do if DOT-PRODUCT(F[i].normal,b) > DotAngTol
          then  Fnew ← Copy-Face(F)
          REVERSE-FACE(F_new)         Point face up.
          STITCH(F_new,support)       Unite face with support.
    supports ← SEPARATE-BODY(support) Separate disconnected regions.
    B ← GET-BODIES(supports)
    for i ← 1 to length[B]
    do VS-PROJECT-LUMP(B[i],supports,b,z_b)
    support ← B[0]                    Put supports into one BODY.
    for i ← 2 to length[B]
    do support ← UNITE(B[i],support)
    temp ← COPY-BODY(body)
    support ← SUBTRACT(temp,support)
    return support
```

Algorithm 3.2 A subroutine used by Algorithm 3.1.

```
VS-PROJECT-LUMP(body,rejects, b,z_b)
    F ← GET-FACES(body)
    base ← NEW-BODY()
    for i ← 1 to length[F]
    do V_F ← GET-VERTICES(F[i])
          for j ← 1 to length[V_F]
          do V[length[V_F] – j + 1] ← V_F[j]
             V[length[V_F] – j + 1].set_z(z_b)  Project point downwards.
    F_new ← MAKE-FACE-FROM-VERTS(V)
    F_b ← GET-FACES(base)
    for j ← 1 to length[F_b]
    do IntGraph ← FACE-FACE-INTERSECT(F_new, F_b[j])
```

```
        if NUM-EDGES(IntGraph) > 2
           then    temp ← UNHOOK-FACE(F[i])
                   reject ← UNITE(temp,reject)
E ← GET-EDGES(body)
for i ← 1 to length[E]
do if      !E[i].coedge.partner
           then    side[0] ← E[i].start_vertex()
                   side[1] ← E[i].start_vertex()
                   side[2] ← E[i].end_vertex()
                   side[3] ← E[i].end_vertex()
                   side[4] ← E[i].start_vertex()
                   side[1].set_z(z_b)
                   side[2].set_z(z_b)
                   F_new ← MAKE-FACE-FROM-VERTS(side)
                   STITCH(F_new ,body)
STITCH(base,bod)
if   reject
     then  temp ← SEPARATE(reject)
           rejects.add(temp)
return
```

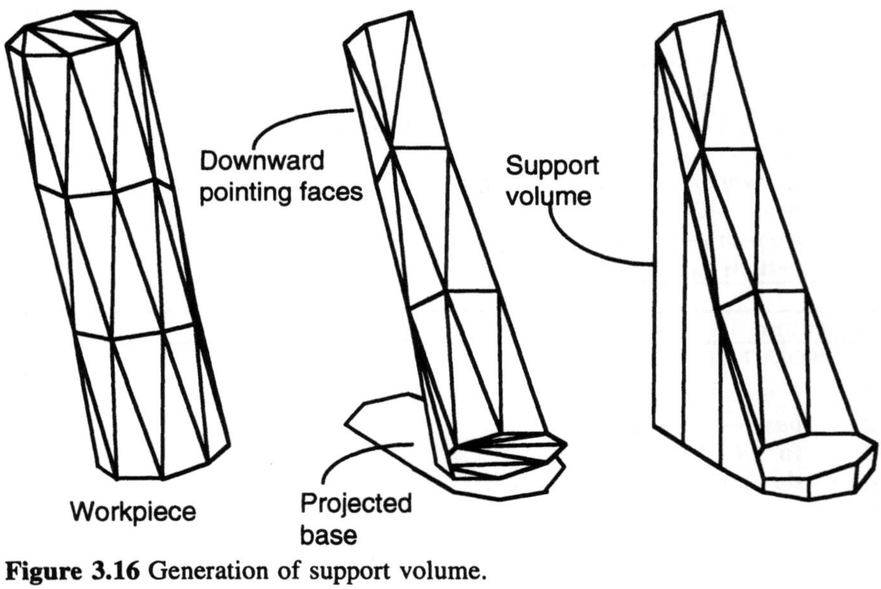

Figure 3.16 Generation of support volume.

3.6 SUMMARY

Computer software is a key enabler of Solid Freeform Fabrication technologies. The scope of information processing in SFF spreads across a range of abstraction levels, from high level CAD software for creating part geometry, to technology-specific processing of geometry, to real-time control of the SFF processes. Clearly without the computer, SFF processes would be difficult or completely impractical to implement. However, many opportunities exist for improving the performance of SFF processes by improving the software, thereby broadening the application of SFF from producing models and prototypes to functional part manufacturing. This chapter provides an outline of some of the issues in software development in the areas of geometric modeling and process modeling and control. To realize the potential of SFF, research must continue simultaneously in all of these areas.

REFERENCES

3D Systems, Inc., "Stereolithography Interface Specification", 3D Systems, Inc., Valencia, CA, June 1988.

Allen, S., and Dutta, D., "On the Computation of Part Orientation Using Support Structures in Layered Manufacturing", *Solid Freeform Fabrication Proceedings*, Austin, TX, August 8-10, 1994, pp. 259-269.

Bartels, K. A., Crawford, R. H., Das, S., Guduri, S., Bovik, A. C., Diller, K. R., and Aggarwal, S. J., "Fabrication of Macroscopic Solid Models of Three-Dimensional Microscopic Data by Selective Laser Sintering", *Journal of Microscopy*, Volume 169, No. 3, 1993, pp. 383-389.

Beaman, J. J., "Machine Issues Associated with Solid Freeform Fabrication", *Solid Freeform Fabrication Proceedings*, Austin, TX, August 3-5, 1992, pp. 309-330.

Bjørke, O., "How to Make Stereolithography into a Practical Tool for Tool Production", *Annals of CIRP*, Vol. 40, No. 1, 1991, pp. 175-177.

Bøhn, J. H. and Wozny, M. J., "Automatic CAD-Model Repair: Shell-Closure", *Solid Freeform Fabrication Proceedings*, Austin, TX, August 3-5, 1992, pp. 86-94.

Carleberg, P., "Product Model Driven Direct Manufacturing", *Solid Freeform Fabrication Proceedings*, Austin, TX, August 8-10, 1994, pp. 270-276.

Chalasani, K., Jones, L., and Roscoe, L., "Support Generation for Fused Deposition Modeling", *Solid Freeform Fabrication Proceedings*, Austin, TX, August 7-9, 1995, pp. 229-241.

Chari, J. K., and Hall, J. L., "Robust Prototyping", *Solid Freeform Fabrication Proceedings*, Austin, TX, August 9-11, 1993, pp. 135-142.

Chen, L., and Woo, T. C., "Computational Geometry on the Sphere with Application to Automated Machining", *Journal of Mechanical Design*, Vol. 114, 1992, pp. 288-295.

Craig, J. I., Fulton, R. E., Stephen, E. R., and Jarnagin, A. B., "Testing CALS Standards for Engineering and Manufacturing Applications", *Engineering Data Management: Key to Integrated Product Development*, proceedings of the 6th annual ASME Database Symposium, T. R. Chase, ed., San Francisco, CA, August 2-6, 1988, pp. 33-41.

Crawford, R. H., Das, S., and Beaman, J. J., "Software Testbed for Selective Laser Sintering", *Solid Freeform Fabrication Proceedings*, Austin, TX, August 12-14, 1991, pp. 21-27.

Cunningham, J. J., and Dixon, J. R., "Designing with Features: The Origin of Features", *Computers in Engineering 1988*, vol. 1, proceedings of the 1988 ASME International Computers in Engineering Conference and Exhibition, San Francisco, CA, July 31-August 4, 1988, V. A. Tipnis and E. M. Patton, eds., pp. 237-243.

Cutkosky, M. R., Tenenbaum, J. M., and Muller, D., "Features in Process-Based Design", *Computers in Engineering 1988*, vol. 1, proceedings of the 1988 ASME International Computers in Engineering Conference and Exhibition, San Francisco, CA, July 31-August 4, 1988, V. A. Tipnis and E. M. Patton, eds., pp. 557-562.

Darrah, J., and Wielgus, M., "A New CAD Model Format for SFF Machines?", *Solid Freeform Fabrication Proceedings*, Austin, TX, August 6-8, 1990, pp. 121-125.

Dolenc, A., and Mäkelä, I., "Slicing Procedures for Layered Manufacturing Techniques",

Computer-Aided Design, Vol. 26, No. 2, 1994, pp. 119-126.

Drake, S. H., "Alpha_1, Remote Manufacturing, and Solid Freeform Fabrication", *Solid Freeform Fabrication Proceedings*, Austin, TX, August 6-8, 1990, pp. 115-120.

Falcidieno, B., and Giannini, F., "Automatic Recognition and Representation of Shape-Based Features in a Geometric Modeling System", *Computer Vision, Graphics, and Image Processing*, vol. 48, no. 1, October 1989, pp. 93-123.

Gadh, R., "A Hybrid Approach to Intelligent Geometric Design Using Features-Based Design and Feature Recognition", Advances in Design Automation, vol. 2, proceedings of the *19th ASME Design Automation Conference*, September 19-22, 1993, Albuquerque, NM, B. J. Gilmore, D. A. Hoeltzel, S. Azarm, and, H. A. Eschenauer, eds., pp. 273-283.

Gadh, R., and Prinz, F. B., "MANUFeature: A Framework for Recognizing CAD Interactions in Design-for-Manufacturing Analysis", *Advances in Design Automation*, vol. 2, proceedings of the 18th ASME Design Automation Conference, September 13-16, 1992, Scottsdale, AZ, D. A. Hoeltzel, ed., pp. 91-103.

Gan, J. G., Woo, T. C., and Tang, K., "Spherical Maps: Their Construction, Properties, and Approximation", *Journal of Mechanical Design*, Vol. 116, 1994, pp. 357-363.

Gilman, C. R., and Rock, S. J., "The Use of STEP to Integrate Design and Solid Freeform Fabrication", *Solid Freeform Fabrication Proceedings*, Austin, TX, August 7-9, 1995, pp. 213-220.

Gursoz, E. L., Weiss, L. E., and Prinz, F. B., "Geometric Modeling for Rapid Prototyping and Tool Fabrication", *Solid Freeform Fabrication Proceedings*, Austin, TX, August 6-8, 1990, pp. 135-145.

Guduri, S., Crawford, R. H., and Beaman, J. J., "A Method to Generate Exact Contour Files for Solid Freeform Fabrication", *Solid Freeform Fabrication Proceedings*, Austin, TX, August 3-5, 1992, pp. 95-101.

Guduri, S., Crawford, R. H., and Beaman, J. J., "Direct Generation of Contour Files from Constructive Solid Geometry Representations", *Solid Freeform Fabrication Proceedings*, Austin, TX, August 9-11, 1993, pp. 291-302.

Henderson, M. R., "Extraction of Feature Information from Three-Dimensional CAD Data", Ph.D. Thesis, Purdue University, 1984.

Hoppe, H., DeRose, T., Duchamp, T., McDonald, J., and Stuetzle, W., "Surface Reconstruction from Unorganized Points", *Computer Graphics*, Vol. 26, No. 2, July 1992, pp. 71-78.

Jacobs, P. F., *Rapid Prototyping and Manufacturing: Fundamentals of Stereolithography*, Society of Manufacturing Engineers, Dearborn, MI, 1992.

Jacobs, P. F., *Stereolithography and other RP&M Technologies: From Rapid Prototyping to Rapid Tooling*, ASME Press, New York, 1996.

Kirschman, C. F., and Jara-Almonte, C. C., "A Parallel Slicing Algorithm for Solid Freeform Fabrication Processes", *Solid Freeform Fabrication Proceedings*, Austin, TX, August 3-5, 1992, pp. 26-33.

Levy, R. A., Guduri, S., and Crawford, R. H., "Preliminary Experience with Selective Laser Sintigraphic (SLS) Models of the Human Temporal Bone", *Solid Freeform Fabrication Proceedings*, Austin, TX, August 3-5, 1992, pp. 161-173.

Joshi, S., and Chang, T. C., "Graph-Based Heuristics for REcognition of Machined Features from a 3D Model", *Computer-Aided Design*, vol. 20, no. 2, March 1988, pp. 58-66.

Kim, Y. S., "Recognition of Form Features Using Convex Decomposition", *Computer-Aided Design*, vol. 24, no. 9, September 1992, pp. 461-476.

Kim, Y. S., "Volumetric Feature Recognition Using Convex Decomposition", *Advances in Feature Based Manufacturing*, J. J. Shah, M. Mäntylä, D. S. Nau, eds., Elsevier Science, Amsterdam, 1994, pp. 39-63.

Mäkelä, I., and Dolenc, A., "Some Efficient Procedures for Correcting Triangulated Models", *Solid Freeform Fabrication Proceedings*, Austin, TX, August 9-11, 1993, pp. 126-134.

Nelson, J.C., S. Xue, J.W. Barlow, J.J. Beaman, H.L. Marcus, and D.L. Bourell, "A model of the Selective Laser Sintering of Bisphenol-A Polycarbonate", *Ind. Eng. Chem. Res.*, 32, 2305-17, 1993.

Pang, T. H., and Jacobs, P. F., "StereoLithography 1993: QuickCast™", *Solid Freeform Fabrication Proceedings*, Austin, TX, August 9-11, 1993, pp. 158-167.

Panzica, C. H., "IGES — Today's Product Data Exchange Technology", *Managing Engineering*

Data: Emerging Issues, proceedings of the 2nd annual ASME Database Symposium, R. E. Fulton, ed., San Francisco, CA, July 31-August 4, 1988, pp. 61-65.

Pavlakos, E. G., Greenfield, I., Cutting, C. B., Wei, C. S., "Conversion of 3D Scan Data to IGES Data Base for Design and Manufacture", *Computers in Engineering 1989*, proceedings of the 1989 ASME International Computers in Engineering Conference and Exposition, Vol. 1, D. R. Riley and T. J. Cokonis, eds., Anaheim, CA, July 30-August 3, 1989, pp. 99-103.

Rajagopalan, M., Aziz, N. M., and Huey, C. O., "A Model for Interfacing Geometric Modeling Data with Rapid Prototyping Systems", *Advances in Engineering Software*, Vol. 23, 1995, pp. 89-96.

Rock, S. J. and Wozny, M. J., "A Flexible File Format for Solid Freeform Fabrication", *Solid Freeform Fabrication Proceedings*, Austin, TX, August 12-14, 1991, pp. 1-12.

Rock, S. J. and Wozny, M. J., "Utilizing Topological Information to Increase Scan Vector Generation Efficiency", *Solid Freeform Fabrication Proceedings*, Austin, TX, August 12-14, 1991, pp. 28-36.

Rock, S. J. and Wozny, M. J., "Generating Topological Information from a 'Bucket of Facets'", *Solid Freeform Fabrication Proceedings*, Austin, TX, August 3-5, 1992, pp. 251-259.

Rogers, W. E., Crawford, R. H., Beaman, J. J., and Walsh, N. E., "Fabrication of Prosthetic Sockets by Selective Laser Sintering", *Solid Freeform Fabrication Proceedings*, Austin, TX, August 12-14, 1991, pp. 158-163.

Rosen, D. W., Dixon, J. R., and Finger, S., "Conversions of Feature-Based Representations via Graph Grammar Parsing", *Proceedings of the 1992 ASME Design Theory and Methodology Conference*, September 13-16, 1992, Scottsdale, AZ, D.L. Taylor and L.A. Stauffer, eds., pp. 83-90.

Ryan, S. A., "PDES/STEP: A Status Report", *Engineering Data Management: Key to Integrated Product Development*, proceedings of the 6th annual ASME Database Symposium, T. R. Chase, ed., San Francisco, CA, August 2-6, 1988, pp. 43-47.

Sachs, E., Curodeau, A., Gossard, D., Jee H., Cima, M., and Caldarise, S., "Surface Texture by 3D Printing", *Solid Freeform Fabrication Proceedings*, Austin, TX, August 8-10, 1994, pp. 56-64.

Sakurai, H., "Decomposing a Delta Volume into Maximal Convex Volumes and Sequencing Them for Machining", Computers in Engineering 1994, vol. 1, *Proceedings of the 1994 ASME International Computers in Engineering Conference and Exposition,* Minneapolis, MN, September 11-14, 1994, K. Ishiii, ed., pp. 135-142.

Sakurai, H., and Chin, C-W., "Definition and Recognition of Volume Features for Process Planning", *Advances in Feature Based Manufacturing,* J. J. Shah, M. Mäntylä, D. S. Nau, eds., Elsevier Science, Amsterdam, 1994, pp. 65-80.

Sakurai, H., and Gossard, D. C., "Recognizing Shape Features in Solid Models, *IEEE Computer Graphics and Applications,* September 1990, pp. 22-32.

Shah, J. J., Rogers, M. T., Sreevalsan, P. C., Hsiao, D. W., Mathew, A., Bhatnagar, A., Liou, B. B., and Miller, D. W., "The ASU Features Testbed: An Overview", *Proceedings of the ASME Computers in Engineering Conference,* vol. 1, Boston, MA, August 5-9, 1990, G. L. Kinzel and S. M. Rohde, eds., pp. 233-242.

Shah, J. J., Shen, Y., and Shirur, A., "Determination of Machining Volumes from Extensible Sets of Design Features", *Advances in Feature Based Manufacturing*, J. J. Shah, M. Mäntylä, D. S. Nau, eds., Elsevier Science, Amsterdam, 1994, pp. 129-157.

Shen, Y., and Shah, J. J., "Feature Recognition by Volume Decomposition Using Half Space Partitioning", *Advances in Design Automation,* vol. 1, *Proceedings of the 20th ASME Design Automation Conference,* September 11-14, 1994, Minneapolis, MN, B. J. Gilmore, ed., pp. 575-583.

Suh, Y. S., and Wozny, M. J., "Adaptive Slicing of Solid Freeform Fabrication Processes", *Solid Freeform Fabrication Proceedings*, Austin, TX, August 8-10, 1994, pp. 404-411.

Suh, Y. S., and Wozny, M. J., "Integration of a Solid Freeform Fabrication Process into a Feature-Based CAD System Environment", *Solid Freeform Fabrication Proceedings* , Austin, TX, August 7-9, 1995, pp. 334-341.

Sun, M. M., Beaman, J. J., and Barlow, J. W., "Parametric Analysis of the Selective Laser Sintering Process", *Solid Freeform Fabrication Proceedings*, Austin, TX, August 6-8,

1990, pp. 146-154.

Sun, M. M., and Beaman, J. J., "A Three Dimensional Model for Selective Laser Sintering", *Solid Freeform Fabrication Proceedings*, Austin, TX, August 12-14, 1991, pp. 102-109.

Tang, K., Woo, T., and Gan, J., "Maximum Intersection of Spherical Polygons and Workspace Orientation for 4- and 5-Axis Machining", *Journal of Mechanical Design*, Vol. 114, 1992, pp. 477-485.

Tseng, Y-J., and Joshi, S. B., "Recognizing Multiple Interpretations of Interacting Machining Features", *Computer-Aided Design*, vol. 26, no. 9, September 1994, pp. 667-688.

Thompson, D. C., *The Optimization of Part Orientation for Solid Freeform Manufacture*, MS Thesis, The University of Texas at Austin, December 1995.

Thompson, D. C., and Crawford, R. H., "Optimizing Part Quality with Orientation", *Solid Freeform Fabrication Proceedings*, Austin, TX, August 7-9, 1995, pp. 362-368.

Turner, G. P. and Anderson, D. C., "An Object-Oriented Approach for Interactive Feature-Based Design for Quick Turnaround Manufacturing", *Computers in Engineering 1988*, vol. 1, proceedings of the 1988 ASME International Computers in Engineering Conference and Exhibition, San Francisco, CA, July 31-August 4, 1988, V. A. Tipnis and E. M. Patton, eds., pp. 551-555.

Vandenbrande, J. H., and Requicha, A. A. G., "Spatial Reasoning for Automatic Recognition of Interacting Form Features", *Proceedings of the ASME Computers in Engineering Conference*, vol. 1, Boston, MA, August 5-9, 1990, G. L. Kinzel and S. M. Rohde, eds., pp. 251-256.

Weissman, E. M., and Hsu, M. B., "A Finite Element Model of Multi-Layered Laser Sintered Parts", *Solid Freeform Fabrication Proceedings*, Austin, TX, August 12-14, 1991, pp. 86-94.

Woo, T.C. "Feature Extraction by Volume Decomposition", *Proceedings of the Conference on CAD/CAM Technology in Mechanical Engineering*, Cambridge, MA, March 1982, pp. 76-94.

Woo, T. C., "Visibility Maps and Spherical Algorithms", *Computer-Aided Design*, Vol. 26, No. 1, 1994, pp. 6-16.

Wozny, M. J., "Systems Issues in Solid Freeform Fabrication", *Solid Freeform Fabrication Proceedings*, Austin, TX, August 3-5, 1992, pp. 1-15.

Wu, Y-J. E., and Beaman, J. J., "Contour Following for Scanning Control in SFF Application: Control Trajectory Planning", *Solid Freeform Fabrication Proceedings*, Austin, TX, August 6-8, 1990, pp. 126-134.

Wu, Y-J. E., and Beaman, J. J., "Solid Freeform Fabrication Laser Tracking Control", *Solid Freeform Fabrication Proceedings*, Austin, TX, August 12-14, 1991, pp. 37-45.

Wu, Y-J. E., and Beaman, J. J., "Laser Tracking Control Implementation for SFF Applications", *Solid Freeform Fabrication Proceedings*, Austin, TX, August 3-5, 1992, pp. 161-173.

4

POLYMERS IN SOLID FREEFORM FABRICATION

With the possible exception of the MIT Soligen process, all commercial SFF techniques currently produce parts from polymeric or oligomeric hydrocarbon materials. The dominance of polymeric materials, relative to metals and ceramics, is largely a consequence of their ease of processing at relatively modest temperatures, generally less than 400°C. This permits lower-cost energy sources to be employed by the shaping or fabricating technology than are required to directly fabricate with metals and ceramics. In addition, polymeric materials (including inorganic glasses) have low surface energies and high melt viscosities, relative to metals and other high surface energy, inorganic crystalline materials. These features eliminate the tendency seen in the metals for the molten material to minimize its energy by forming spherical droplets that can cause poor surface quality. Polymeric materials are especially advantageous in terms of the relative ease with which their chemical structures and molecular sizes, characteristics which affect their mechanical and rheological properties, can be modified and optimized for each specific SFF technique and need.

The discussion of relationships between SFF processes and polymeric material requirements is limited, below, to three processes; the Stratasys, Inc., Fused Deposition Modeling (FDM) process that builds shapes by extruding and placing filaments of thermoplastic material, the 3-D Systems, Inc., Stereolithography (SLA) process that builds shapes by photopolymerizing and crosslinking a suitable monomer with a UV laser, and the DTM Corporation Selective Laser Sintering (SLS) process that builds shapes by fusing polymeric powder with an IR laser. The almost infinite flexibility of design of polymeric materials permits development of materials that are optimized to meet the material property needs for each of these three very different processes.

The following discussion of polymeric materials focuses on some of the issues that are important to each of these three processes. A fundamental property of polymers is presented in each subsection, followed by a paragraph that relates the significance of this property to each of the three processes. It is hoped that this organization will, at least, emphasize the strong interactions between process and material that necessarily exist in all material processes, including SFF. Numerous excellent texts, including Billmeyer (1962), Stevens (1990), Tadmor and Gogos (1979), Progelhof and Throne (1993), and Allcock and Lampe (1990), can be consulted to learn about the many details of polymer synthesis and properties that are not discussed here.

4.1 POLYMER MATERIAL PROPERTIES

4.1.1 Molecular Weight

Polymers are large, chain-like, molecules with variable molecular weight that are synthesized from simple, lower molecular weight, chemical repeat units called monomers. Unlike metals and simple organic and inorganic compounds, polymers do not have a precise molecular weight. Instead, the chain lengths within a given polymer batch are found to be distributed about a mean value. Terms such as *number average molecular weight, M_n,* and *weight average molecular weight, M_w,* are used to describe averages in the distribution (Billmeyer, 1962). The molecular weight averages of the polymer are determined by the details of its specific chemical synthesis and can be controlled to any values between 5000 (below which polymers are termed oligomers) and infinity. As suggested by Figure 4.1, alteration and control of the average molecular weight permits the formation of many products which have widely differing mechanical and rheological properties behavior yet the same chemical repeat unit. For example, 5000 molecular weight cis-polybutadiene (cis-PBD) is a viscous, but easily poured liquid at room temperature, however 2 million molecular weight cis-PBD is a non-pourable rubbery product that, when further reacted, is a major constituent in automobile tires.

Significance to Processes

To achieve functionally strong prototypes, all three processes need to work with polymers that have molecular weights as high as possible. As is discussed below in some detail, high molecular weight can lead to difficulty in placing and forming the material because of the high viscosities involved. One solution to this problem is to build the molecular weight after placing a lower molecular weight material. This is the essence of the SLA process. However, the need to control chemistry and shrinkage make operation of the SLA process somewhat more difficult than the FDM and SLS processes which do not employ chemical reactions.

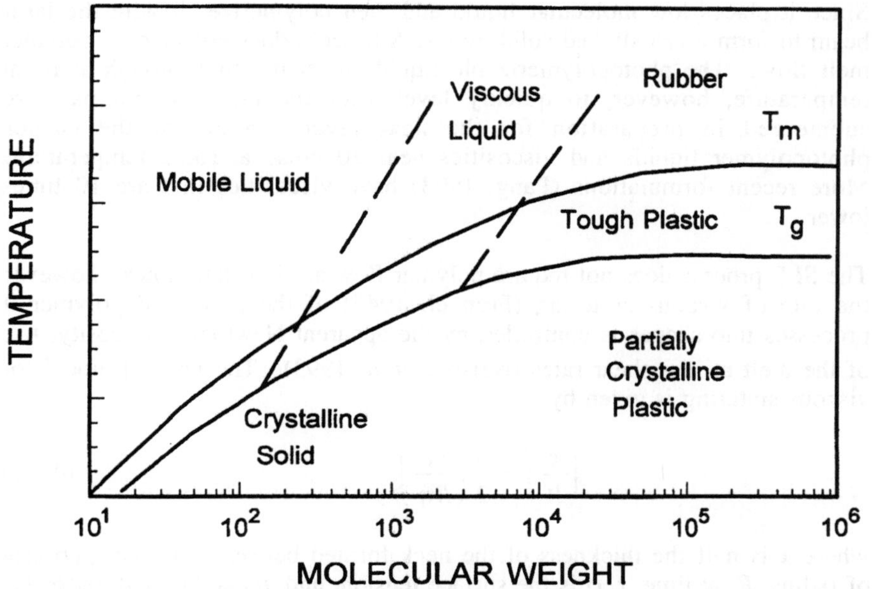

Figure 4.1 Effects of Molecular Weight on Polymer Characteristics (Billmeyer, 1962).

4.1.2 Rheology

Figure 4.2 shows some melt viscosities for typical injection molding grade polymers at various melt temperatures and shear rates. Clearly, their viscosities are high, 10^2 - 10^4 poise relative to that for water, 10^{-2} poise, and in contrast to the behavior of small molecules which usually show constant Newtonian viscosities, the viscosities of polymers decrease with increasing flow rate or shear rate. Both characteristics are very important to polymer processing, generally. The viscosity is shown in Figure 4.2 to become nearly constant at very low shear rates. This apparent Newtonian viscosity plateau, η_o, is known to be related to the molecular weight of the polymer by the relationship (Graessley, 1974),

$$\eta_o = k'(M_w)^{3.4} \tag{4.1.1}$$

The large exponent, 3.4, indicates that η_o is a very sensitive function of molecular weight. The viscosity is also a fairly sensitive function of temperature, see equation (4.2.2) below, and will double when the temperature is reduced by roughly 15-25°C (Tadmor and Gogos, 1979). This suggests that temperature control may be very important in processes where polymer flow is involved.

Significance to Processes

Since it places low molecular liquid and then polymerizes it with the laser beam to form a crosslinked solid, the SLA process does not involve polymer melt flow. The photopolymerizable liquids must be fluid enough at room temperature, however, to quickly level after the elevator and part are submerged in preparation for the next layer. Some of the earlier photopolymer liquids had viscosities near 20 poise at room temperature. More recent formulations (Pang, 1993) have viscosities that are 10 times lower.

The SLS process does not require polymer flow at high shear rates, however the rate of viscous sintering (Frenkel, 1945) of the powdered polymer it processes into shapes is controlled by the apparent Newtonian viscosity, η_o, of the melt at low shear rates (Nelson, *et al*, 1993). The Frenkel model for viscous sintering is given by,

$$\left(\frac{x}{R}\right)^2 = \frac{3}{2}\left(\frac{\sigma t}{R\eta}\right) \tag{4.1.2}$$

where x is half the thickness of the neck formed between adjacent particles of radius, R, at time, t, σ is the surface tension, and η_o is the melt viscosity. For the sintering of pure polymers, material viscosities in the range indicated by Figure 4.2 can be run successfully in the SLS machine. Full density parts can be made from nylons and waxes, materials which have very low viscosities in the melt. Porous parts with acceptable strengths for many applications are made with higher melt viscosity materials such as injection molding grades of glassy polymers such as ABS (SAN plus dispersed rubber), Polycarbonate, PVC, and PMMA.

The polymer flow from the extruder tip in the FDM process is not a simple shear flow, but rather a combination of shear and elongational flow. The viscoelastic nature of the polymer fluid strongly affects the ability of the material to be drawn as a filament from the extruder tip (Tadmor and Gogos, 1979), and the resistance to stretching that is incorporated in the *elongational viscosity* is used to characterize the phenomena (Bird, *et al*, 1977). The elongational viscosity is similar in magnitude to the apparent Newtonian viscosity at low shear rates, however it is not sensitive to shear rate. Melt strength is, however, sensitive to molecular entanglement (Graesley, 1974). Polymers with molecular weight averages below that required for entanglement may have insufficient melt strength for successful use in the FDM process.

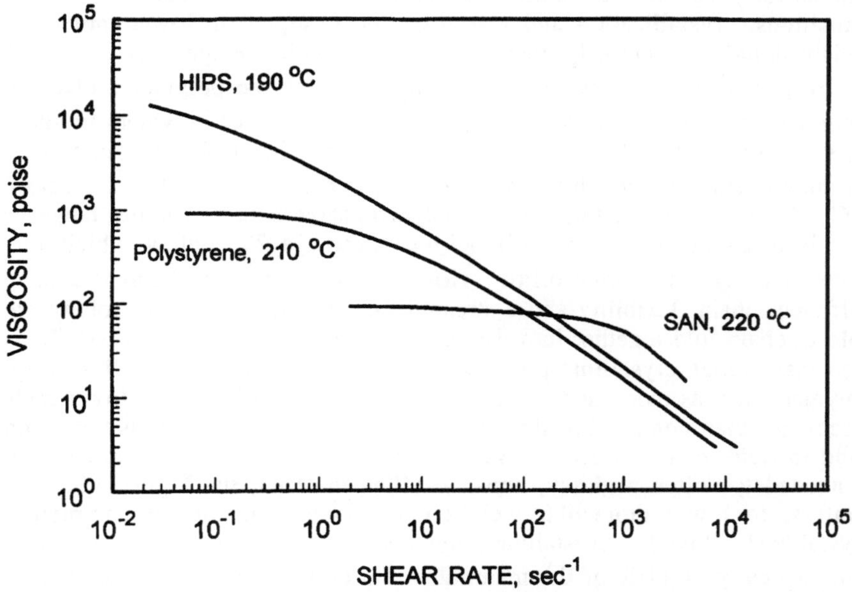

Figure 4.2 Melt Viscosities of Various Polymers (Tadmor and Gogos, 1978)

4.1.3 Mechanical Properties

All three SFF processes produce plastic parts by thermal and chemical processing techniques that are considerably different than conventional injection molding, compression molding, and extrusion casting processes. The stresses and material orientations that are built into parts by each of these SFF processes can be considerably different than those that are generated by conventional techniques. For example, flow orientation in the direction of filament placement may well be a feature of the FDM process. This can cause higher moduli and strengths in the placement direction that could be considerably different from these same properties in an injection molded part. Similarly, the SLS process builds objects from successive layers of fused powder, and SLS parts can show anisotropy in thermal expansion and sintering as a result of the incomplete fusion of layers (Nelson, *et al*, 1993, 1995, Deckard and Claar, 1993, Vail, *et al*, 1993). Consequently, parts from these SFF processes are not to be used to qualify materials or otherwise infer a suitability for use of a design that is intended for use by conventional processes. Similarly, characterization of materials that are formed by conventional casting techniques may be inappropriate for predicting the behavior of parts made by these SFF processes. Despite these problems, the basic properties of polymers, reviewed below, are of general importance to each of the SFF processes being reviewed here.

Polymer melts are simply high molecular weight liquids, and as such they can solidify as their temperatures are lowered through the same phase transitions, crystallization and glassification, observed in lower molecular weight liquids. Figure 4.1 suggests that the transition temperatures, T_m and T_g, respectively, are elevated to plateau values as the polymer molecular weight is increased. Consequently, it is not unusual for polymers to show T_g's in excess of 100°C while those of ordinary chemical compounds and the monomers from which the polymers are constructed are often far below 0°C. Figure 4.1 also infers that all polymers show both transitions, however that is often not the case. All polymers show a T_g, below which the material is rigid and often brittle. However, only polymer chains that have sufficient chain flexibility and a streamlined, regular, chain morphology are able to chain-fold together into the small ordered regions, called crystallites, that characterize crystalline polymers. Because of higher chain flexibility, polymers such as nylon and waxes that are crystalline often show lower melt viscosities than comparably-sized amorphous polymers. The transition from solid to fluid melt also occurs over a much narrower temperature range than is needed to soften a glassy, non-crystalline material, see Figure 4.3. In contrast to lower molecular weight compounds that can be completely crystallized below T_m, crystalline polymers are actually only semicrystalline with typically 50-60% of the material in the higher density crystallites and the remainder in a lower density, disordered or amorphous state. Since for most polymers, $T_m \approx 2T_g$, on an absolute temperature scale, semicrystalline polymers can be designed so that their use temperature lies between the T_g and T_m transitions. These materials can be tough because their amorphous regions are in the rubbery state, yet strong and rigid because their crystallites are rigid and serve to connect the amorphous regions.

The presence in crystalline polymers of two phases with different densities causes transmitted light to scatter. Consequently, crystalline polymers, such as nylon, polyethylene, waxes, and polypropylene, appear to be opaque. Likewise, amorphous, non-crystalline, polymers, such as polystyrene, cis-PBD, epoxies, polycarbonate, poly(methyl methacrylate) or PMMA, and poly(vinyl chloride) or PVC are transparent unless a second phase, filler or rubber impact modifier, is deliberately added to modify the material properties. Likewise, crosslinked materials such as the SLA photopolymers are transparent glasses because chain folding is not usually possible in crosslinked networks, see below.

As suggested by the above discussion, polymers that are amorphous are generally used for structural purposes at temperatures below their T_g's. Above T_g, these materials will show reduced moduli and may flow under the influence of stress. Figure 4.3 summarizes the effects of temperature, polymer molecular weight, crystallinity, and degree of crosslinking on tensile modulus. The moduli of polymers that do not crystallize and are not crosslinked are seen to drop dramatically, indicating substantial creep and material flow, as their temperatures are raised through their T_g. The moduli of semi-crystalline materials also decrease with increasing $T > T_g$, however the rate of decrease is much reduced due to the presence of the unmelted crystallites that physically link the chains. Once the temperature is raised above T_m, the crystallites melt, and substantial flow and creep again occurs. Systems that are crosslinked consist of chemically-coupled chains, see Figure 4.4. Crosslinked systems show some softening or modulus reduction for $T > T_g$, usually do not show T_m due to the constraints to motion associated with crosslinking, and show moduli plateau that are largely temperature independent until polymer degradation occurs. Relative to thermoplastics, crosslinked materials show much reduced creep and stress relaxation, and they show no fusion and material flow after the crosslinking process has occurred.

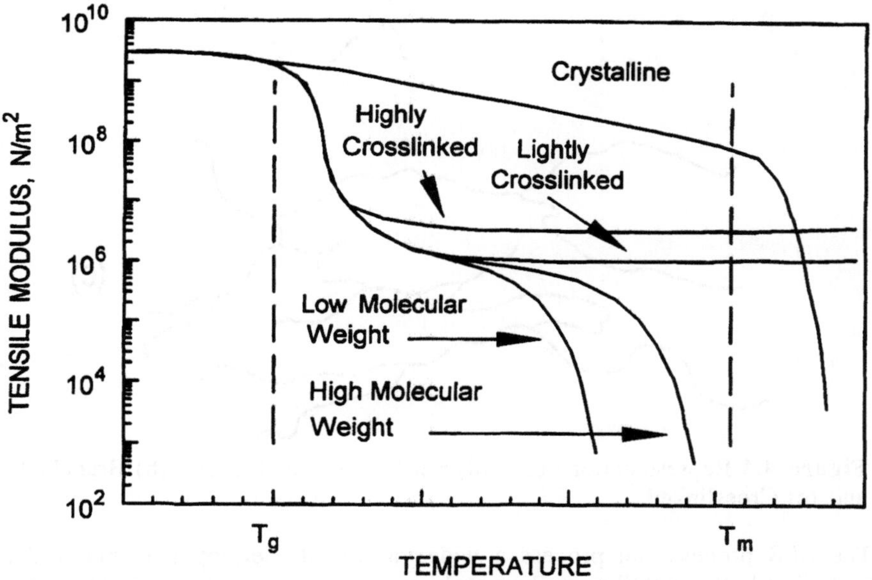

Figure 4.3 Effect of Temperature on Tensile Modulus of Various Materials (taken from Progelhof and Throne, 1993)

Significance to processes

The SLS and FDM processes are designed to handle thermoplastic materials, either linear or branched, Figure 4.4, that can be repeatedly heated to melt or soften and cooled to freeze or solidify as the temperature passes through the physical transitions signified by T_g and/or T_m. The SLA process employs layers of low molecular weight multifunctional liquids that are locally crosslinked by the application of geometrically directed UV light.

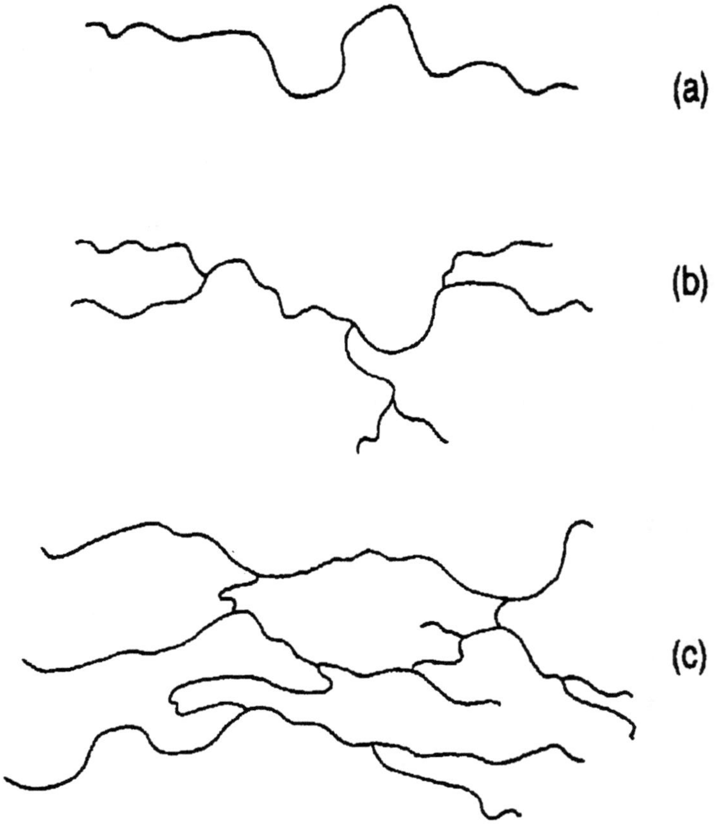

Figure 4.4 Representations of Polymer Types: (a) Linear , (b) Branched, and (c) Crosslinked.

The SLS process can process a wide variety of thermoplastic materials, including both crystalline and amorphous polymers, to achieve parts with a wide variety of properties. However, the properties of the part will be probably lower than the properties of the cast material comprising it because of layered porosity. Strengths of porous metal and plastic parts typically follow expressions of the form (German, 1989),

$$\sigma = C\sigma_0\rho^m \tag{4.1.3}$$

where σ and σ_0 are the strengths of the porous and solid forms of the material, respectively, C and m are empirical constants, and ρ is the fractional relative density of the porous object. Assuming the presence of a critical strain at break and linear elastic mechanical behavior, the modulus is related to density and material properties in a manner similar to equation (4.1.2). Depending on the material and processing conditions used, ρ can vary from little more than the density of the powder bed, in the case of polymer/ceramic composite powders (Nelson, *et al*, 1995) to nearly 1.0 for waxes and nylon. Because of their lower melt viscosities and higher sintering rates, equation (4.1.2), the crystalline polymer powders can give part properties that approach cast values. Higher densities can be achieved with amorphous materials by reducing polymer molecular weight to improve rates of sintering, however this may not lead to improved strength if the molecular weight is reduced below the critical value (Joseph, *et al*, 1982, Gardner and Martin, 1979).

The FDM process is currently limited to four materials: two waxes, an ABS, and a nylon (Comb and Priedeman, 1993). Except for the ABS, all are semicrystalline polymers with moduli near 80,000 psi, tensile strengths less than 2,000 psi, and elongations at break under 7%. Particularly the low elongations are evidence for low molecular weights, relative to conventional injection molding materials. Low molecular weights are required to keep the pressure low in the liquefier tip. The pressure is limited, in turn, by the force at which the filament drive wheels (the polymer filament is the "piston" for extruding molten polymer from the liquefier) slip on the filament surface, as well as the buckling force for the filament piston. Temperature control is critical; low extrudate temperatures result in poor bonding, possible delamination, and low material flow rates, while high extrudate temperatures lead to surface rippling.

The SLA process uses photopolymerizable oligomers that can be crosslinked to form transparent polymeric systems with essentially infinite molecular weights. New, low viscosity, photopolymerizable epoxy resins show cast and cured tensile moduli near 390,000 psi and elongations at break near 10% with factor of 10 improvements in impact strengths (to 30 kJ/m^2). These properties may not be achieved in actual SLA parts for a variety of reasons that are related to operation of the SLA machine, some of which are discussed below.

4.2 DIMENSIONAL ACCURACY

The ability of a SFF process to produce accurately-shaped geometry is critical to its overall acceptance in the market place. All three SFF processes can achieve, with appropriate adjustments to process parameters, highly accurate parts with linear dimensions that are accurate to within an uncertainty less than ±0.005 in/in, however it is fair to state that this accuracy is influenced by both material and choice of process parameters and may not be achieved on the first part processed from a new data file.

Shrinkage and related warpage are probably the biggest generic problems faced by the SFF processes under discussion. If shrinkage were uniform, the problem would be trivial; one would simply scale the part geometry to accommodate the shrink rate. Unfortunately shrinkage is often not uniform, locally, and part warpage can result.

Material shrinkage during temperature and phase changes (FDM and SLS), morphology changes (SLS), and chemical changes (SLA) contribute to inaccurate part formation and can be an important cause of warpage and curl in the SFF produced part.

Due to formation of crystallites, semicrystalline polymers typically show 4% to 8% decrease in volume relative to the amorphous rubber state at the freezing point, taken as T_m, see Figure 4.5. Amorphous materials, by contrast, show only a reduction by roughly a factor of approximately two in the rubber linear thermal expansion coefficient, from $\alpha_R \approx 5 \times 10^{-4}$ in/in°C to $\alpha_G \approx 5 \times 10^{-5}$ in/in°C as the material cools through the T_g. Consequently, an amorphous material with T_g of 110°C could be expected to show a linear shrinkage of 0.008 in/in between a melt temperature of 150°C and 30°C while a crystalline material with T_m of 150°C would show a linear shrinkage approaching as much as $\approx$ 0.039 in/in between the same temperature boundaries. The higher shrinkage in crystalline polymers often precludes their use where high precision injection molded parts are required, and can be a contributing cause to inaccuracy in parts formed by SLS and FDM as well.

Nearly all polymerizing systems, including those used by the SLA process, show a decrease in volume that is related to the formation of new chemical bonds. As suggested by Figure 4.6 (Bajaj, 1995), opening of vinyl bonds can cause substantial volumetric shrinkage that is non-linearly proportional to the concentration of vinyl bonds in the monomer for monomers as diverse as vinyl chloride liquid (34.4 vol% shrink) and di-vinyl-bi-benzyl (6.3 vol% shrink). Monomers such as methyl methacrylate (MMA), that have vinyl concentrations in the bulk monomer near 10 M (gmols/liter) typically show a 21 vol% polymerization shrinkage. For MMA, this translates to about 0.07 in/in linear shrinkage in each principle direction due to polymerization, a value that is roughly twice the expected shrinkage of a thermally processed crystalline polymer and nearly nine times that expected for the thermally processed amorphous example above.

Jacobs (1992) points out that the relationship between SLA processing conditions, shrinkage, and related curl of the part is not as simple as suggested by Figure 4.6, however he notes that monomers that show high polymerization shrinkage generally also show higher degrees of warpage. Part of the solution to the shrinkage problem in SLA is to develop complex monomers that have lower vinyl concentrations and low viscosity yet

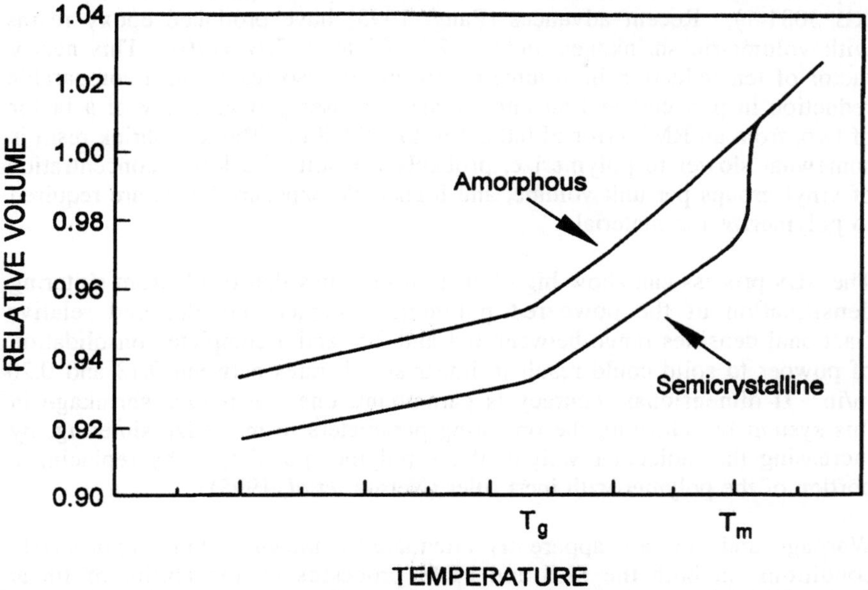

Figure 4.5 Volume vs Temperature for Crystalline and Amorphous Materials. (Billmeyer, 1962)

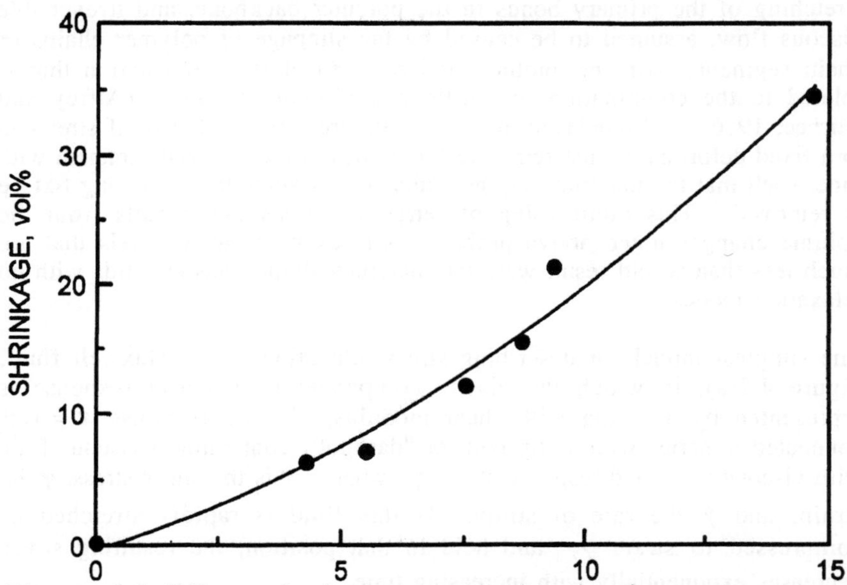

Figure 4.6 Effect of vinyl concentration in the monomer on polymerization shrinkage (Bajaj, 1995).

react rapidly and completely enough to form well crosslinked networks. Typical early SLA resins showed shrinkages in the range 4-6 vol% (Cibatool XB 5081-1). Recent advances (Pang, 1993) have produced epoxy resins with volumetric shrinkages under 0.7% (Cibatool XB 5170). This nearly factor of ten reduction in volumetric shrinkage also results in a comparable reduction in part curl and an improvement in user part accuracy of a factor of two, from an RMS error of 0.0053 in to 0.0028 in. The low shrink resin is somewhat slower to polymerize, probably a result of a lower concentration of vinyl groups per unit volume, and higher uV exposure levels are required to polymerize the material.

The SLS process can show high linear shrink rates that result from sintering densification of the powdered polymers. Typical powder bed relative fractional densities range between 0.4 and 0.6, and a complete consolidation of powder to solid could result in linear shrink rates between 0.13 and 0.20 in/in. If dimensional accuracy is paramount, one can reduce shrinkage in this system by adjusting the operating parameters to minimize sintering, by increasing the molecular weight of the polymer powder, or by replacing a portion of the polymer with inert filler (Nelson, *et al*, 1995).

Warpage and curl are apparently attenuated somewhat, under appropriate conditions, in both the SLS and FDM processes by the ability of linear thermoplastic materials to stress relax. Thermoplastic materials are viscoelastic materials whose total deformation is a combination of instantaneous elastic deformation, assumed to be caused by the bending and stretching of the primary bonds in the polymer backbone, and irreversible viscous flow, assumed to be caused by the slippage of polymer chains or chain segments past one another, and retarded elastic deformation that is related to the combination of elastic and viscous responses (Alfrey and Gurnee, 1956). Viscoelastic materials can stress relax; that is, if stretched to a fixed deformation, the retractile forces within the material diminish with time, such that the material may not fully retract when the stretching fixture is removed. This diminishing of retractile forces that results from the volume changes noted above probably reduces warpage to levels that are much less than would result were the materials simply elastic solids with no relaxation modes.

The simplest model for describing stress relaxation is the Maxwell fluid, Figure 4.7(a), in which the elastic component of material response is represented by a spring with shear modulus, G, and response $\tau = G\gamma$, connected in series with a slip unit, or "dashpot", containing a viscous fluid with viscosity, η_o, and response $\tau = \eta_o\dot{\gamma}$, where τ is the shear stress, γ the strain, and $\dot{\gamma}$ the rate of strain. If this fluid is rapidly stretched or compressed to strain γ_o and held in that position, the resulting stress decreases exponentially with increasing time,

$$\tau = G\gamma_0 exp(-t/\lambda)$$
$$\lambda = \eta_0/G \qquad (4.2.1)$$

where λ is a time constant for the response. Because the strain is fixed in this experiment, one can equivalently define the time dependent relaxation modulus of the material to be,

$$G(t) = G_0(-t/\lambda) \qquad (4.2.2)$$

Since λ is proportional to η_o and since η_o decreases with increasing temperature, T, according to,

$$\eta_0 = Aexp(\Delta E/RT) \qquad (4.2.3)$$

where ΔE is the activation energy for viscous flow, and A a prefactor (McKelvey, 1962), it follows that one can obtain the same value of $G(t)$ by adjusting either time, t, or temperature, T. This is the basis for the famous "time-temperature superposition principle" (Ferry, 1961) that permits data gathered at short times and high absolute temperatures to be used to predict material behavior at longer times and lower temperatures. For our immediate purposes, we note that increasing temperature reduces viscosity, reduces the time constant, and hastens the relaxation.

Actually, polymer dynamic response is too complicated to be accurately explained by one time constant, λ. What is usually done is to assemble an n-element parallel array of Maxwell units, Figure 4.7(b), to describe the fluid (Tobolsky, 1960). A curve fit to experimental data by polynomials of exponential terms of the form,

$$G(t) = \sum_{i=1}^{n} G_i(t-t/\lambda_i) = \int_{-\infty}^{+\infty} G(\lambda)exp(-t/\lambda)d\lambda \qquad (4.2.4)$$

where $G(\lambda)$ is a continuous relaxation spectral function, is usually done. Equations that are similar in form to equation (7) can also be written to describe the time dependent compliance, $J(t)=(1/G(t))$ and other modulus functions for Maxwell and related Voigt fluids.

Figure 4.8 shows the tensile relaxation moduli, $E(t)$, for polystyrenes with differing molecular weights at 15°C above its T_g (Tobolsky, 1964). The time for the modulus to decrease from 107 to 105 dynes/cm2 increases from about 10 sec for 18,600 molecular weight material to 9.5 x 10^5 sec (11 days) for 315,000 molecular weight material. According to equations (1) and (4), a time ratio for these molecular weights should be (315/18.6)3.4 or 1.5 x 10^4 as opposed to the observed ratio of 9.5 x 10^4. The prediction is of the correct order of magnitude, and we can conclude that most of the relaxation of the lower molecular weight materials appears to be related to their lower viscosities, as suggested by the Maxwell fluid models.

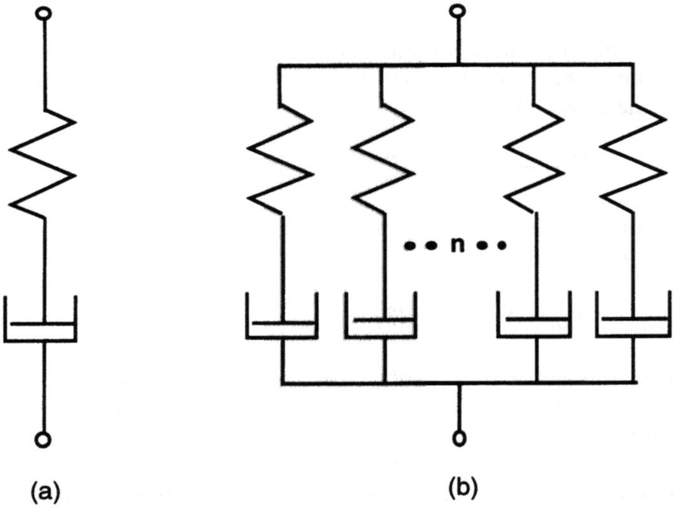

Figure 4.7 Mechanical Analogs of Viscoelastic Liquids

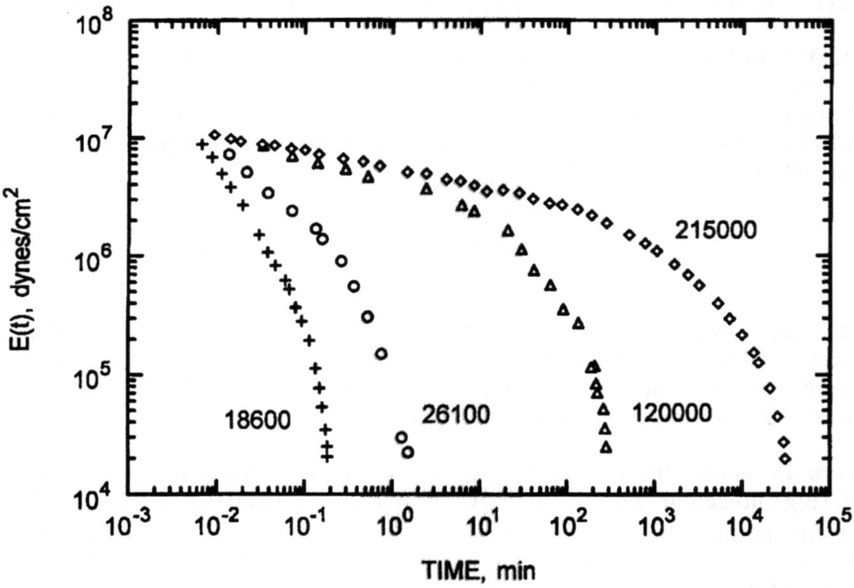

Figure 4.8 Effect of Molecular Weight on Relaxation Moduli of Polystyrene at 115°C (Tobolsky, 1964).

Significance to Processes

The crosslinked materials that are formed by SLA will not stress relax to the same extent as linear polymers, and one should not expect stress relaxation to greatly alter curl and warp in this system. The SLA process reduces warpage by partly polymerizing the monomer through its WEAVE™ and STAR-WEAVE™ laser build patterns (Jacobs, 1992) to set the shape. The shape is then removed to an oven where further polymerization, hopefully under more uniform conditions, can be carried out.

To reduce warpage, the Stratasys FDM process deposits material on a removable foam foundation that is firmly anchored to the bed of the machine. The foam lowers the cooling rate of the extruded filaments, which permits the viscoelastic polymer to relax out some of the stresses that are induced by crystallization. The melt temperature and surrounding air temperature are carefully controlled to reduce the cooling rate of the extrudate, thereby permitting longer time for relaxation of shrinkage stresses, which cause warpage and delamination in the part.

In the SLS process, warping and curling are generally reduced by operating the powder bed temperature at as high a temperature as possible, consistent with the spreading and blocking characteristics of the powder. This temperature, in the case of amorphous materials, is normally near the polymer T_g , and depending on powder size and polymer molecular weight, may actually exceed T_g .

Warpage and shrinkage stresses in semicrystalline nylon are also controlled by carefully regulating the thermal environment in the present SLS process to reduce the rate of cooling to a level that is low enough to permit stress relaxation phenomena to occur (Forderhase, 1994). This strategy is necessary in working with nylon due to its large inherent shrinkage. Because nylon's melt viscosity and relaxation time constants, at temperatures just below the melting point, are quite small, it is possible to build full density, warp free, nylon parts from powder despite these large shrinkages. In earlier work with wax powders, a "superbase" of beeswax was placed in the part cylinder and wax anchors were attached between the part base and the superbase to produce flat parts (Forderhase and Corden, 1993). They found that flat parts could also be produced by reducing the laser energy per unit area so as to reduce the degree of sintering and associated shrinkage, at the cost of reduced density and strength.

4.3 CHEMISTRY CONSIDERATIONS

Those who are interested in SFF processes that employ thermoplastic materials need not be overly concerned about the chemistry of formation of these materials. Such processes simply heat the materials above T_g and T_m

to cause them to flow and fuse to the previous layer, then cool the materials to form load bearing elements in the prototype. These are physical, not chemical, transitions, and a wide variety of thermoplastic materials may be used with often only minor modifications to process parameters. Because of its potential impact on recycling and cost, polymer degradation by repeated exposure to heat and oxidants, primarily atmospheric oxygen, is the primary chemical concern in thermoplastics processing, and is discussed below. Processes, such as SLA, that employ thermosetting materials are actually using chemical reactions to build molecular weight and properties. Somewhat greater attention to those parameters that control the details of thermosetting chemical reactions are consequently required by these processes.

Primarily because their discussion covers the widest range of materials, only radical-based polymerization and degradation mechanisms are considered in this section. By so doing, the author understands that an important class of acid-base polymerizations that lead to polyesters, polyamides, and other condensation polymers is neglected as are the related degradation reactions, associated with hydrolysis. The interested reader is referred to Stevens (1990) and Allcock and Lampe (1990) for further details on these systems.

4.3.1 Free Radical Polymerizations

Monomers that contain vinyl groups, designated CH_2=CHX where X is a pendant substituent that may be polymeric and may contain additional vinyl groups, C is chemical symbol for carbon, and H that for hydrogen, can rapidly polymerize by a free radical reaction to high molecular weight polymers. The free radical polymerization process is a three-step process of radical initiation through decomposition of an initiator by addition of heat or light energy, propagation during which molecular weight is increased by the addition of monomer units, and termination during which radicals are destroyed and the reaction stops. Figure 4.9 illustrates these steps for a benzoyl peroxide-initiated reaction where the peroxide is designated as initiator with concentration, [I]. During initiation, the initiator thermally decomposes to form radical products, with concentration [R•], that react with the vinyl double bond in the monomer, at concentration [M], to produce a vinyl radical, M_1•. During propagation, the vinyl radicals rapidly react with more monomer to form vinyl radical chains of increasing molecular weight. Termination occurs when two radicals interact by either coupling or disproportionation mechanisms. Because the rate constants for termination, k_t, are at least 10,000 times larger than those for initiation, k_d, it is reasonable to assume that shortly after the reaction begins, the formation and destruction of radicals occur at the same rates; hence the overall, "steady state" concentration of radicals, $[M\bullet] = \Sigma_x [M_x\bullet]$, is given by,

$$\text{rate of initiation} = R_i = 2fk_d[I] = 2k_t[M\bullet]^2 = R_t = \text{rate of termination} \quad (4.3.1)$$

which gives,

$$[M\bullet] = \sqrt{\frac{fk_d[I]}{k_t}} \tag{4.3.2}$$

where f is an efficiency factor to account for the fact that not all radicals lead to the formation of vinyl radicals. The rate of polymerization is the rate of monomer consumption,

$$R_p = -\frac{d[M]}{dt} = k_p[M][M\bullet] = k_p[M]\sqrt{\frac{fk_d[I]}{k_t}} \tag{4.3.3}$$

and is proportional to the square root of the initiator concentration and to the first power of monomer concentration. A measure of the average molecular weight is the *kinetic average chain length*, v_o, as the average number of monomer units polymerized per chain initiated, a ratio that is equal to the rate of propagation per rate of initiation (or termination at steady state), or

$$v_0 = \frac{R_p}{R_t} = \frac{k_p[M]}{2\sqrt{fk_tk_d[I]}} \tag{4.3.4}$$

The degree of polymerization, DP_o, a term that is defined as the average molecular weight of the polymer divided by the molecular weight of the repeat unit, is related to the kinetic average chain length by

$$DP_0 = v_0(1+y) \tag{4.3.5}$$

Where y is the ratio of the rate of termination by coupling to that by all mechanisms. If coupling predominates, DPo is twice v_o. In some polymerizations, the polymer molecular weight is controlled by the addition of chain transfer agents. These chemicals act as additional termination sites during polymerization to reduce the molecular weight via the relationship,

$$\frac{1}{DP} = \frac{1}{DP_0} + C_S\frac{[S]}{[M]} \tag{4.3.6}$$

where C_S is a coefficient particular to the chain transfer agent and polymerization employed, [S] is the concentration of agent, and [M] that of the polymerizing monomer.

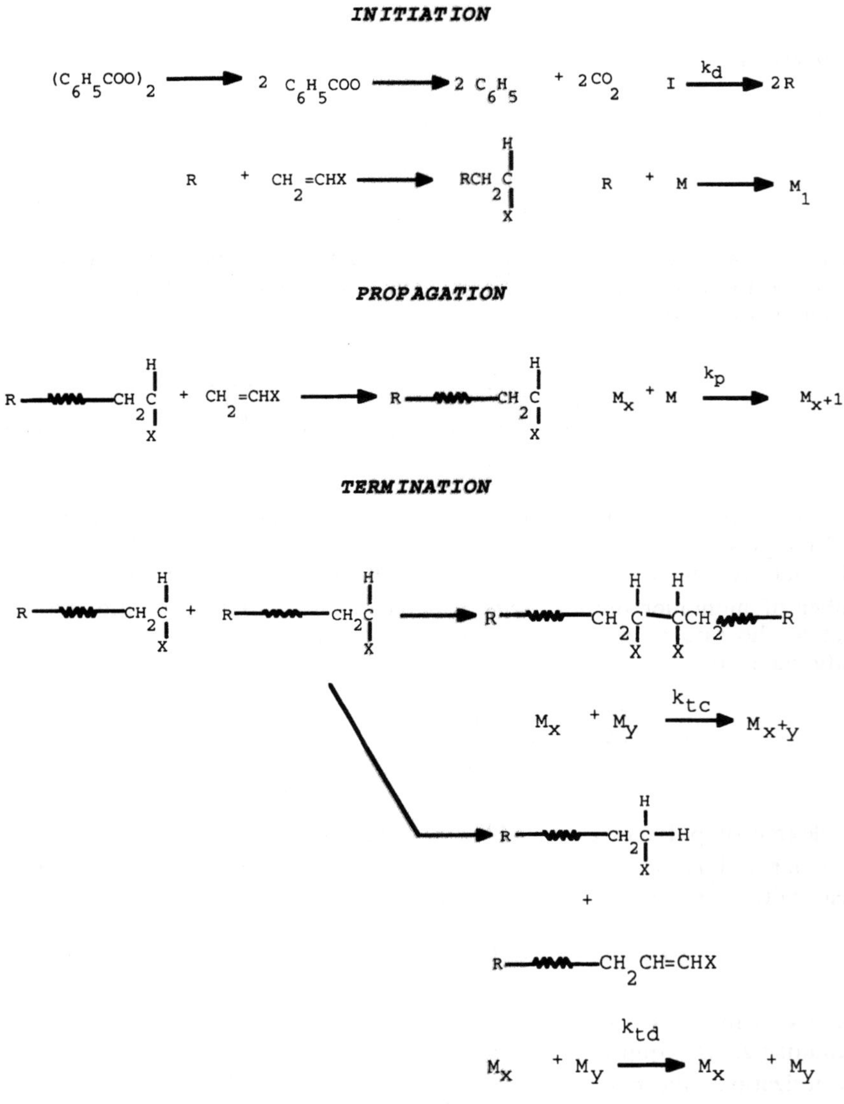

Figure 4.9 Schematic of Free Radical Polymerization

Equations (4.3.3) and (4.3.4) correctly suggest the major compromise that exists between desire for high molecular weight and the economic or process need for high polymerization rates when selecting initiator concentrations. Doubling the initiator concentrations increases the rate by a factor of 1.4 but decreases the average molecular weight by the same factor. Raising the reaction temperature or increasing the efficiency of the initiation process also raises the rate of polymerization and lowers the average molecular weight. As suggested by equations (4.3.5) and (4.3.6), numerous deviations from equations (4.3.3) and (4.3.4) can occur. These include a shift in termination mode from predominantly coupling (y=1) to predominantly disproportionation (y=0), equation (4.3.5) to transfer of radicals to a non-propagating solvent molecule, equation (4.3.6). Examples of the use of chain transfer agents are presented in the discussion of binders for SLS materials, later in this chapter, see text that is related to Figure 4.16.

4.3.2 Photopolymerization

Any monomer that can be polymerized by free radical addition can be polymerized by photopolymerization or photosensitized polymerization. With reference to Figure 4.9, all that changes is the nature of the initiation step. Few vinyl monomers absorb light in the near ultraviolet, UV, wavelength range, 250-500 nm, that is most convenient. Photosensitizers, chemicals that absorb light in the desired wavelength range and dissociate to free radicals, are consequently required to initiate most photopolymerizations. Often, photosensitizers such as azoisopropane,

$$Me_2CHN{=}NCHMe_2 + h\nu \rightarrow 2Me_2CH\bullet + N_2$$

$$I + \text{light} \xrightarrow{k_d} 2R\bullet + \text{nitrogen gas} \qquad (4.3.7)$$

can be found that do not thermally dissociate over wide ranges of temperature yet readily photodissociate when irradiated with uV light at wavelengths between 300 and 400 nm (Allcock and Lampe, 1990). Such materials permit the practice of StereoLithography™ (SLA) in which the surface of a vat of photopolymerizable monomer is locally polymerized with a UV laser to build a shape layer. The thermal stability of the monomer /photosensitizer system must be excellent to permit vat storage of monomer for extended periods near room temperature.

Comparison of equation (4.3.7) and the free radical thermal initiation scheme in Figure 4.9 suggests that $R_i = 2k_d[A]$ should describe the rate of initiation from photoinitiator with concentration [A]. Actually, the photoinitiation behavior is a bit more complicated because the photoenergy or power is not uniform throughout the liquid but instead follows Beer's Law (Daniels and Alberty, 1961),

$$Log_{10}\left[\frac{P}{P_o}\right] = -\varepsilon[\mathrm{A}]x = \frac{1}{A_\lambda} \qquad (4.3.8)$$

where P_o is the beam radiant power per unit area entering the system boundary at the surface, $x = 0$, P the power per unit area leaving the system at distance x from the surface, ε is the molar absorptivity, and A_λ the absorbance of the system at the particular frequency or wavelength of radiation employed. Equation (4.3.8) can be rewritten as,

$$P = P_0 exp(-2.303\varepsilon[\text{A}]x) \tag{4.3.9}$$

which clearly shows an exponential decrease of photopower with distance, x, under the surface. The rate of initiation, R_i, is related to the absorbed photopower per unit volume, dP/dV, and the quantum yield, ϕ,

$$R_i = 2\phi(dP/dV) \tag{4.3.10}$$

where ϕ is defined as the number of chains initiated per quantum of light absorbed, a value that lies between 0 and 1, and the factor 2 is associated with the assumption that two radicals are generated, via equation (4.3.7). By considering the energy balance on a thin slice, dx, of fluid at distance x from the surface, one can easily show that,

$$\frac{dP}{dV} = \frac{AP_x - AP_{x+\delta x}}{A\delta x} = \alpha P_0 exp(-\alpha x) \tag{4.3.11}$$

where A is the area of the volume element through which the radiant power is passed, and $\alpha = 2.303\varepsilon[\text{A}]$. The rate of initiation is then given by combining equations (4.3.10) and (4.3.11),

$$\begin{aligned} R_i &= 2\phi P_0 \alpha exp(-\alpha x) \\ &= 4.6\varepsilon[\text{A}]\phi P_0 \quad \text{at } x \approx 0 \end{aligned} \tag{4.3.12}$$

The remaining terms in the photopolymerization process are the same as for the thermally initiated process, and the steady-state assumption for radical population still holds. Consequently, the propagation rates and kinetic average chain lengths are given by,

$$\begin{aligned} R_p &= -\frac{d[M]}{dt} = k_p[\text{M}][\text{M}\cdot] = k_p[\text{M}]\sqrt{\frac{2.303[\text{A}]\phi P_0}{k_t}} \\ v_0 &= \frac{R_p}{R_t} = \frac{k_p[\text{M}]}{2\sqrt{2.303k_t\varepsilon[\text{A}]\phi P_0}} \end{aligned} \tag{4.3.13}$$

in the limit of small values of αx. Some monomers, such as alkyl vinyl ketones and vinyl bromide, can directly photoreact to give radicals that initiate polymerization. For such systems, the monomer is also the photosensitizer, and equations of the form given in (20) can still be used by simply replacing [A] by [M].

Rigorous calculation of the gel point, the conversion at which the reacting mass solidifies, or equivalently, calculation of the cure depth is a difficult task for free radical reactions of thermosetting, multifunctional, monomers and is an on-going research effort (Bajaj, 1995, Flory, 1953, Mikos, 1986). If all vinyl groups are isoreactive and intramolecular cyclizations can be neglected, the fractional conversion, θ_C , at the gel point is given by (Flory, 1953),

$$\theta_C = \frac{[M_0]-[M_C]}{[M_0]} = \frac{1}{\rho_0 \bar{y}_w} \quad (4.3.14)$$

where $[M_o]$ and $[M_C]$ are the initial and gel vinyl bond concentrations, respectively, $\bar{y}_w$ is the weight average degree of polymerization of the chains and ρ_o is the fraction of double bonds that belong to the bi-unsaturated molecules in the mixture. Unfortunately in many systems, equation (4.3.14) does not generally predict gelation very well (Walling, 1945), and more complex treatments are required (Mikos, 1986) to link kinetic expressions such as equations (4.3.13) to prediction of gelation.

Significance to Processes

Of the three processes presently under discussion, the kinetics of thermal and photopolymerization are of greatest concern to the SLA process. This process actually builds a crosslinked polymer in the shape of the desired part and care must be taken to adjust and operate the process to ensure that the part, once made, will have the appropriate properties. Equations (4.3.13) show that a compromise exists between the conflicting desires for high polymerization rates and high product molecular weights. Higher laser power leads to higher polymerization and part-build rates, however both linear and nonlinear polymeric products may become brittle due to the lower molecular weight, of primary and crosslinking chains, respectively, that result.

The cure depth strongly affects SLA part quality. If the cure is not deep enough, the layers do not adhere to each other and delamination can occur. If the cure depth is too great, curl will occur, causing distortion and part inaccuracy. Since the curing characteristic of each batch of resin is slightly different (variations in [A], for example), Jacobs (1992) points out that the WINDOW PANE™ technique has been developed to calibrate the SLA process to the particular resin being photopolymerized. Basically, laser power and scan speed are controlled to build test parts from the resin, parts are post-processed, and cure depths are measured. A plot of the "working line" is made of cure depth, C_d, vs the natural logarithm of the laser beam exposure at the beam centerline, E_{max} , and the cure characteristics of the resin are established from the slope (D_p) and the intercept where cure is zero (E_c), according to,

$$C_d = D_p \ln(E_{max} / Ec) \tag{4.3.15}$$

Recognizing that the maximum beam exposure, E_{max}, for a Gaussian beam is related to the beam power, P, beam speed, V, and a diameter of influence, W, by,

$$E_{max} = \frac{\sqrt{\frac{2}{\pi}}\, P}{WV} \tag{4.3.16}$$

Equation (4.3.15) is simply a restatement of Beer's Law, equation (4.3.9).

4.3.3 Degradation Mechanisms

The molecular weights and associated properties of polymers used in rapid prototyping applications can degrade by chemical oxidation, by thermal processes including random chain scission and depolymerization, and by radiation. While polymer degradation is usually considered to be undesirable, several applications of the SLS rapid prototyping technology to produce metal or ceramic parts require removal of a polymer in subsequent thermal processing steps. The success of these methods heavily depends on choosing polymers that have suitable thermal degradation characteristics, see below.

The oxidative degradation of an organic polymer generally proceeds through free radical reactions. At temperatures below 100°C, purely thermal degradation is very slow and degradation often appears to involve the reaction of oxygen with easily abstractable hydrogens on the polymer chain to form hydroperoxides, -OOH. The hydroperoxides can decompose causing a reduction in molecular weight and generation of free radicals, see Figure 4.10. The radicals thus formed then react with oxygen to yield more additional peroxides and hydroperoxides through a set of reactions that have initiation, propagation, and termination steps in much the same manner as described for the free radical polymerization process. That is, oxidation occurs under conditions where reaction propagation is much more rapid than initiation, and the chain oxidations are terminated by bimolecular reactions to nonradical products (Shalaby, 1979). Consequently, steady state radical populations are assumed, and overall oxidation rates can take the form, by analogy with the radical chain reactions, Figures 4.9 and 4.10, and equation (4.3.1),

$$R_{OX}=\frac{d[O_2]}{dt}=k_P\sqrt{\frac{R_i}{2k_t}}\,[\mathrm{H}] \tag{4.3.17}$$

where the uptake of oxygen per unit mass per unit time is taken as the oxidation rate, R_{ox}, and [H] is the concentration of easily abstractable hydrogen atoms per unit mass polymer. As suggested by Figure 4.10, the rate of initiation, R_i, is proportional to the oxygen concentration, however the proportionality may not be linear as a consequence of limitations of oxygen diffusion into the polymer. Through the term, [H], equation (4.3.17) suggests that polymers with high concentrations of easily abstractable hydrogen atoms will oxidize more rapidly. Generally, aliphatic polymers with branch points, such as the material illustrated in Figure 4.10 and polypropylene, or with heteroatoms in their backbones, such as polyethylene oxide, or with unsaturated functional groups that can activate hydrogens on adjacent carbons, such as polybutadiene, are prone to rapid oxidation. Polymers with aromatic backbone structures, such as polycarbonate and PET, are oxidatively more resistant than aliphatic materials, because hydrogens on the aromatic rings are unreactive to peroxy free radicals.

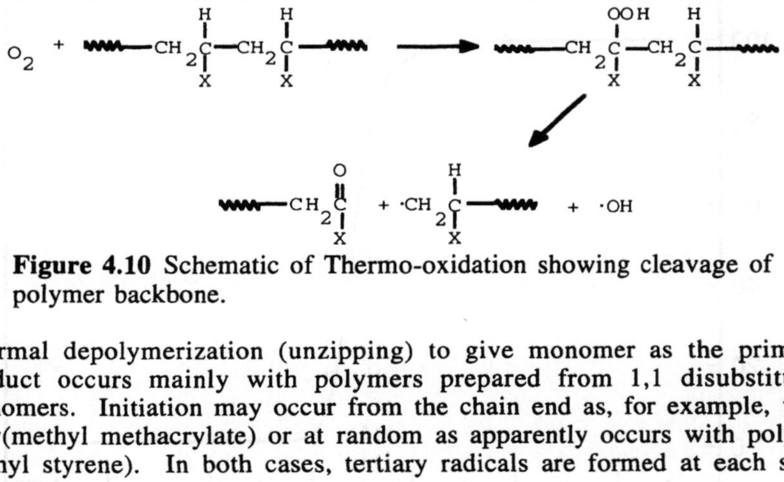

Figure 4.10 Schematic of Thermo-oxidation showing cleavage of polymer backbone.

Thermal depolymerization (unzipping) to give monomer as the primary product occurs mainly with polymers prepared from 1,1 disubstituted monomers. Initiation may occur from the chain end as, for example, with poly(methyl methacrylate) or at random as apparently occurs with poly(a-methyl styrene). In both cases, tertiary radicals are formed at each step, according to

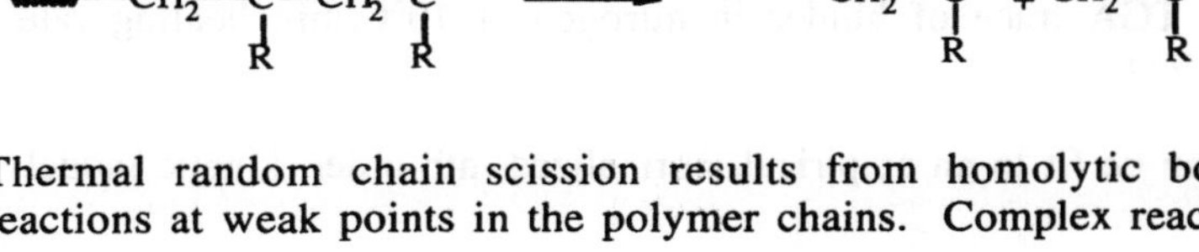

Thermal random chain scission results from homolytic bond cleavage reactions at weak points in the polymer chains. Complex reaction mixtures

with a wide range of molecular weights, including gel, can result. Random chain scission occurs in all polymers to varying degrees, but it occurs less with increasing substitution of the backbone, and depends as well on temperature. Radiation may similarly cause both crosslinking and degradation, depending on the dosage, polymer structure, and temperature. Ultraviolet and light causes 1,1-disubstituted materials to degrade exclusively to monomer at elevated temperatures and to degrade and crosslink at room temperature (Shalaby, 1979).

Because they are deliberately designed to depolymerize as an aid to their thermal removal in post-SLS processing, polymer binders that are based on copolymers of n-butyl methacrylate (BMA) and methyl methacrylate (MMA) have been observed to also depolymerize during SLS processing (Vail, *et al*, 1994, 1996). Figure 4.11 shows a typical Thermo-Gravimetric-Analysis (TGA) trace that shows the copolymer binder to decompose completely, leaving less than 0.5% residue. The depolymerization process begins near 275 °C and is completed by 400 °C when heated at 10 °C/min. TGA traces can also be done isothermally to obtain useful kinetic information about the depolymerization process.

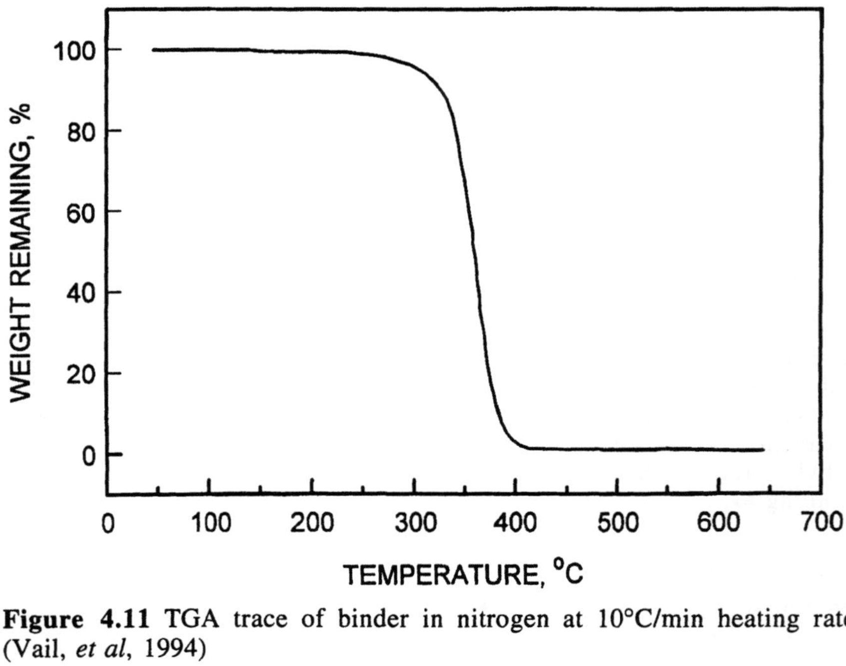

Figure 4.11 TGA trace of binder in nitrogen at 10°C/min heating rate (Vail, *et al*, 1994)

These data can be fit to an empirical, normalized, nth-order, kinetic model,

$$\frac{d}{dt}\left(\frac{w}{w_o}\right) = k_d\left(\frac{w}{w_o}\right)^n \qquad (4.3.17)$$

employed to express the mass loss rate dependency on mass, where w_o is the initial mass and w the mass remaining at time t. The rate constant, k_d, is assumed to follow the Arrhenius relationship,

$$k_d = A\exp\left(-\frac{\Delta E}{RT}\right) \qquad (4.3.18)$$

where A and ΔE are the frequency factor and the activation energy for the decomposition reactions, respectively, R the gas constant and T the absolute temperature. The pseudoreaction order, n, can be determined from the rearranged integral form of equation (4.3.17),

$$\left(\frac{w_o}{w}\right)^{n-1} - 1 = (n-1)k_d t \qquad (4.3.19)$$

by simply finding the value of n that best fits all the data. Figure 4.12 shows a typical construction for the copolymer that suggests that the reaction is approximately 3/2 order over the temperature range 290-350 °C and to conversions of the polymer to gas as high as 97%. This result contrasts with the first order response observed by Inaba, *et al*, (1988) for PMMA to conversions of about 65% over the temperature range 312-360 °C.

There is no compelling reason to believe that n-butyl methacrylate degrades in a manner that is profoundly different from that of methyl methacrylate, and the difference in reaction order may simply be a consequence of the changing molecular weight and resulting volatility, of the polymer residue during the course of the decomposition. Kashiwagi, *et al*, (1985) have shown that chain scission occurs before substantial depolymerization by unzipping so that the average zip length of PMMA in nitrogen declines by a factor of 4 as polymer conversion to gas increases from zero to about 80%. These same authors also show that the average zip length also declines by a factor of two as the decomposition temperature is increased from 258 to 324°C.

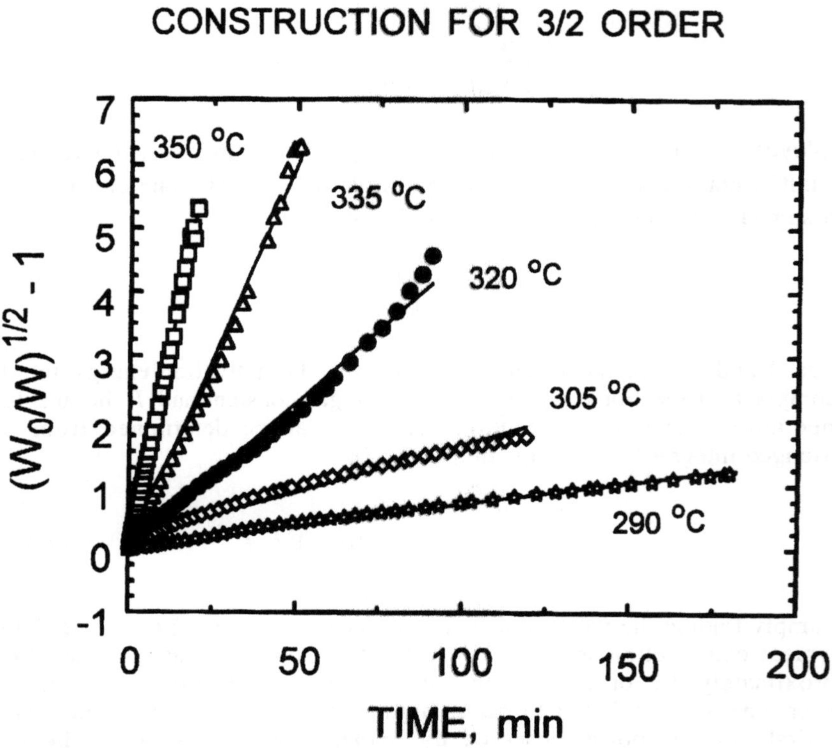

Figure 4.12 Construction for n=3/2 order (Vail, *et al*, 1996)

Figure 4.13 shows the construction to determine the Arrhenius constants for the copolymer degradation rate that were obtained from the line slopes in Figure 4.12 for 3/2 order. A fairly good straight line results which yields,

$$\Delta E = 44.1 \pm 6 \text{ kcal/mol}$$
$$A = 2.907 \pm 2.45 \text{ x } 1013 \text{ sec-1} \qquad (4.3.20)$$
$$n = 1.5$$

Significance to Processes

Most polymer forming processes hold the polymer at high temperatures for only a few minutes, and oxidation of the polymer is usually minimal under these circumstances. Antioxidants, typically hindered phenols and bis-phenols that preferentially react with free radicals, may be added to the polymer to protect it for the usual 15-30 min exposure to melt processing temperatures. The materials employed by the Stratasys filament extrusion process for SFF may be protected in this manner.

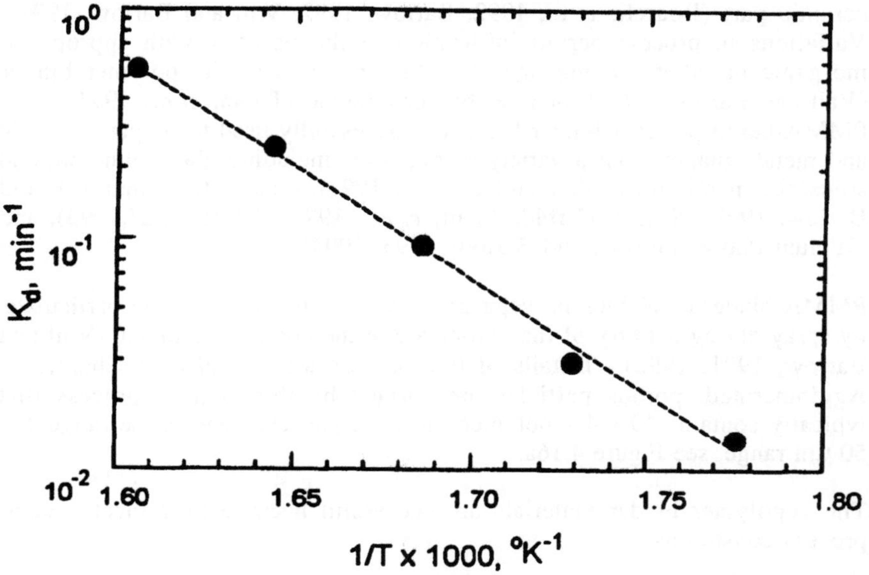

Figure 4.13 Effect of temperature on depolymerization rate constant (Vail, *et al*, 1996).

As suggested by Figure 4.10, elimination of oxygen effectively eliminates thermo-oxidation. For example, the SLS process holds polymer powder at 20-50°C below its fusion temperature for extended periods of time, 6-10 hours, and the unfused powder, typically half of total volume of parts and powder, is recycled so that some substantial portion of powder has seen considerable thermal history. Virtually no degradation in molecular weight has been observed in the SLS process, despite tens of hours of exposure to high temperature. The primary processing reason that degradation has not been a problem is the careful control of oxygen concentration to below 2% in the part-making chamber of the Sinterstation™.

The significance of the other polymer degradation issues to SLS processing and post-processing are discussed are in Chapters 6 and 5, respectively.

4.4 POLYMER BINDERS

A class of polymer binders, based on copolymers of MMA and BMA monomers have been developed at the University of Texas to permit the formation of metal and ceramic parts by the SLS process. Essentially, the metal or ceramic powder is coated with plastic binder, the coated powder is processed in the Sinterstation™, using the same IR laser that is normally used for polymers and waxes, to make a porous green shape. The porous green shapes are then processed further to thermally remove the PMMA-based binder and to sinter the polymer-free shape, thus creating a metal or ceramic part (Bourell, *et al*, 1992, Barlow, 1992, Vail and Barlow, 1991). Variations in process permit infiltration of the porosity with appropriate inorganic or other binding agents, before removal of the polymer binder (Vail and Barlow, 1992), or after binder removal (Tobin, *et al*, 1993). The PMMA-based polymer binder has been successfully used to prepare ceramic and metal shapes from a variety of materials including glass, alumina, and silica/zircon mixtures (Vail and Barlow, 1991), copper (Badrinarayan and Barlow, 1992), Silicon Carbide (Vail, *et al*, 1993, Nelson, *et al*,1995), and Calcium Phosphate (Lee and Barlow, 1993, 1994).

PMMA binder emulsions in water are coated on to the inorganic particulate by spray drying a slurry of the particulate in the polymer emulsion (Vail and Barlow, 1991, 1992). Details of this process are provided in Chapter 5. Agglomerated, porous particles are formed by this coating process that typically contain 20 vol% polymer and have particle sizes in the range 10-50 μm range, see Figure 4.16a.

The copolymer binder materials are successful because they meet several process constraints:

The binders should not be water soluble. This requirement is essential to produce a polymer-coated powder whose flow characteristics are independent of relative humidity.

Water insolubility of the polymer binder is also a key feature of post-SLS processes for producing ceramic and bioceramic parts. In these post-processes, water-borne ceramic cementing agents are infiltrated into the polymer bound part that has been previously shaped by SLS. Ease of handling and maintenance of shape of the green part throughout the cementing process require that the polymer binder must not dissolve or otherwise lose its strength when exposed to the water-borne cementing agent. The cementing agent is dried and set, and the polymer burned out to produce an all-ceramic part that has very low dimensional changes, typically <1% linear shrink, relative to the green part.

Binders must be applied to inorganic particles from an aqueous vehicle to yield a free flowing powder that is suitable for processing into shapes by SLS. This is a very practical issue that relates to technologies for coating the inorganic particulate with polymer. The polymer must be distributed so as to cover as much of the surface of the inorganic particulate as possible. This could be accomplished, in principle, by dissolving the polymer in a suitable organic liquid solvent to achieve a low concentration of polymer, depositing the solution on the surface of the inorganic, then evaporating the solvent to deposit the polymer. The problems of solvent-recovery cost and potential environmental harm associated with the use of organic solvents largely preclude such a process. These problems can be avoided, however, if water is used as the vehicle. Since the polymer can not be water soluble, the need for a water vehicle requires that the polymer binder material be highly dispersible in water, or water-emulsifiable.

Many polymers can be made by a technique known as *emulsion polymerization*. In this technique, liquid monomers are emulsified in water with the aid of an appropriate emulsifying agent. Water-borne, ionic initiators are typically incorporated to polymerize the monomer thereby forming the desired emulsified polymer. Typical emulsion particles are quite small, about 100 nm ($4x10^{-6}$ in) in diameter, so that the polymer is well distributed throughout the water vehicle with typical concentrations in the range 2-$5x10^{14}$ particles per cubic centimeter (Bovey, *et al*, 1955), 40-50% solids, and the emulsion is mechanically stable to the effects of gravity. Emulsion polymerization is generally known to produce high molecular weight products at high polymerization rates and is the method of choice for preparing the polymer binder that is commonly used in "water-based acrylic paint" systems.

The polymer binder must soften and flow at temperatures between 40°C and 100°C. Despite the similarity of some of its ingredients and of its manufacture, the polymer binder systems that have been developed for SLS processes are distinctly different from water-based acrylic paint formulations in several ways. First of all, the polymer coating must be rigid and non-tacky at room temperature to prevent "blocking" or premature fusing of the polymer-coated powder during storage and to provide some structural permanence (low creep rates and low flexibility at ambient temperature) to composite parts made from this powder by SLS. Paint formulations, on the other hand, require that the particles of emulsion polymer fuse together at ordinary temperatures to form a flexible protective film.

The binder fusion temperature is a function of its molecular constituents. This temperature can be characterized by the polymer's glass transition temperature, T_g, if the polymer is amorphous, or by its melting temperature, T_m, if it is semi-crystalline. The SLS binder application requires constituents that lead to amorphous polymers, copolymers, and terpolymers with T_g's above 40°C. This contrasts with typical paint formulations that have T_g's near -20°C. The 100°C upper-bound to the softening temperature is set by the normal pressure boiling point of the water vehicle in the emulsion. Regardless of the process used for coating the inorganic particles with polymer binder, the surface temperature of the particle will be limited to 100°C as long as a water film is present. To prevent "fines" that can cause powder spreading problems in the SLS process, see Figure 4.17, it is desirable for polymer flow, film formation, and wetting of the inorganic surface to occur simultaneously with water evaporation. This can not occur if the binder softening temperature greatly exceeds the normal boiling point of water.

One way to adjust the softening temperature is to copolymerize two monomers. Depending on the composition, the glass transition temperature, T_g, will lie between the two T_g's of polymers formed from the pure monomers. For example, poly(methyl methacrylate) has a T_g near 110 °C which is a little too high for water-based processes. Poly(n-butyl methacrylate) has a T_g near 20°C which is too low to provide the required storage and mechanical stability. The T_g of the MMA/BMA copolymer (Vail, *et al*, 1994) is seen in Figure 4.14 to follow copolymer composition according to the well known Gordon-Taylor Equation (Olabisi, *et al*, 1979),

$$T_g = \omega_1 T_{g1} + \omega_2 T_{g2} \tag{4.4.1}$$

where ω_i is the mass fraction of component i in the copolymer and T_{gi} is the glass transition temperature for the homopolymer that is formed from component i.

The polymer or copolymer glass transition temperature, T_g, is known (Wu, 1981) to decrease with number average molecular weight via

$$T_g = T_{g\infty} - \frac{K'}{M_n} = T_{g\infty} - \frac{K''}{M_w} \tag{4.4.2}$$

where M_n and M_w are number average and weight average molecular weights, respectively, and are related by a constant for constant polymerization conditions, and $T_{g\infty}$ is the glass transition temperature at infinite molecular weight. The average molecular weight decreases in linear proportion to an increasing [S]/[M] ratio in the polymerization, where [S] and [M] are the concentrations of chain transfer agent and monomer, respectively (Vail, *et al*, 1994). Consequently, one expects that the T_g should decrease linearly with increasing [S]/[M]. Figure 4.15 confirms this expectation.

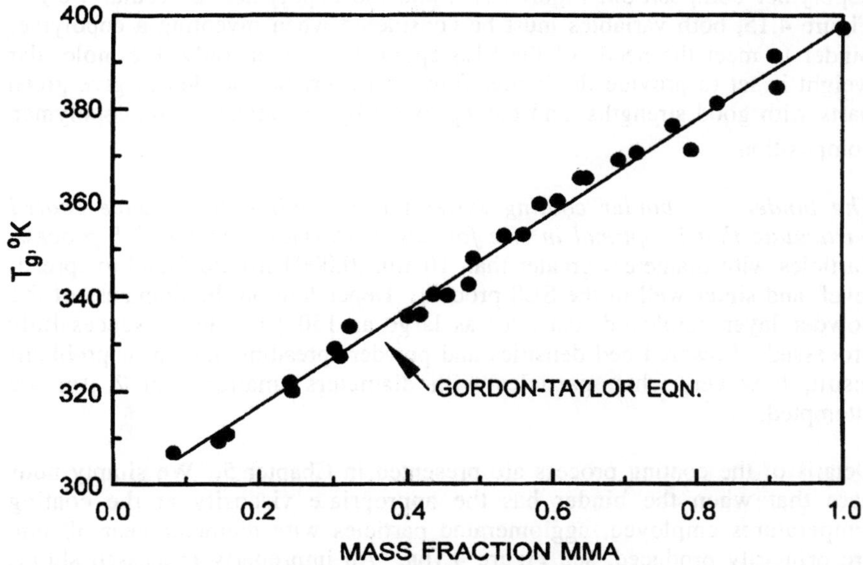

Figure 4.14 Comparison of experimental glass transition temperatures with various predictions (Vail, *et al*, 1994).

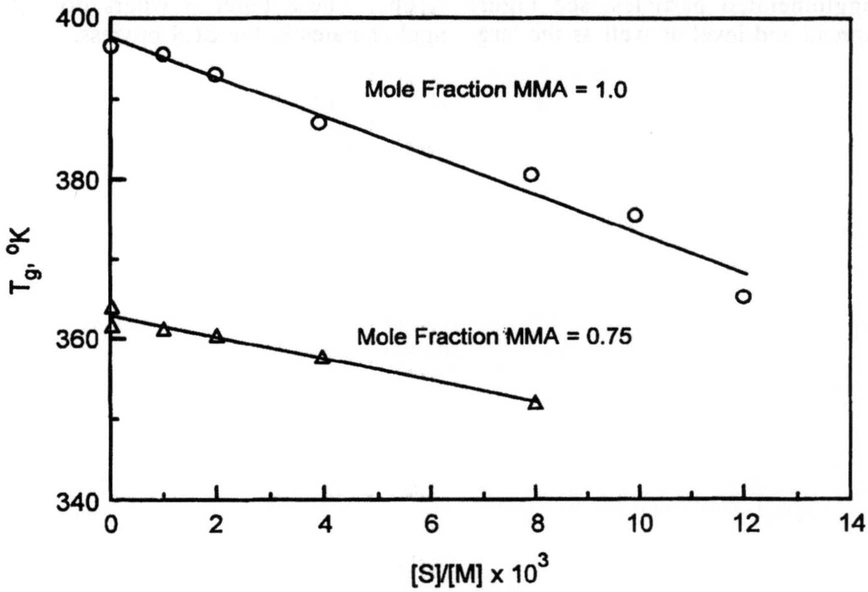

Figure 4.15 Effect of Chain Transfer Agent on Glass Transition Temperature (from Vail, *et al*, 1994).

Since the copolymer glass transition temperature is a function of both the copolymer composition, Figure 4.14, and the copolymer molecular weight, Figure 4.15, both variables must be considered when inventing a copolymer binder to meet the needs of the SLS application. Generally, the molecular weight is set to provide the binder flow characteristics needed to give green parts with good strengths, and the T_g is set by the adjusting the copolymer composition.

The binder and binder coating systems are required to produce coated particulate that is optimal in size for best performance in the SLS process. Particles with diameters greater than 10 μm (0.0004 in) are found to spread, level, and sinter well in the SLS process. Depending on the thickness of the powder layer employed, particles as large as 150 μm can be successfully processed. Lowered bed densities and powder spreading and shear problems result, however, when particles with diameters smaller than 2 μm are attempted.

Details of the coating process are presented in Chapter 5. We simply note here that when the binder has the appropriate viscosity at the coating temperatures employed, agglomerated particles with diameter near 50 mm are primarily produced, see Figure 4.16a. An improperly processed slurry, where the processing temperature is too low relative to the polymer fusion temperature or where the binder viscosity is too high at the processing temperature employed may produce a large quantity of smaller, non-agglomerated particles, see Figure 4.16b. These latter powders do not spread and level as well as the larger agglomerates in the SLS process.

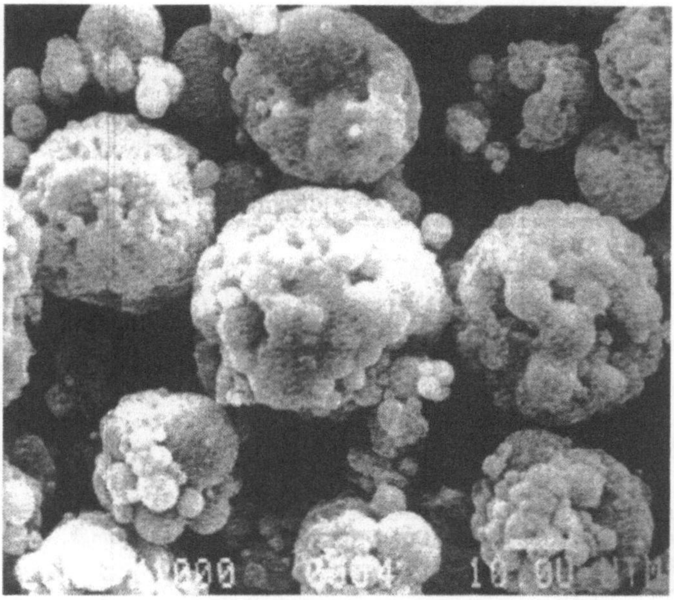

(**a**) Low Viscosity (PMMA) (Vail and Barlow, 1991)

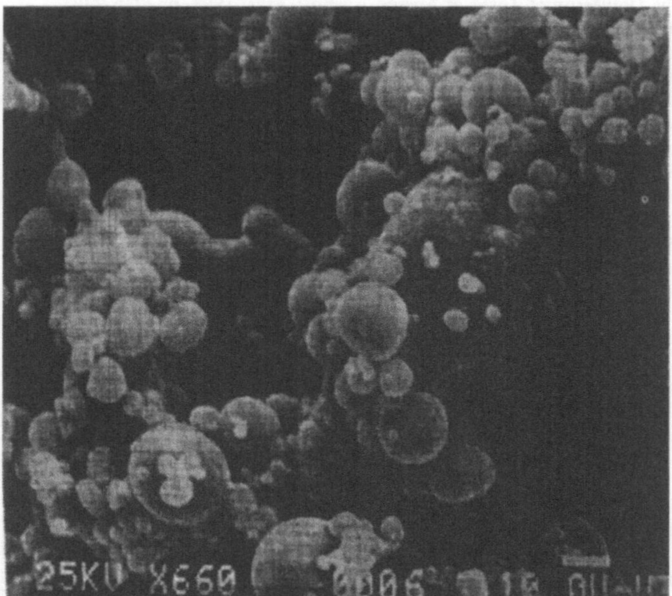

(**b**) High viscosity (UCAR 430) (Vail and Barlow, 1990)

Figure 4.16 Effect of binder viscosity on formation of spray-dried agglomerates from glass powder.

REFERENCES

Alfrey, Jr., T., and Gurnee, E. F., "Dynamics of Viscoelastic Behavior," in Eirich, E. R., Ed., *Rheology*, Vol. 1, Academic Press: New York, 1956.

Allcock, H. R., Lampe, F. W., *Contemporary Polymer Chemistry*, 2nd Edition, Prentice Hall: Englewood Cliffs, New Jersey, 1990.

Badrinarayan, B., and Barlow, J. W., "Metal Parts from Selective Laser Sintering of Metal-Polymer Powders," *Solid Freeform Fabrication Symposium Proceedings*, The University of Texas at Austin: Austin, Texas, 1992, pp. 141-146.

Bajaj, R., "Synthesis and Analysis of New Monomers for Microelectronics Packaging Applications, " Ph.D. Dissertation, The University of Texas at Austin: Austin, Texas, 1995.

Barlow, J. W., "Metallic and Ceramic Structures from Selective Laser Sintering of Composite Powders," *3rd Int. Conference on Rapid Prototyping*, University of Dayton: Dayton, OH, 1992, pp. 73-76.

Billmeyer, F. W., Jr., *Textbook of Polymer Science*, Interscience Publishers: New York, 1962.

Bird, R. B., Armstrong, R. C., and Hassager, O., *Dynamics of Polymeric Fluids*, Vol. 1, Wiley: New York, 1977.

Bourell, D. L., Marcus, H. L., Beaman, J. J., and Barlow, J. W., "Selective Laser Sintering of Metals and Ceramics," *Int. J. Powder Metallurgy*, 1992, 28, pp. 369-381.

Bovey, F. A., Kolthoff, I. M., Medalia, A. I., and Meehan, E. J., *Emulsion Polymerization*, Interscience: New York, 1955.

Comb, J. W., and Priedeman, W. R., "Control Parameters and Material Selection Criteria for Rapid Prototyping Systems," *Solid Freeform Fabrication Symposium Proceedings*, The University of Texas at Austin: Austin, Texas, 1993, pp. 86-91.

Daniels, F., and Alberty, R. A., *Physical Chemistry*, Wiley: New York, 1961.

Deckard, L. and Claar, T. D., "Fabrication of Ceramic and Metal Matrix Composites from Selective Laser Sintered Preforms," *Solid Freeform Fabrication Symposium Proceedings*, The University of Texas at Austin: Austin, Texas, 1993, pp. 215-222.

Ferry, J. D., *Viscoelastic Properties of Polymers*, J. Wiley: New York, 1961.

Flory, P. J., *Principles of Polymer Chemistry*, Cornell University Press: Ithaca, New York, 1953.

Forderhase, P., McAlea, K., Michalewicz, M., Ganninger, M., and Firestone, K., "SLS Prototypes from Nylon," *Solid Freeform Fabrication Symposium Proceedings*, The University of Texas at Austin: Austin, Texas, 1994, pp. 102-109.

Forderhase, P. and Corden, R., "Reducing or Eliminating Curl on Wax Parts Produced in the Sinterstation 2000 System," *Solid Freeform Fabrication Symposium Proceedings*, The University of Texas at Austin: Austin, Texas, 1993, pp. 94-101.

Frenkel, J., "Viscous Flow of Crystalline Bodies under the Action of Surface Tension," *J. Phys.* (USSR), 1945, 9, pp. 385-396.

Gardner, R. J., and Martin, J. R., "Humid Aging of Plastics: Effect of Molecular Weight on Mechanical Properties and Fracture Morphology of Polycarbonate", *J. Appl. Polym. Sci.*, 1979, 24, pp. 1269-1280.

German, R. M., *Particle Packing Characteristics*, Metal Powder Industries Federation: Princeton, New Jersey, 1989.

Graessley, W. W., "The Entanglement Concept in Polymer Rheology," *Advances in Polymer Science*, 1974, 16, pp. 1-179.

Inaba, A., Kashiwagi, T., and Brown, J. E., "Effects of Initial Molecular Weight on the Thermal Degradation of Poly(methyl methacrylate): Part 1.," *Polym. Degradation and Stability*, 1988, 21, pp. 1-12.

Jacobs, P., *Rapid Prototyping and Manufacturing Fundamentals of Stereolithography*, Soc. of Manufacturing Eng.: Dearborn, MI, 1992.

Joseph, E. A., Paul, D. R., and Barlow, J. W., "Boiling Water Aging of a Miscible Blend of Polycarbonate and a Copolyester, " *J. Appl. Polym. Sci.*, 1982, 27, pp. 4807-4819.

Kashiwagi, T., Hirata, T., and Brown, J. E., "Thermal and Oxidative Degradation of Poly(methyl methacrylate): Molecular Weight," *Macromolecules*, 1985, 18, pp. 131-142.

Lee, G., and Barlow, J. W., "Selective Laser Sintering of Biomaterials for Implants," *Solid*

Freeform Fabrication Symposium Proceedings, The University of Texas at Austin: Austin, Texas, 1993, pp. 376-380.

Lee, G., and Barlow, J. W., "Selective Laser Sintering of Calcium Phosphate Powders," *Solid Freeform Fabrication Symposium Proceedings*, The University of Texas at Austin: Austin, Texas, 1994, pp. 191-197.

McKelvey, J. M., *Polymer Processing*, J. Wiley: New York, 1962.

Mikos, A. G., "Kinetic Modeling of Copolymerization/Crosslinking Reactions," *Macromolecules*, 1986, 19, pp. 2174-2182.

Nelson, J. C., Xue, S., Barlow, J. W., Beaman, J. J., Marcus, H. L., and Bourell, D. L., "A Model of the Selective Laser Sintering of Bisphenol-A Polycarbonate," Ind. Eng. Chem. Res., 1993, 32, pp. 2305-2317.

Nelson, J. C., Vail, N. K., Barlow, J. W., Beaman, J. J., Bourell, D. L., and Marcus, H. L., "Selective Laser Sintering of Polymer-Coated Silicon Carbide Powders," *Ind. Eng. Chem. Res.*, 1995, 34, pp. 1641-1651.

Olabisi, O., Robeson, L. M., and Shaw, M. T., *Polymer-Polymer Miscibility*, Academic Press:New York, 1979.

Pang, T., "StereoLithography Epoxy Resin Development: Accuracy and Dimensional Stability," *Solid Freeform Fabrication Symposium Proceedings*, The University of Texas at Austin: Austin, Texas, 1993, pp. 11-26.

Progelhof, R. C. Throne, J. L., *Polymer Engineering Principles*, Hanser Publishers: New York, 1993, chapt. 3.

Shalaby, S. W., "Radiative Degradation of Synthetic Polymers: Chemical, Physical, Environmental, and Technical Considerations," *Macromol. Rev.*, 1979, 14, pp. 419-458.

Stevens, M. P., *Polymer Chemistry*, 2nd Edition, Oxford University Press: New York, 1990.

Tadmor, Z. and Gogos, C. G., *Principles of Polymer Processing*, Wiley-Interscience: New York, 1979.

Tobin, J. R., Badrinarayan, B., Barlow, J. W., Beaman, J. J., and Bourell, D. L., "Indirect Metal Composite Part Manufacture using the SLS Process," *Solid Freeform Fabrication Symposium Proceedings*, The University of Texas at Austin: Austin, Texas, 1993, pp. 303-307.

Tobolsky, A. V., *Properties and Structure of Polymers*, J. Wiley: New York, 1960.

Tobolsky, A. V., "Aspects of Viscoelasticity," Chapt. IE. in Rosen, B. Ed., *Fracture Processes in Polymeric Solids*, Interscience: New York, 1964.

Vail, N. K., Badrinarayan, B., Barlow, J. W., and Marcus, H. L., "A Thermal Model of Polymer Degradation during Selective Laser Sintering of Polymer-Coated Ceramic Powders," *Rapid Prototyping J.*, 1996, 2(3), in press.

Vail, N. K., Barlow, J. W., Beaman, J. J., Marcus, H. L., and Bourell, D. L., "Development of a Poly(methyl methacrylate - co - n-butyl methacrylate) Copolymer Binder System, *J Appl. Polym. Sci.*, 1994, 52, pp. 789-812.

Vail, N. K., Barlow, J. W., and Marcus, H. L., "Silicon Carbide Preforms for Metal Infiltration by Selective Laser Sintering," *Solid Freeform Fabrication Symposium Proceedings*, The University of Texas at Austin: Austin, Texas, 1993, pp. 204-214.

Vail, N. K., and Barlow, J. W., "Ceramic Structures by Selective Laser Sintering of Microencapsulated, Finely Divided Ceramic Materials," *Solid Freeform Fabrication Symposium Proceedings*, The University of Texas at Austin: Austin, Texas, 1992, pp. 124-130.

Vail, N. K. and Barlow, J. W., "Effect of Polymer Coatings on Sintering of Ceramic Parts," *Solid Freeform Fabrication Symposium Proceedings*, The University of Texas at Austin: Austin, Texas, 1991, pp. 195-204.

Walling, C., "Gel Formation in Addition Polymerization," *J. Am. Chem. Soc.*, 1945, 67, 441-447.

Wu, S., *Polymer Interface and Adhesion*, Marcel Dekker: New York, 1981.

5

INDIRECT FABRICATION OF METALS AND CERAMICS

Various nomenclatures are in use to describe SFF processes that create metal and ceramic objects. At the University of Texas and in this book, the three major processing methods are called *transfer methods*, *indirect methods*, and *direct methods*. In simple terms, transfer methods use patterns, either castings or spray metal patterns. Indirect methods[1] (which could also be called low-density matrix methods) involve making relatively low-density objects and then infiltrating or post-sintering these objects to high density. Direct methods (which could also be called high-density methods) create high density structures without a secondary processing step. There is already commercial application of the indirect methods; the direct methods are all still in the research stage. Transfer and indirect methods are the subject of this chapter. A discussion of direct methods is deferred until Chapter 7.

5.1 TRANSFER PROCESSES

Most rapid prototyping processes produce plastic-based parts. It is possible, however, to produce metal parts from plastic RP patterns using a number of traditional transfer processes. The most common processes are investment casting and sand casting, although a number of other routes, including plaster casting and spin casting of low temperature alloys in silicone rubber molds have also been explored. In most cases, these pattern-based transfer processes are desirable when a limited number of prototype parts are required. The impact of rapid prototyping on the two most important transfer processes, investment casting and sand casting, is discussed below.

[1]This terminology is by no means universal. For example, the term direct tooling is used to describe indirect methods of creating metal tooling to differentiate it from "indirect" transfer methods.

Figure 5.1 A polycarbonate pattern made with the SLS process and the resulting casting. (Courtesy of DTM)

5.1.1 Investment Casting.

In the investment casting process, a pattern is assembled with a feeding system - sprues, runners, and gates - into what is referred to as a "tree". The assembly is then dipped into a ceramic slurry, dried, and additional layers added until a relatively thick shell has been formed. The pattern is then melted and/or burned out of the shell before the metal is cast. In investment casting, both the pattern and the mold are expendable. Since the pattern does not have to be removed from the mold, investment casting is well suited for producing complex geometries. It is also considered to be more accurate than many other casting processes.

Traditionally, patterns for investment casting are made by injection molding wax. In principle, however, any material which can be melted and/or burned out of the shell without damaging it can be used as a pattern. Typically, shell damage is caused by excessive thermal expansion of the pattern or chemical attack of the shell face coat by the pattern material. Residual ash can also cause defects in the cast part. Today, nearly all rapid prototyping technologies - SLS (DTM Corporation, 1995, Venus, 1995), SLA (Pang and Jacobs, 1993), FDM (Comb *et al*, 1994), and LOM (Warner, 1995) - can produce patterns which can be used for investment casting with some degree of success. In the case of the SLS and FDM processes, it is possible to build patterns directly from investment casting waxes. A number of plastic-based material systems which can be used as patterns have also been developed. Typically, these plastic systems have some degree of porosity which prevents excessive pattern expansion and shell cracking during the pattern removal process. In the case of SLA, this has been achieved with epoxy resins and a special build style. In the case of SLS, a number of plastic materials (e.g., polycarbonate) which sinter to less than full density can also be used. An example of a SLS pattern and the resulting casting is shown in figure 5.1.

Investment casting of parts using rapid prototype patterns is typically economical when only a limited number of parts are required. Much recent attention has focused on the use of rapid prototyping patterns to make cast metal tooling for injection molding plastics.

5.1.2 Sand Casting.

Sand casting materials are made from several different refractories (SiO_2 and zircon) and a variety of bonding agents including clay, silica gel, and resins. Sand molds are typically made by pouring the free flowing sand material around a pattern, compacting the sand, and then hardening the binder using a catalyst, heat, or a combination of both. The sand is held in a container referred to as a flask; the pattern is normally split at a parting line and the two halves are called the cope and drag halves. Sand cores for the mold are formed separately in molds called core boxes and placed in the mold after the pattern has been removed. Sand molds are expendable and the sand is generally recycled. The patterns - which can be made of wood, plastic, or metal - are used to make multiple sand molds. An exception to this is a process in which a pattern made from a material such as a low density polystyrene foam is used as an expendable pattern. In this case, the pattern is left in the mold and burned up during the casting process. The gaseous degradation products are able to escape through the porous sand structure.

Parts made from SLS (Venus, 1995), SLA (Young, 1994), and LOM (Warner, 1995, Bowman, 1993) have been used as semi-permanent patterns for sand casting molds. The number of molds which can be made from a RP pattern depends on the particular material used, but is typically 100 or less. There is also an example of a porous polycarbonate part made by SLS used as an expendable pattern as described above. It is also possible to make sand molds and cores directly using the SLS process. In this case, the thermal energy from the laser is used to melt the binder on a resin coated sand material. The part is then cured in an oven for a short time at a temperature of approximately 200 °C. An example of a SLS sand mold and metal casting is shown in Figure 5.2. This process is particularly attractive for making cores since the need for producing a core box mold is eliminated.

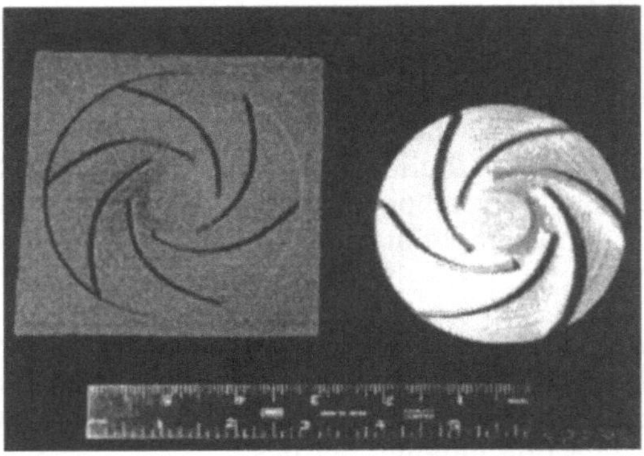

Figure 5.2 A sand mold made directly in the SLS process and the resulting casting. (Courtesy of DTM)

5.2 INDIRECT METHODS

Several processing routes have been developed to produce metallic and ceramic parts which are neither direct processes (Chapter 7) nor strictly a conversion technology (Chapter 5.1). These indirect methods involve production of a part containing metal or ceramic material bound by a wax or polymer binder. Secondary or post processing is required to remove the binder and to instill a degree of structural integrity to the remaining material. This section is a brief description of several of these processes. The next section deals more broadly with issues of post processing.

5.2.1 FDC and Reactive Stereodeposition.

Fused Deposition of Ceramics (FDC) is being developed for the fabrication of functional ceramic components (Agarwala *et al*, 1995). It is based on existing Fused Deposition Modeling (FDM) technology for processing of polymers and waxes. FDC is being developed to create ceramic components using injection molding type ceramic-polymer feedstocks. These feedstocks are extruded into filaments, which are then used as feed material for fabrication of three-dimensional green ceramic objects using a commercial FDM system (Walters, 1992, Comb and Priedman, 1993). During FDC processing, the polymer/wax acts as a carrier and binder for ceramic particles as the material flows out of the heated extruder head. The green ceramic object thus created is then subjected to conventional binder removal and sintering processes, to produce fully dense structural ceramic components.

The process has been demonstrated using silicon nitride, alumina, and zirconia (Agarwala *et al*, 1995, Crockett *et al*, 1995). Typical is FDC of silicon nitride. Here, commercially available Si_3N_4 was mixed with proprietary binder in volume ratios of the order of 60% ceramic-40% binder. Filaments were produced by capillary extrusion, and part production followed the usual FDM path. Post processing involved a two-stage binder removal process similar to post processing in ceramic injection molding. First, the majority of the binder is removed by embedding the part in pulverized charcoal or alumina and wicking the viscous binder by heating to 250°C to 450°C. The second stage is a traditional high-temperature burn out which thermally dissociates the remaining binder, followed by sintering to induce solid-state bonding of contacting ceramic particles. Figures 5.3a and 5.3b shows FDC parts as processed and after binder removal/sintering, respectively.

Injection SFF and reactive stereodeposition are also well known SFF techniques capable of indirectly fabricating ceramic components. These techniques were developed at the University of Arizona and Advanced Ceramics Research Inc., and entail the CAD computer controlled, sequential deposition of liquid monomer slurry layers onto a heated platen surface where they subsequently polymerize to form the desired green ceramic component. Final densification of the component is accomplished through binder pyrolysis followed by ceramic sintering.

A wide variety of alumna, zirconia and silicon nitride ceramic components have been successfully injection solid freeformed from slurries composed of high loadings (50-56 volume percent solids concentration) of ceramic powders dispersed in a liquid organic monomer vehicle (Stuffle *et al*, 1993a). These high ceramic solids loadings impart a strongly thixotropic rheology to the SFF deposited slurry, enabling accurate deposition with negligible layer slumping/spreading. The high solids loading also enables the green bodies to be pressureless sintered to 95-99 percent theoretically dense components that exhibit mechanical properties similar to conventionally processed ceramics. Injection SFF slurry vehicles are typically composed of 20-35 weight percent multifunctional acrylate monomers (i.e., 1,6-hexanedioldiacrylate & trimethylolpropanetriacrylate) dissolved in low volatility/low toxicity solvents (i.e., dimethyl glutarate/succinate). Small amounts (less than 1 weight percent) of dibenzoyl peroxide are also present in the slurry vehicle which thermally initiates the free radical addition polymerization of the acrylate monomers present in the injection SFF deposited slurry layers.

A significant advantage of the injection SFF compared to other techniques is that it is not equipment intensive due to the fact that the monomer slurries have low viscosities and can be readily dispensed using low pressures (less than 100 psi). Injection SFF techniques are also not limited to ceramic materials processing and have been successfully used in the fabrication of various metal/ceramic functional gradient materials combinations, including tungsten-alumina and inconel-zirconia. Current research efforts are underway to extend this technique for the fabrication of pure refractory metal components.

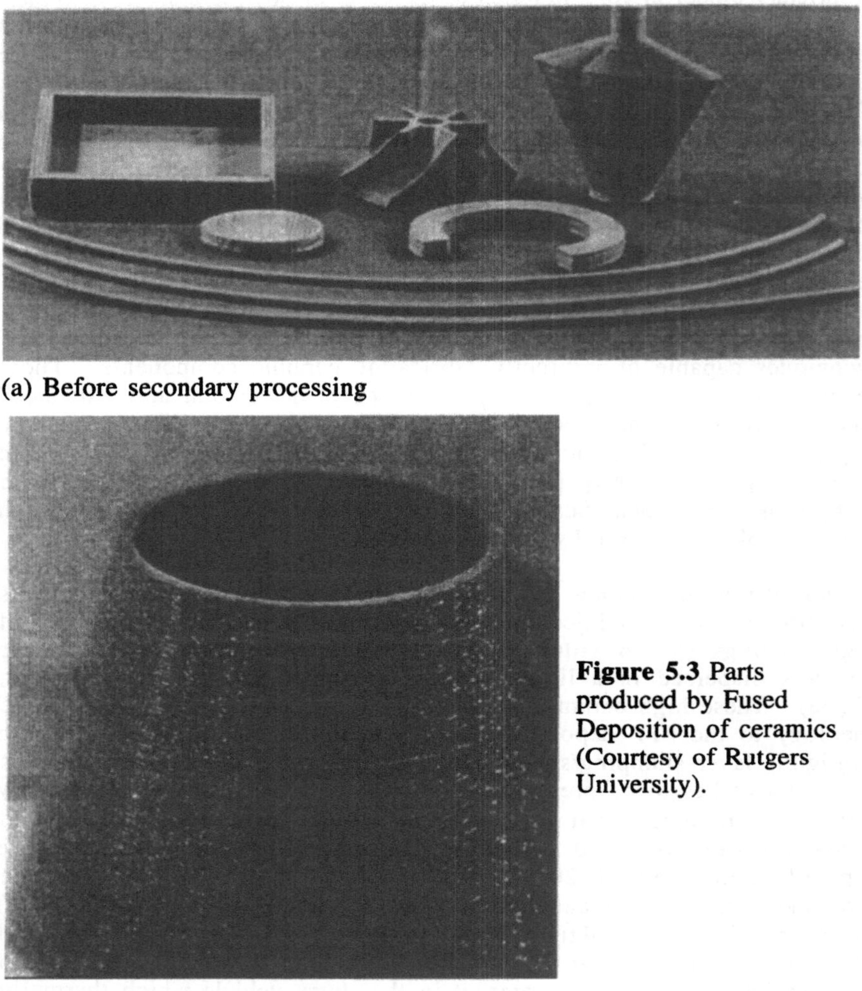

(a) Before secondary processing

(b) After secondary processing

Figure 5.3 Parts produced by Fused Deposition of ceramics (Courtesy of Rutgers University).

Reactive stereodeposition (RS) is similar to injection SFF in that both techniques involve the controlled deposition and layerwise chemical curing of a low viscosity slurry. A major difference between the former and the latter is that reactive stereodeposition involves an inorganic silicate binder-material which is not completely pyrolyzed like the polyacrylate binder present in the injection SFF green ceramic bodies. RS is also limited to fabricating siliceous ceramic materials due to the fact that most other ceramic precursor metal alkoxides are too reactive and shelf stable slurries are not able to be formulated with them. RS also suffers from the disadvantage that its tetraethoxysilane/silica slurries have a tendency to prematurely gel and clog the IRS apparatus deposition needle.

This technique has been used to produce alumina and silicon nitride parts from slurries containing ceramic powders or inorganic fibers dispersed in liquid acrylic monomers (Stuffle *et al*, 1993b). A low-pressure stereodeposition system has been used to produce silica, mullite, and fiber-reinforced borosilicate glass objects. In this case, the extruded slurry contains inorganic powder in solvent plus liquid silicon alkoxide.

The process employs ethyl silicate, a metal alkoxide compound, as a binder. Ethyl silicate is a partially condensed oligomer of tetraethoxysilane (TEOS). The oligomer used in this work has an average molecular weight of 610 g/mol and will condense to 40 wt% silica upon complete densification. TEOS, on the other hand, condenses to only 29 wt% silica upon densification. The oligomer is used as opposed to the monomer to reduce shrinkage due to loss of chemically bound solvent and because ethyl silicate has previously proven useful in binding refractory powders through slip casting (Emblem, 1983, Biddle *et al*, 1977).

Ethyl silicate reacts with water in the presence of an acidic or basic catalyst to form a continuous inorganic network or gel. In using ethyl silicate as a binder, the resulting Si-O-Si network remains in the object throughout the drying and sintering processes, in practice contributing ~3-10 vol% of amorphous silica to the ceramic matrix. This results in a higher effective green body loading, produced by a material which promotes either viscous or liquid-phase sintering and thus helps enable pressureless sintering of the ceramic object. Additionally, because gelation is a room temperature curing process, there is less chance for warpage due to localized temperature gradients. There are also fewer burnout problems during the sintering phase, as there is less material actually leaving the object.

To create a ceramic slurry, dry oxide powders are added to a carrier solvent (ethanol) which contains the ethyl silicate. The volume ratio of these components is defined by a tradeoff between ease of dispensing and rate of slurry rigidization. Reactive Stereodeposition requires the slurry to be fluid enough to deposit at low pressures (5-20 psi), yet harden sufficiently once on the surface to support the weight of successive layers. Additionally, it is desirable to achieve the highest particulate loading possible for improved densification during sintering. Thus the slurry's rheological properties, and especially the impact of shear thinning, become driving criteria for slurry development.

5.2.2 LOM Pre-Ceramic Sheet

Engineers at Lone Peak Engineering have developed a laminated object manufacturing (LOM) route for production of ceramic and metal parts

(Griffin *et al*, 1994). The ceramic process was initially demonstrated using thin sheets of alumina filled polymer. These sheets were prepared by tape casting using a proprietary process. The powder was mixed with an organic binder system using a ball mill. The resulting slurry was cast into thin tapes approximately 0.015 mm thick. Squares 10 by 15 cm were cut from the tape-cast roll. The system was similar to commercial LOM, excepting that a heated plate and pressure were used to assist layer lamination and parts were laid up by hand. After part generation, the binder was removed by thermal degradation in air at 600°C, followed by sintering at 1550°C for 2 hours in air. Parts produced using the LOM technique possessed essentially identical structure, physical and mechanical properties to conventional powder processed alumina, Figure 5.4. Stainless steel metal components have been prototyped by Lone Peak Engineering using a similar process. After part generation, the binder was removed by thermal degradation and the prototypes were infiltrated with bronze. Infiltration creates fully-dense parts, Figure 5.5.

Figure 5.4 Alumina part produced in LOM process. (Courtesy of Lone Peak Engineering)

Figure 5.4 Metal and ceramic parts produced in LOM process (Courtesy Lone Peak Engineering)

5.2.3 SLA Slurries

Another approach to the creation of ceramic parts is to charge a UV-curable resin with particulate prior to stereolithography. As a UV laser scans the slurry, the resin is cured, forming a ceramic-polymer composite, layer by layer. The part is a ceramic-polymer composite which, after building, is heated in a furnace to remove the polymer binder and sinter to produce the final ceramic. For investment casting and biomedical applications, researchers at the University of Michigan (Griffith and Halloran, 1994, Griffith *et al*, 1995) have developed a water-based resin of acrylamide with added crosslinking monomer methylene bis-acrylamide (Janney, 1990, Young *et al*, 1991). Ceramic materials investigated include silica, a refractory material for shells and cores in the investment casting industry; alumina, a high strength structural ceramic; and hydroxyapatite, an artificial bone material for custom ceramic implants. For silica and alumina, the stereolithography cure depth decreased with increasing solids loading and particle size; however the cure depth was sufficient for stereolithography use. The cure depth behavior is strongly dependent upon refractive index difference between the ceramic and the UV-curable solution. Based on X-ray diffraction, the hydroxyapatite structure was preserved after curing and binder burnout.

5.2.4 3D Printing

This process was introduced in Chapter 2. It involves spreading powder in a thin layer followed by placement of a binder using technology similar to ink-jet printing. Soligen, Inc., located in Northridge CA, was the first company to market ceramic parts, namely, cores for sand casting.

Parts from both metallic and ceramic powder have been produced using 3D Printing. Metals include 316L stainless steel, iron, tungsten, a tungsten heavy-metal alloy, and a tungsten carbide/cobalt alloy (Michaels *et al*, 1992). The general approach was demonstrated for stainless steel (Sachs *et al*, 1995). A powder bed was prepared from metal powder. The binder material was an aqueous acrylic copolymer emulsion which self cross linked upon drying in air (HA-16 from Rohm and Haas) Initial efforts required preparation of the powder bed surface with a salt coating to prevent disruption of the powder bed during 3D Printing. After a brief, low temperature curing cycle, parts were debound in argon at 400°C and sintered at 1000°C for nominally 6 hours. Further firing produced parts with densities ranging between 78% and 92% of theoretical density.

Ceramics studied using 3D Printing include both undoped and MgO-doped alumina(Yoo *et al*, 1993) and silicon nitride (Cima *et al*, 1994). Binders were diluted acrylic copolymers and latex binders. For alumina, after 3D Printing, parts were cured and dried to remove residual moisture, cold or warm isostatically pressed, debound at 450°C and fired at 1650°C for 4 hours. Densities greater than 95% were achieved with undoped alumina; for doped material, the density exceeded 99%. Particle flow was improved by spray drying the alumina to produce a polyacrylic acid dispersant coating (Cima *et al*, 1995). In the same report, fired alumina was infiltrated with glass to produce a high density (93% to 99%) glass-ceramic composite.

More recently, ink-jet binders have been modified to incorporate fine ceramic powder. These slurry binders may contain as much as 40% ceramic. Both alumina and zirconia have been successfully deposited. A significant advantage is the increase in green part density associated with use of slurry binders. An alternative process involves spraying an isopropanol-based alumina slurry onto the powder bed prior to printing with a solution of polymethyl methacrylate dissolved in acetone.

5.2.5 Selective Laser Sintering

The commercial Selective Laser Sintering process uses a rastered, 25 W, CO_2 laser to selectively fuse thermoplastic powders into the desired shape. This laser power is generally too low to permit the fusion of ordinary metal and ceramic powders, although some success with mixtures of powdered metal and braze compounds, for example (Bunnell, *et al*, 1995), has been achieved. Generally laser power near 100-300 W is required for this task as are high bed temperatures and reducing atmospheres (Zong, *et al*, 1992).

SLS Binders

The manufacture of green parts by techniques such as die compaction and powder injection molding (PIM) is well established (Jones, 1960, Lenel, 1980, Hausner and Mal, 1982, German, 1994). The first process uses high compaction pressures and produces relatively high density parts with some anisotropy and internal stresses, while the latter process uses low compaction pressure and produces lower density parts which have less anisotropy and reduced internal stresses. In the SLS process, no compaction pressure is applied and so the green parts are relatively low density and nearly isotropic and free from internal stresses. The latter two characteristics are desirable. If the green density is too low, however, the parts can exhibit distortion during the furnace process. As a general rule, green part density should be at least 50% of the theoretical value to avoid such problems. In addition, it is important to achieve a minimum green strength so that parts can be handled and cleaned without loss of definition. Experience with SLS green parts has indicated that the strength should be at least 250 psi (1.7 MPa).

The intrinsic bonding capability between the polymer binder and a given particle surface is dominated by the interfacial characteristics of the polymer. It is possible in some cases to modify the polymer interfacial properties to improve adhesion between the two materials. If the intrinsic bonding capability is enhanced, it may be possible to reduce the amount of polymer without a reduction in green strength. It is desirable to decrease the amount of binder since a decrease in binder is typically accompanied by an increase in green density.

The impact of the final three binder variables - viscosity, composition, and morphology - must be considered together. There are two extreme powder/binder morphologies - the uniformly coated system and the mixed system - as shown in Figure 5.6.

Considerable work has been done, described in Chapter 4, to develop thermoplastic, water-based emulsion binders for inorganic materials. The binder is coated on inorganic particulate by fluidized drying techniques, described below, to produce polymer-coated particulate agglomerates. Coating the particulate with binder is preferred over simply mixing the two materials because the mixed materials can segregate by density, a situation that could lead to highly variable powder and part properties. The uniformly coated system produces parts with little variability in properties and requires only limited binder flow at contact points to develop strength. Hence, viscosity control is not critical to the performance of such a system. On the other hand, because such systems have no particle-particle contacts, the coating thickness must be kept small to produce an acceptable particle density and minimize the shrinkage which occurs when the binder is removed in the furnace process. The mixed system will often produce greater variability in part properties and requires substantial binder flow to develop strength. Hence, control of binder viscosity is more critical for such systems. Since much of the binder resides in the powder pore space, however, the particle density and furnace shrinkage are less sensitive to the amount of binder than they are for the uniformly coated system. In practice, it is possible to make high quality green parts utilizing either type of binder/particle morphology or an intermediate morphology.

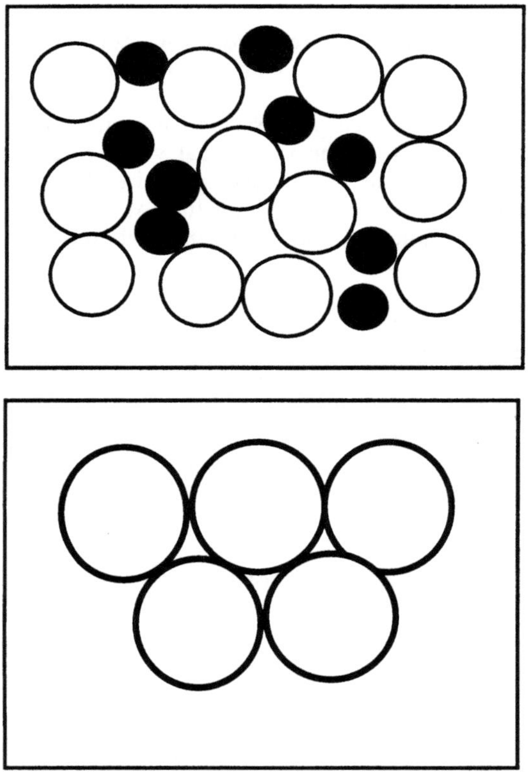

Figure 5.6 Two extreme binder/particle morphologies. In the first case, the binder is mixed with the particles while in the second, the binder is uniformly coated on the particles. Intermediate morphologies are also possible.

The strength of green parts made from coated particulate is usually higher than that of parts made using a powder mixture at the same polymer content (Vail and Barlow, 1991, Badrinarayan, 1995). Figure 5.7 shows a typical comparison of strengths achieved for green parts made from PMMA coated copper powder and from the same PMMA emulsion that is spray dried separately to form a powder that is mixed with the copper powder.

The agglomerates can be processed by existing SLS technology to form green shapes of polymer-bound metal or ceramic powder. The polymer is subsequently removed, usually by thermal processes, and the polymer-free part is then sintered or further processed to form a load bearing structure with geometric precision that is comparable to that of the green shape. This general approach has been used to prepare parts from a variety of materials including copper (Badrinarayan and Barlow, 1994), glass (Vail and Barlow, 1991), alumina (Vail and Barlow, 1994), silicon carbide (Vail, *et al*, 1993, Deckard and Claar, 1993), iron (Badrinarayan and Barlow, 1994), and calcium phosphate (Lee and Barlow, 1994), and steel.

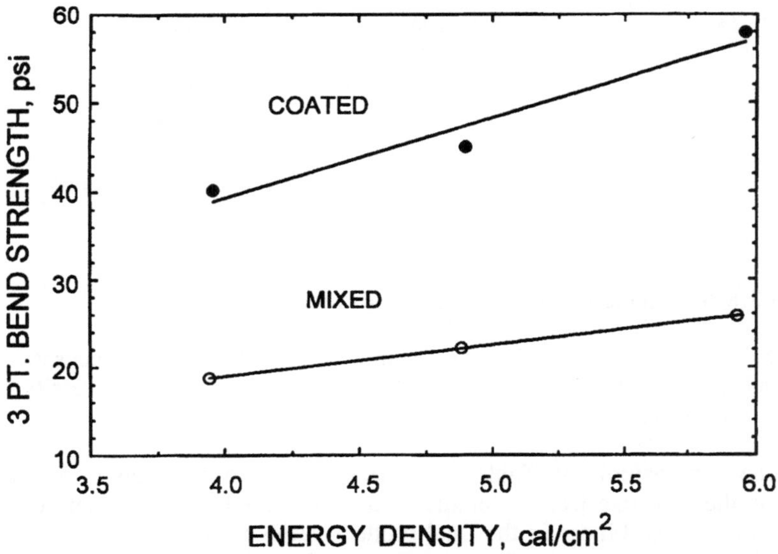

Figure 5.7 Comparison of bend strengths of SLS processed parts made from mixed copper/PMMA powders and from PMMA-coated copper powder.

Spray drying can be used to prepare polymer-coated materials. The spray drying process, shown schematically in Figure 5.8, is best for coating relatively small particles, with 2-20 μm diameter, with relatively large quantities of binder, typically 20 vol% on a dry basis. The process is limited to particles with diameters less than 40 μm by particulate settling and clogging in slurry feed lines. Slurries that contain particles greater than 40 μm or particles with specific gravity greater than 4.0 require viscosity enhancers to prevent settling (Badrinarayan, 1995). Typically, raising the water viscosity to 2-3 cps by the addition of less than 0.1% Xanthan gum to the water is found to reduce settling sufficiently to permit processing. Further increases in viscosity can prevent proper atomization of the fluid by the rotary atomizer.

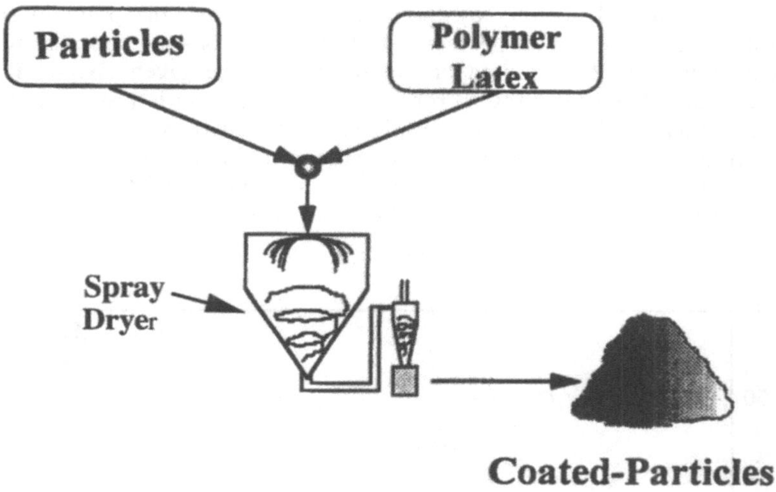

Figure 5.8 Schematic of the Spray Drying Process.

Table 5.1 shows some feed slurry compositions that were successfully processed in the Anhydro Spray Dryer that is available at the University of Texas (Vail, 1994).

All of the compositions in Table 5.1 were designed to coat about 25 vol% binder on the substrate from an emulsion that contained 40% polymer solids. The weight percent binder in the final product varies from 11.8% to 5.3% as the density of the substrate increases. The weight percent solids (binder plus substrate) in the slurry feed can be quite high, again depending on the density of the substrate, however it is usually kept in the range shown to facilitate slurry transport.

Table 5.1 Spray Dryer Feed Slurry Compositions

Substrate	Slurry Composition g/g substrate			Slurry Solids, %	Binder %
	Emul-sion (0.4 Solids)	Water	Viscosity Enhancer		
Glass, 5 μm	0.335	0.933	0	50	11.8
SiC, 13.4 μm	0.297	0.302	0	70	10.6
Tool Steel, 11 μm	0.135	0.184	0.53*	85	5.3

Some Slurry Compositions (*g/g water)

The atomizer disperses the mixture of the inorganic particulate and binder emulsion into a flowing stream of hot air (Masters, 1985). Heat from the air evaporates the water and fuses the polymer binder to create a free flowing powder of agglomerated particulate. The particulate is then separated continuously from the gas stream by a cyclone separator, see Figure 5.8. For the binders described in Chapter 4, the spray dryer is typically operated at a air inlet temperature above 200 °C and an exit temperature near 140 °C. The air exit temperature is controlled by the slurry feed rate used, and care must be taken to ensure good drying and fusion by adjusting the feed rate to keep the air exit temperature above 100 °C. When coating metal particulate, the need to maintain the air exit temperature may require raising the inlet air temperature to near 250 °C or preheating the slurry feed.

Commercial SLS Secondary Processing

For the first three years of commercial use, the SLS process was used primarily to produce durable nylon-based prototype parts and patterns for making silicone rubber molds. Rubber molds are used to vacuum cast durable plastic prototype parts from liquid urethane or epoxy systems. The direct production of nylon-based parts is generally economical when a limited number of parts are needed, while vacuum casting is attractive when more (up to 20 for a single mold) parts are required, particularly if the parts are larger. While nylon-based and cast prototype parts are useful for some level of functional evaluation, it is necessary to have prototype parts made with the production material and process if more rigorous testing and analysis are required. As a result, there has always been a great deal of interest in using rapid prototyping processes to produce durable prototype molds for injection molding plastic parts. A number of the options discussed in Section 5.1 as well as spray metal and investment cast molds have been explored by the rapid prototyping community in recent years. In all of these examples, a rapid prototyping process is used to make a plastic pattern which is then converted to a mold using a secondary process. While these options have some attraction, particularly as the quality of rapid prototyping patterns has improved, they all suffer some limitations associated with lead time, accuracy, durability, or some combination of these factors.

The first commercial metals-based rapid prototyping process (the RapidTool Process) was introduced in late 1995 to address these limitations and offer SLS users a more direct route to durable prototype molds. Powders for the indirect process are typically coated or mixed with a small amount of polymer as described in the previous section. In the SLS process, the laser is used to melt the polymer and produce a green part in which the particles are bound together by polymer-polymer bonds. This green part is processed in a high temperature furnace to remove the polymer and sinter the particles, creating metal-metal bonds. It is possible to produce full density parts using furnace sintering and other consolidation processes such as hot isostatic pressing (HIP). Typically, however, SLS green parts have a large amount of pore space and so full density consolidation is accompanied by large shrinkages. To reduce total shrinkage, a partially sintered part can be infiltrated with a lower temperature material to produce a fully dense composite structure.

The commercial process has three distinct steps. In the first step, green mold parts are made from a carbon steel powder using the SLS process. Next, these parts are placed in a reducing atmosphere furnace, the polymer is removed, the steel is lightly sintered together, and the resulting structures are infiltrated with copper. Finally, the fully dense composite metal mold parts are finished before using them to mold plastic parts. A flow diagram for the process is shown in Figure 5.9 and discussed in more detail below.

The optimized powder for the commercial process is a 1080 carbon steel (RapidSteel metal) with a mean volume average particle size of 0.0022 inch (55 μm). This powder is uniformly coated with 0.8 weight % (approximately 5 volume %) polymer. The softening temperature of this polymer coating allows the steel powder to be processed in the SLS machine without heating the feed and part beds. The density of the powder in the part bed and the density of green parts is 0.16 lb/ft^3 (4.3 g/cm^3) which is 55% of the theoretical density of the steel. There is negligible shrinkage during the SLS process. The strength of green parts is approximately 400 psi (2.8 MPa). This strength is sufficient for parts to be handled as well as for features as small as 0.020 inch (0.5 mm) to be built and cleaned. A SEM micrograph of a green part fracture surface showing the development of polymer bonds or "necks" between particles is shown in Figure 5.10.

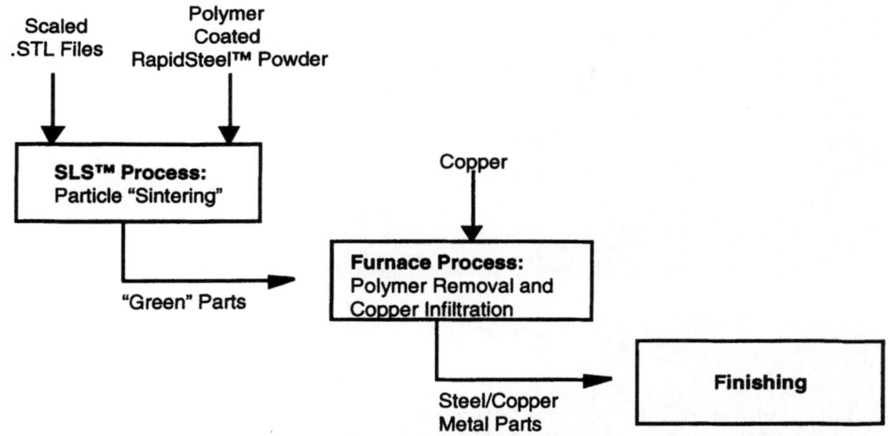

Figure 5.9 The process steps for producing metal molds using an indirect SLS metals approach

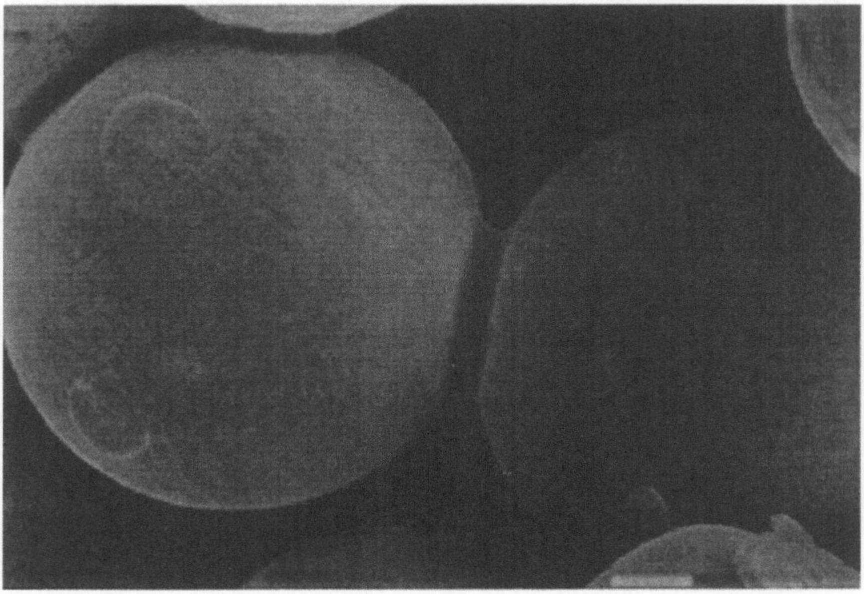

Figure 5.10 A SEM micrograph of the fracture surface of a green part showing the development of polymer-polymer bonds or "necks" between particles (55 µm particles).

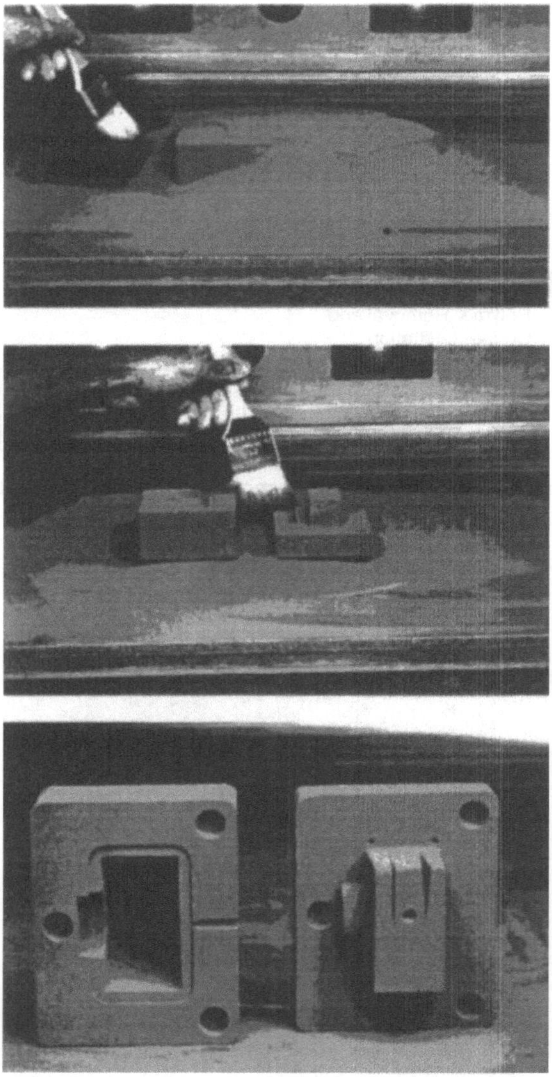

Figure 5.11 The "breakout" process for a green core and cavity. The loose powder is pushed back into the feed cartridges and reused. The final green part cleanup is performed with small brushes and compressed air.

Nearly all the unmelted coated powder is easily removed from parts with brushes and can be completely recycled to make additional parts. The "breakout" process for a green cavity and core is shown in Figure 5.11. The final part cleanup is accomplished outside the SLS machine with compressed air and small brushes.

In the portion of the furnace cycle after the coating softens, but before it is removed (between approximately 60°C and 400°C), gravity induced creep of the lubricated particle structure can cause feature distortion as well as variable rates of settling in different regions of the part. This phenomenon can be eliminated by cross-linking the polymer binder. To accomplish this, the green parts are soaked in a dilute (3 -10 wt% solids) aqueous acrylic emulsion bath which contains a cross-linker. The saturated parts are then dried at approximately 100°F (50°C) in a laboratory oven until nearly all the water is removed. It is also possible, in principle, to incorporate a cross-linker directly into the polymer coating and eliminate this step.

Currently, the time to prepare a furnace-ready core and cavity is typically 3 days. The SLS build rate can potentially be increased by using a shorter wavelength energy source such as a Nd:YAG laser which heats the metal powder more efficiently. A substantial fraction of CO_2 wavelength radiation is reflected from metal surfaces (Carter and Jones, 1993). By increasing the SLS build rate and by eliminating the cross-linker infiltration, it should be possible to produce furnace-ready green parts in approximately 1 day.

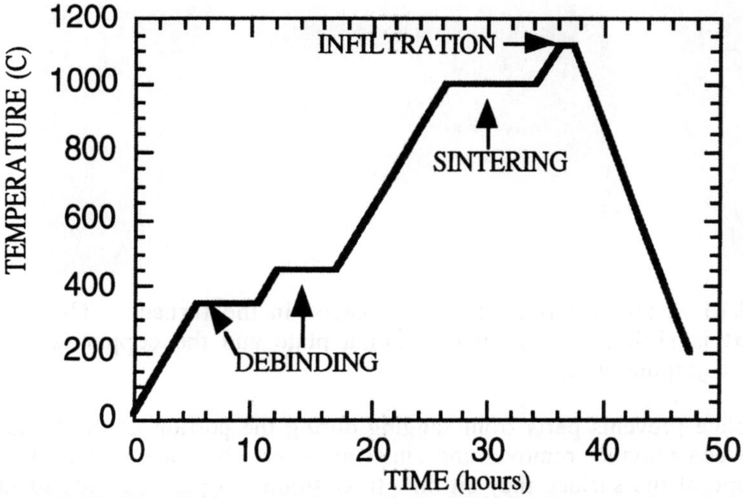

Figure 5.12 The furnace cycle used to debind, sinter, and copper infiltrate green parts made with the SLS process.

In the commercial process, a furnace atmosphere of 30% hydrogen and 70% nitrogen is used. The hydrogen reduces oxides on the steel surface, while the nitrogen helps maintain carbon levels in the steel and prevent copper-induced swelling (Krantz, 1969). The polymer is removed during two temperature holds: one at 570°F (350°C) for five hours and the second at 840°F (450°C) for five hours. At these temperatures, the polymer decomposes and the degradation products are removed from the furnace by the gas mixture, which is circulated through the chamber at a fixed flow rate. The sample is then heated to a temperature of 1832°F (1000°C) and held for 8 hours. At this temperature, the steel powder begins to sinter at a significant rate. In the temperature region between polymer removal and sintering, the porous steel part is held together by friction between particles. The steel skeleton is infiltrated with copper at a temperature of 2048°F (1120°C) with a hold time of 1 hour. The complete furnace cycle is shown in Figure 5.12. The infiltration is performed in a single cycle. To accomplish this, the green part is placed on a flat (less than 0.005 inch or 0.12 mm distortion) alumina plate inside a graphite crucible as shown in Figure 5.13.

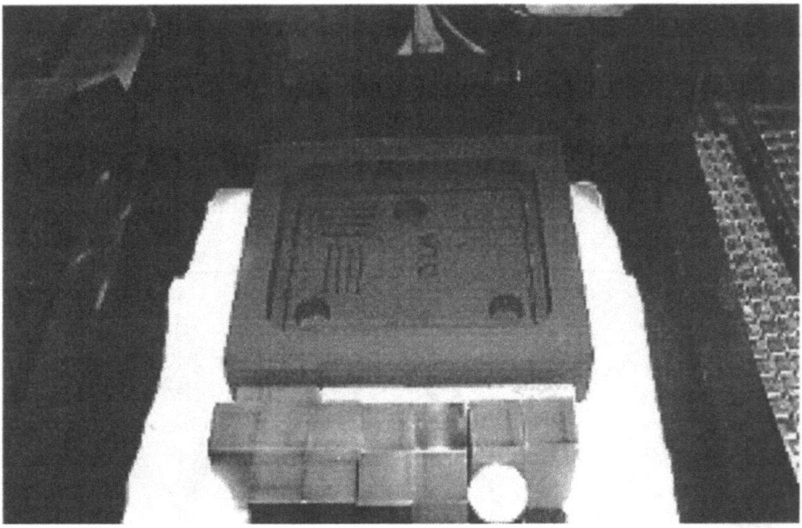

Figure 5.13 A green part ready to be placed in the furnace. The green part is sitting on a ground alumina plate and the copper is sitting on a graphite wedge.

The flat plate prevents parts from sagging during the portion of the furnace cycle between polymer removal and sintering where the parts will conform to the shape of the surface they sit on. In addition, copper does not adhere to the alumina surface and so the cool, infiltrated part can easily be removed from the plate. The copper is placed on an angled graphite wedge so that there is a component of gravity assisting the infiltration. The amount of copper used corresponds to 80% of the green part weight.

The total furnace process time is two days and the process shrinkage is 3.2% in the x and y dimensions and 4.5% in the z (gravity) dimension. As discussed in Section 6.4, the tolerances which can be achieved today for these parts are +/- 0.010 inch (+/- 0.25 mm) within a 6 inch by 6 inch by 2 inch (152 mm by 152 mm by 51 mm) build area. It is anticipated that with improvements to the material system and the SLS and furnace processes that the tolerances can be improved to +/- 0.005 inch (+/- 0.125 mm). A copper infiltrated part is shown in Figure 5.14.

Figure 5.14 A copper infiltrated steel part after it is removed from the furnace.

The surface finish of the as-infiltrated parts is 400 μin (10 μm) Ra. To finish mold parts, it is necessary to correct critical dimensions and polish display surfaces. Typically, metal-to-metal contact surfaces such as parting lines and shutoffs are machined. There is 0.020 inch (0.5 μm) of additional material added to parting line and shutoff surfaces in the CAD file to account for this correction. Mold polishing is performed with standard techniques. In some cases, "stair-steps" on curved surfaces which result from the layer-wise build method can be removed at the green part stage by light sanding. Additional mold finishing operations such as chemical etching and plating are also possible with the steel/copper composite material.

The infiltrated part is 60% steel and 40% copper. The mechanical and thermal properties of this composite material are compared to two common mold materials - 7075 aluminum alloy (T6 temper) (ASM Handbook Committee 1990) and P20 steel (Menges and Mohren, 1986) in Table 5.2. The steel/copper composite has a modulus and coefficient of thermal expansion similar to the steel, but a strength, hardness, and thermal conductivity more like aluminum than steel.

Table 5.2 Mechanical and Thermal Property Comparison: Copper Infiltrated Steel (SLS), 7075 Aluminum, and P20 Steel

Property	Copper Infiltrated Steel (SLS)	Al 7075-T6	P20 Steel
0.2% Yield Strength (ksi)	37 (ASTM E8)	73	109
Tensile Strength (ksi)	69 (ASTM E8)	83	138
Elongation (%)	15 (ASTM E8)	11	20
Young's Modulus (Msi)	30 (ASTM E8)	10	30
Hardness (HRB)	75 (ASTM E18)	81	180-210
Thermal Conductivity (BTU/hr ft°F)	107 (ASTM E 457) (212°F)	76	17
Coefficient of Thermal Expansion (μin/in/°F)	7.99 (ASTM E 831) (-60°F to 450°F)	13.1	9.4

Figure 5.15 A copper infiltrated steel mold made with the SLS process. More than 10,000 polypropylene parts were produced with the mold

The time required to finish a typical core and cavity mold set is approximately 5 days. As a result, the total time to prepare a finished core and cavity mold set today is approximately 10 days - 3 days to prepare the green parts, 2 days for the furnace process, and 5 days for the finishing work.

A finished mold which has been used to produce in excess of 10,000 parts is shown in Figure 5.15. No deterioration in the quality of the parts was noted during this run. It is anticipated that is should be possible to produce at least 50,000 parts from most polymers using this mold material. For abrasive materials, such as filled polymers, it would be necessary to plate or harden the mold surfaces to achieve a similar mold life.

Epoxy Molding via SLS Indirect Processing.

The engineering reasons for building a prototype mold are several and include evaluation of the processability of candidate materials in the mold and of the mechanical behavior and molded geometry of candidate parts, prior to committing to a costly production tool. For the evaluation to be truly valid, molding temperatures, pressures and cycle times should be as similar to those intended for production as possible (Progelhof and Throne, 1993; Tadmor and Gogos, 1979). Consequently, engineers have often concluded that even prototype molds should be made of the same materials as production molds to ensure a proper evaluation. When faced with the time and production costs associated with this conclusion, most engineers have decided not to build prototype tools, sometimes with very expensive consequences. The cost and time savings inherent in modern rapid prototyping methods and materials can help to solve this dilemma.

The University of Texas has developed a rapid prototyping process for preparing mold inserts that are suitable for injection molding a limited quantity of polymeric materials (Badrinarayan and Barlow, 1994; Tobin, *et al*, 1993; Badrinarayan, 1995; Barlow, *et al*, 1994). In this process, Selective Laser Sintering is used to form green mold cavity inserts from metal powder that is coated with fusible thermoplastic binder. In subsequent steps, the binder is thermally removed and the metal powder is oxidized to form a porous metal/ceramic cavity that shows little shrinkage and generally excellent retention of geometry, relative to the green part. The cavity is then strengthened and sealed by infiltration and cure of an epoxy tooling resin. This mold-production process has been called Rapid Mold (RM) by the inventors. Among the advantages of the RM system are ease and speed of manufacture, low-cost post treatment in a simple air oven, and good hardness, thermal conductivity, and thermal expansivity, relative to filled-epoxy tooling. The RM system is named so as to distinguish it from the RapidTool mold-making system that was recently introduced by DTM Corporation, Austin, Texas. Both RM and RapidTool may use the same polymer-coated metal powder feed stock, and the preparation of green shapes is identical. The processes deviate from one another in the post-SLS processing steps. One interesting possibility is to use the RM process to prototype the plastic part, then to switch to the RapidTool process to build the production mold using the same powder and cavity part file.

Table 5.3 shows a comparison of the material properties achieved with the RM system to that of a tool steel that is typically used to make injection molds. The properties in Table 1 result from a feed stock comprising 40 vol% of a copolymer that contains 20 mol% butyl methacrylate and 80 mol% methyl methacrylate (Badrinarayan, 1995) and 60 vol% -325 mesh iron powder (Hoeganaes ANCOR ATW 230). The SLS-produced test bars were then oxidized in an air oven, according to a proprietary temperature schedule and subsequently infiltrated with epoxy resin (Dow DER 331). DTM Corporation, Austin, Texas, has recently announced a polymer-coated metal powder, Rapid Steel, that uses a smaller amount of a similar copolymer binder and a larger metal particle size than were used in this study. The properties obtained with Rapid Steel in the RM process should be similar to those in Table 5.3.

The thermal conductivity of the RM material is seen in Table 5.3 to be substantially lower than that of steel; however, it is still about seven times higher than the typical 0.2 W/m-°C seen for unfilled polymeric materials (Jurand, 1989). Consequently, the limiting resistance to heat transfer in many circumstances will remain that associated with the polymer melt.

Some estimates of the relative thermal performances of molds made of steel and of the RM material system were recently made (Barlow, *et al*, 1996) by solving the simultaneous partial differential equations that describe the transient heat transfer from polymer melt, through the mold material, to the cooling water. Two general conditions were considered: (1) injection of polymer melt, one time, into a cold mold at $T(x)$ = 25°C, and (2) injection of polymer melt into the mold with a repetitive cooling time until "steady state" is achieved. For this case, the mold temperature, T(x) is determined by the mold temperature profile from the previous injection cycle at the prescribed cooling time.

Figures 5.16 and 5.17 show calculated temperature profiles at the centerline of a 1/8 inch thick polymer part and at the cavity wall, respectively, for the case where the mold temperature is initially 25°C and the mold conduction length is 0.9375 inches. Very little difference is seen in the performance of the two molds, tool steel and RM, in terms of the time required for the centerline to reach some required value. For example, if the polymer center needs to reach 100°C, $(T-T_0)/T_0 = 3$, before the part can be ejected, the required cooling time in the steel mold is only 9 sec shorter, 24 versus 33 sec, than that in the RM produced mold.

Table 5.3 Material Properties (* Compression failure is actually in failure in shear for RM material.)

Material Property	**Rapid Mold System**	**Conventional Tool Steel**
Tensile Modulus, 10^6 psi	0.40	30
Tensile Strength, 10^3 psi	10	150
Compress. Strength, 10^3 psi	24	----
Shear Strength*, 10^3 psi	12	75 (est.)
Est. Poisson's Ratio	0.45	0.35
Glass Transition Temp., °C	130-150	-----
CTE, ppm/°C	26	12
Thermal Diffusivity, in^2/min	0.121	1.181
Specific Heat, cal/g	0.16	0.11
Density, g/cm^3	1.72	7.86
Thermal Conductivity, W/m-°C	1.47	50

While this difference could be important for production tooling, it is probably not a great issue when prototyping. Figure 5.16 follows the cavity surface temperature versus time after injection into the 25°C mold. The surface temperature of the steel rises to only 41°C vs 75°C for the Rapid Mold material even though the initial melt temperature in contact with it is at 275°C, because of the much higher thermal diffusivity of both mold materials relative to that of the polymer.

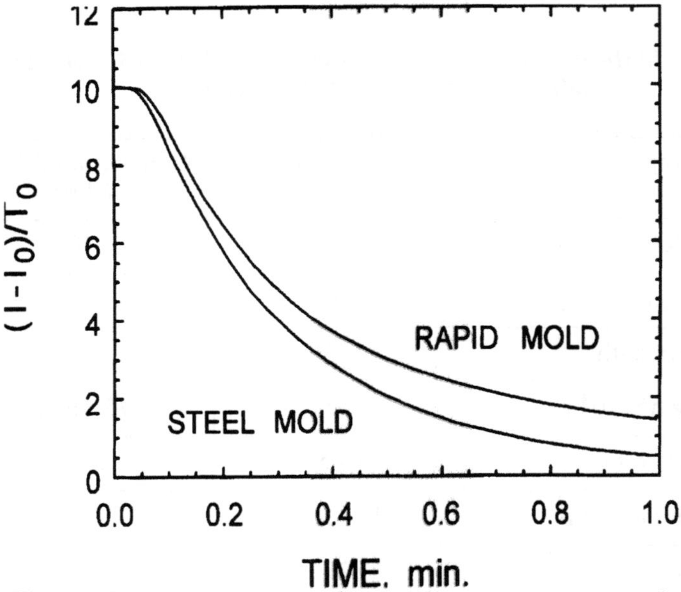

Figure 5.16 Comparison of centerline temperatures of 1/8 in. thick polymer parts injection molded at 275°C into a 25°C mold (Barlow, *et al*, 1996).

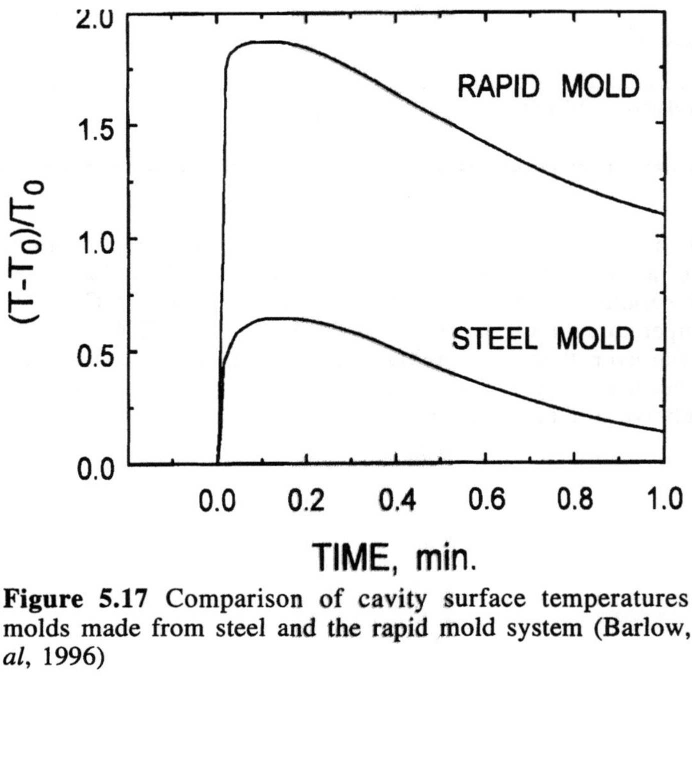

Figure 5.17 Comparison of cavity surface temperatures in molds made from steel and the rapid mold system (Barlow, *et al*, 1996)

Clearly from Figure 5.17, the RM material conducts less heat than steel. This suggests that the temperature could build with repeated use of the mold on, for example, a 40 sec cycle. This does appear to occur (Barlow, *et al*, 1996). A design solution to the mold-overheating problem is to simply reduce the conduction length from the cavity surface to the cooling water. Figure 5.18 shows the calculated maximum temperatures on a 40 sec cycle when the conduction length is 0.4375 in. Steady state for this case is reached after about three part cycles, the plastic part centerline temperature stays below 100°C, and the maximum surface cavity temperature is below 110°C. This mold is probably acceptable, and parts molded from it will probably be quite similar to those obtained from conventional steel molds. Typical cycle times for 1/8 inch thick parts molded in steel are usually close to 40 sec, with 5 sec of the cycle associated with melt injection and part ejection and 30-35 sec associated with cooling the part (Bernhardt,1959).

The room temperature mechanical properties of the RM system in Table 5.3 are quite similar to those of the epoxy infiltrant used to construct the test parts (Badrinarayan and Barlow, 1994). Below the epoxy T_g, 130-150°C, the coefficient of thermal expansion, CTE, of the RM system is substantially lower than the 60 ppm/°C of the unfilled epoxy, although it is still higher than that of steel by about a factor of two. As discussed above, injection molds are temperature controlled to temperatures below 100°C, and these differences in CTE between the RM material and conventional steels are not expected to pose insurmountable problems for the designer.

The tensile strength of the RM material is substantially lower than that of steel. For this reason, it is not recommended for construction of the entire mold. Instead, a composite mold design in which the cavity geometry is provided by the RM system and mold strength is provided by surrounding the cavity with steel has been successfully demonstrated and is recommended (Apelskog-Killander, 1996; Barlow, *et al*, 1996). The design details of this approach, in which stresses are computed as a function of cavity pressure in two dimensions (stresses in the third being compensated by the mold clamp) are discussed in Barlow, *et al*, 1996. To summarize, they found that a 3 in x 3 in square cavity in a 10 in diameter mold made entirely of the Rapid Mold material could handle cavity pressure near 3100 psi but would fail in tension at the cavity corner at injection pressures near 6000 psi, a situation that they described as "not robust" because injection pressures often exceed 10,000 psi in many molding applications. Their solution was to constrain the deformation of Rapid Mold material by inserting it into a steel pocket, or as illustrated in Figure 5.19, a steel ring. The maximum stresses now occur in the steel ring, not in the RM material. Using a factor of safety of two, and the steel mechanical properties in Table 5.3, the maximum cavity pressure for shear failure in the ring is 21,000 psi and that for tensile failure in the ring is 27,800 psi for the 3 in x 3 in square cavity constrained within a 10 in outer diameter, 1 in thick steel ring.

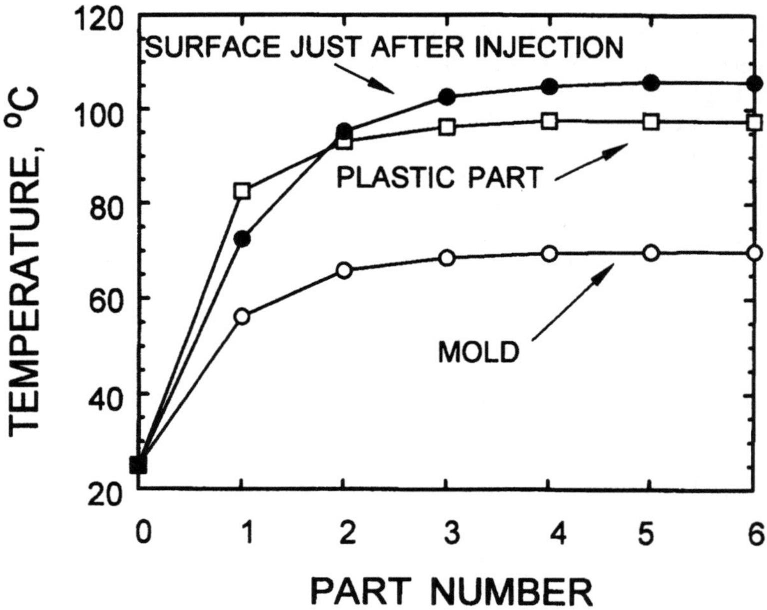

Figure 5.18 Maximum temperatures after injection of 275°C melt into 1/8 inch part cavity on a 40 sec cycle (Barlow *et al*, 1996).

The deformation of the RM material is greatly constrained by the presence of the steel ring, and, as a result, the shear stress in the corners of the cavity is reduced to 0.2 P. Using the working shear stress for the RM material, 6000 psi above, the working cavity pressure with the steel ring in place would be about 30,000 psi, as opposed to 6000 psi without the reinforcing ring. With the ring in place, x-directed tensile stress is limited to $-P$ in the RM material, indicating its fluid-like compression. The compressive strength of the RM material is about 24,000 psi and seems to follow the behavior seen in many materials, including concrete and filled epoxy systems, which show compression strengths that are 2-10 times as great as their tensile or bending strengths, respectively (Marin and Sauer, 1954). Again using a factor of safety of two, the limiting cavity pressure for failure by compression of the composite is approximately 12,000 psi, well within the usual working range for injection molding. Failure of this particular design therefore results from compression failure in the RM material. In actuality, cavity pressure can probably be raised above this conservative estimate. However, it should not exceed approximately 24,000 psi for the RM material, in particular where stress concentration may occur.

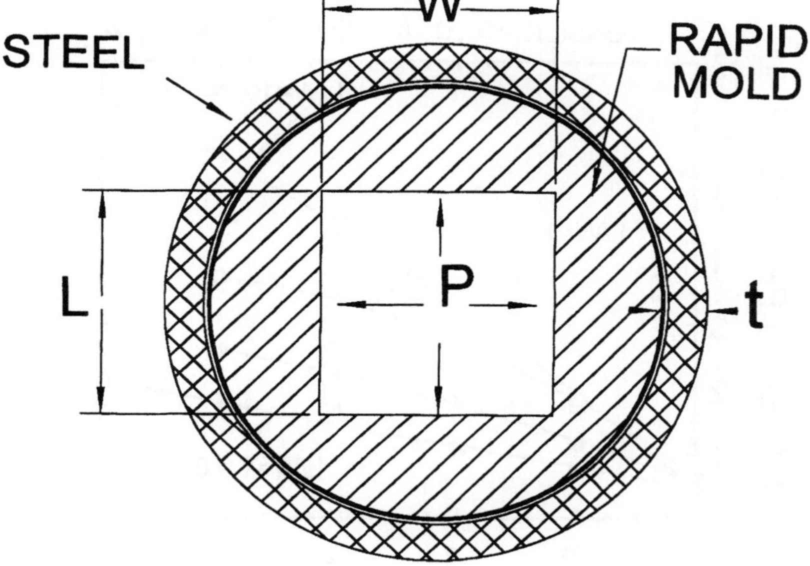

Figure 5.19 Schematic of mold cross section.

Of course, the use of different cavity and ring geometry than the example above will influence the permitted cavity pressure. As a design aid, the relationships between maximum tensile and shear stresses in the steel ring, cavity pressure, and cavity geometry are summarized in Figure 5.20. No safety factors are provided for these calculations. The tensile stress in the ring increases as the ratio of cavity edge L to ring thickness t is increased. For example, a square cavity, L/W=1, five inches on a side, constrained with an 10 inch outer diameter, one inch thick, steel ring would be limited to a cavity pressure of (75,000/3.7) = 20,300 psi. For this design, shear failure in the steel limits the cavity pressure to 37,500/2.6 = 14,400 psi, while consideration of the shear strength and tensile strength of the Rapid Mold material again limit the cavity pressure to about 24,000 psi. Tensile stress in the ring also increases as the cavity becomes less square, due to additional moments on the ring. For example, a 5 inch by 0.5 inch cavity (L/W=10) constrained by the same steel ring (L/t=5) is limited to a cavity pressure of (75,000/5.7) = 13100 psi by the design tensile stress in the ring and to (37,500/1.85) = 20,300 psi by the design shear strength of the ring. As in the previous cases, the allowed cavity pressure of 24,000 psi based on the compressive strength of the RM material would be the limiting pressure in this case as well. Of course, reducing the steel thickness or further increasing the cavity L/W ratio would cause the cavity pressure to be limited by the properties of the steel ring and would result in design cavity pressures less than 24,000 psi. Little difference is seen between L/W= 10 and greater cavity *L/W* ratios.

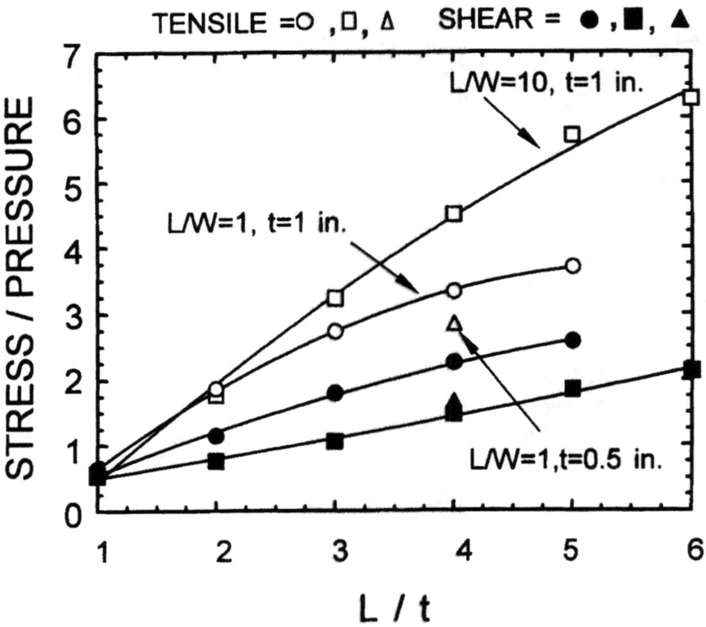

Figure 5.20 Effect of cavity geometry on maximum shear and x-directed tensile stresses in a 10 in outer diameter steel ring (Barlow, *et al*, 1996)

Because even rapid prototype molds are expensive to build, only three have been constructed and tested. Consequently, there is very little information with which to check the veracity of the predictions above. The first mold was built to test the material, not the mold making process (Tobin, *et al*, 1993). As pictured in Figure 5.21, a simple slab mold insert was constructed by machining a slab that had been cast from the mixed iron and binder powders in an air oven, then oxidized and infiltrated with epoxy in the usual way. The machined insert was fastened to a standard MUD mold base and knock-out pin assembly with two bolts and was further constrained on two opposing boundaries by close fitting metal plates. The cavity was only 0.125-0.187 inches deep. This mold was used to prepare a total of 176 parts with no mold cooling and a one minute cycle from four different plastics whose molding conditions ranged from 200-300°C melt temperature and from 8000-35,000 psi injection pressure. Two small failures, a hairline after shot number 125 and a small chip breakout around an ejector pin at shot number 123, were observed, but no catastrophic failure resulted. The mold was checked for wear, and none was found. Two lessons emerged from this study: (1) the RM material is machinable. It can be drilled, tapped, and surface ground and behaved much like cast iron during these operations; (2) one should probably install sleeves for the ejector pins if one needs more than 50-100 parts.

Figure 5.21 Mold assembly with Rapid Mold insert (Tobin, *et al*, 1993).

A small mold was built for injection molding polymer-ceramic slurries (Badrinarayan and Barlow, 1994). This mold was unconstrained and failed, prompting the present study.

L. Apelskog-Killander of the Swedish Institute of Production Research (SIPR), Stockholm, Sweden, recently prepared a rather complex mold insert with the DTM Rapid Steel powder and the University of Texas RM process. A photograph of the green insert half is shown in Figure 5.22. The insert was constrained in a pocket cavity and produced about 50 parts before failing as a consequence of a pre-existing and known defect. More details about insert preparation, geometric accuracy, and failure are presented in a recent article (Apelskog-Killander, 1996).

5.3 SECONDARY/POST PROCESSING

As noted in the previous section, several freeform processes produce parts consisting of a mixture of a metal or ceramic with a binder. Generation of a completely metal or ceramic part requires several secondary or post-processing steps, including binder removal and sintering of the remaining metal or ceramic phase. Secondary processing may also include infiltration of the porous part. The key variables are: the debinding atmosphere, the degree of sintering, the amount of infiltrant (if any), and the method of infiltration. The atmosphere must be controlled for several reasons. First, oxides must be removed from the metal surfaces in order to achieve efficient sintering, complete infiltration, and good mechanical properties. A reducing gas or vacuum is generally used for this purpose. In addition, metallurgical reactions which cause part swelling can occur during the infiltration process. Such problems can often be controlled by modifications to the furnace atmosphere.

Figure 5.22 S.I.P.R. mold insert, prior to firing and infiltration steps (Courtesy of L. Apelskog-Killander).

5.3.1 Binder Removal

Binder removal is prerequisite to part sintering/densification. Process control is critical. If the binder content is high, thermal dissociation produces gaseous phases which induce distortions. Modeling of debinding is therefore highly desirable.

A model is presented here which is conservative in two respects. It assumes that none of the polymer binder flows out of the green part by the combined mechanisms of gravity and wicking into the packing material. We know that some polymer removal does occur by these mechanisms in reality. The model further assumes that polymer depolymerization occurs in a nitrogen inert gas atmosphere. Since oxidation can severely interfere with subsequent sintering and infiltration steps in metal systems, this is a suitable assumption in metal powder systems where oxidation of the metal powder during debinding is to be avoided. Kinetic rate studies (Kashigawi, *et al*, 1985, Badrinarayan, 1995) show that polymer binders of the type used decompose 2-3 times faster in air than in nitrogen. This is apparently caused by chain scission due to the presence of oxygen radicals. Consequently, ceramic systems that can better tolerate oxidative debinding may be binder-free in less time than is estimated by the model presented here.

The model assumes that each part is uniformly surrounded by a 2 cm layer of alumina packing and that the oven walls, also made of alumina, completely surround the bed, see Figure 5.8. It is customary to place the packed part in a cold oven, to heat the oven to 200°C, and to hold at this temperature for 3-4 hours to permit polymer to "drain" from the part into the packing material, prior to further heating the oven at 100°C/hr to higher hold temperatures. Since no depolymerization of binder occurs at 200°C, the model considers all binder to exist at this temperature. No account is taken of binder that may have left the part boundaries by gravitational flow.

The unsteady state, one dimensional, energy balance is given by

$$\begin{aligned} &@\ (L-2)>z>2\ \{\text{cm}\} \\ &\rho_P C_P \frac{\partial T}{\partial t} = \frac{\partial}{\partial z}\left(k_P \frac{\partial T}{\partial z}\right) + G(z,t) \\ &@\ 0<z<2 \text{ and } (L-2)<z<L\ \{\text{cm}\} \\ &\rho_A C_A \frac{\partial T}{\partial t} = \frac{\partial}{\partial z}\left(k_A \frac{\partial T}{\partial z}\right) \end{aligned} \tag{5.3.1}$$

where the boundary conditions are defined as

$$\begin{matrix} @z=0,\ L\ \{\text{cm}\},\ t>0 & @z=2,\ L-2\ \{\text{cm}\},\ t>0 & @z=\frac{L}{2},\ t>0 \\ -k_A \frac{\partial T}{\partial z} = -h(T-T_0) & k_A \frac{\partial T}{\partial z} = k_P \frac{\partial T}{\partial z} & \frac{\partial T}{\partial z}=0 \end{matrix} \tag{5.3.2}$$

and ρ is the density, C is the heat capacity, $T(t,z)$ is the temperature, k is the thermal conductivity, subscript A denotes packing material, subscript 0 denotes the oven wall, and subscript P denotes part material. The linear distance from the bed surface is given by z, L is the bed thickness, and t denotes time.

The parameter $G(t,z)$ represents the heat sink due to depolymerization of the polymer. This term could be an important part of the energy balance on the part because the heat of depolymerization, estimated as the negative of the heat of polymerization, 13.8 kcal/mol (Kine and Novak, 1985), is quite large. Its consideration potentially requires the simultaneous solution of equations 5.3.1 and 5.3.2 and

$$G(z,t) = -\frac{d(w/w_0)}{dt}\frac{\Delta H_R}{M}\phi_{PO}\rho_P \tag{5.3.3}$$

where M is the molecular weight of the repeat unit of the polymer, ϕ_{po} is the initial weight fraction of polymer in the part, and where $-d(w/w_o)/dt$ is the rate of disappearance of polymer, ΔH_R is enthalpy change of depolymerization. For the typically 20 vol% binder loading used in this study, ϕ_{po} ranges between 0.03 and 0.10 depending on the density of the substrate, and G(z,t) turns out to be quite small. That is, nearly all of the absorbed energy is used to heat the substrate not to depolymerize the polymer. Consequently, equation 5.3.1 can be solved directly using $G(z,t) = 0$ without concern that the heat of depolymerization is distorting the unsteady state temperature profiles. The polymer decomposition can then be independently calculated from equation(5.3.3) and the time and temperature history at each node. For the sake of completeness, however, $G(z,t)$ was retained as shown in equation 5.3.1 and solved simultaneously as part of the complete energy balance. The simulations, presented in Figures 5.23 and 5.24, use ϕ_{po}=0.03 for the initial mass fraction polymer in a green part made of iron powder.

The parameter h in equation 5.3.2 is the combined radiation and convection heat transfer coefficient that describes heat transfer between the oven wall temperature at temperature T_o and the packing surface. Assuming that both the oven and packing are gray bodies, the radiation heat transfer portion, h_R, is defined (Pitts and Sisson, 1977) by

$$h_R = \frac{\sigma(T_0^4 - T^4)}{(T_0 - T)\left[\frac{A_A}{A_0\varepsilon_0} + \left(\frac{1}{\varepsilon_A} - 1\right)\right]} \tag{5.3.4}$$

where A_A/Ao is the ratio of the surface area of the packed bed to that of the oven interior walls, taken in the simulation to be 0.1, ε_o and ε_a are the emissivities of the oven wall and packed bed surfaces, respectively, and σ is the Stefan-Boltzmann constant for radiation heat transfer, 1.355 x 10^{-12} cal/cm^2-sec-K^4 (Bird, Stewart, and Lightfoot, 1960). Although the emissivity of alumina decreases with increasing temperature (Rohsenow, *et al*, 1985), the values used in the simulation, ε_o = 0.5 (Pitts and Sisson, 1977) and ε_a = 0.8 (Sih and Barlow, 1995) are assumed constant. The

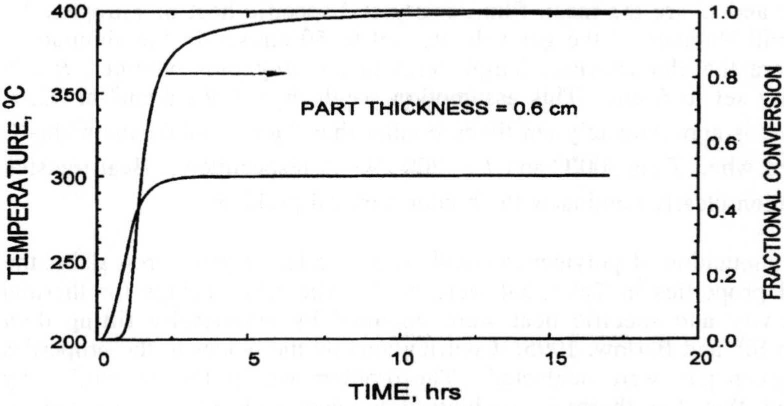

Figure 5.23 Calculated centerline temperature and binder content for a 0.6 cm thick part, surrounded by 2 cm of alumina.

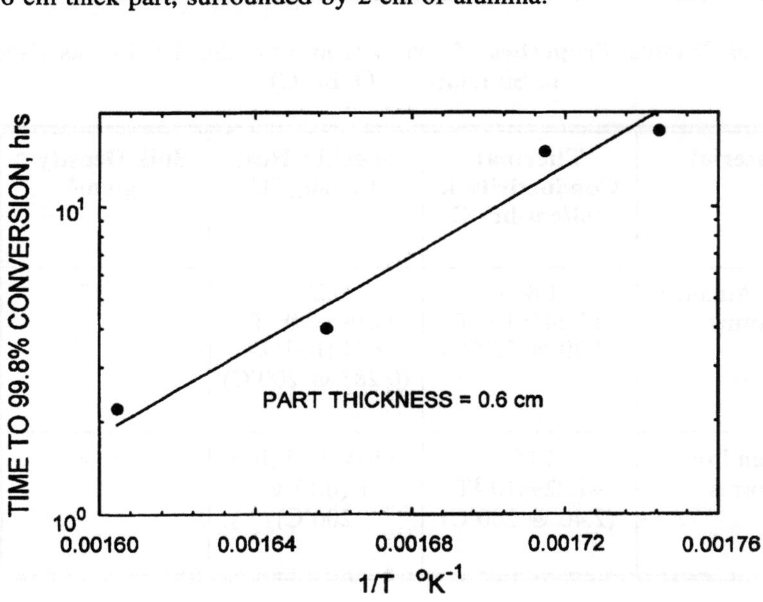

Figure 5.24 Effect of oven hold temperature on the calculated time to achieve a 99.8% binder conversion to gas at the centerline.

convection portion of the heat transfer coefficient, h_c, is estimated by the Pohlhousen Equation (McAdams, 1954),

$$h_c = \rho C\, V\, Pr^{-2/3} Re^{-1/2} \qquad (5.3.5)$$

where ρ and C are the mean film density and specific heat of nitrogen, Pr the Prandtl Number, V the gas velocity, set to 50 cm/sec in the simulation, and where the characteristic length term in the Reynolds Number, Re, is arbitrarily set to 6 cm. This assumption yields $h_C \approx 0.9$ cal/cm^2-hr-°C, a value that is approximately ten times smaller than $h_R = 9.5\text{-}8.0$ cal/cm^2-hr-°C evaluated when T_o is 300°C and T is 200-295°C, respectively. Heat transfer by radiation clearly dominates the binder removal problem.

For the simulation of polymer removal from a series of green iron slabs, the physical properties in Table 5.4 were used. The relationships for thermal conductivity and specific heat were obtained by empirically fitting data found in Sih and Barlow, 1995. Contributions by the binder to the properties of the green part were neglected. The problem was further simplified by assuming that the thermal conductivities were constant at the values determined at 200°C. The thermal diffusivity ratios, $k/\rho C$, of the two powders are quite similar, 7.1 cm^2/hr for alumina versus 5.9 cm^2/hr for iron.

Table 5.4 Physical Properties of Green Iron and Alumina Forms Used in Simulation. (T in °C)

Material	Thermal Conductivity k cal/cm-hr-°C	Specific Heat C, cal/g-°C	Bulk Density ρ g/cm^3
Green Alumina Forms	1.840 $+2.245\times10^{-3}$T (2.29 @ 200°C)	0.277 $+2.995\times10^{-5}$T - 8.31×10^{-11}T^2 (0.283 @ 200°C)	1.13
Green Iron Forms	2.151 $+1.229\times10^{-3}$T (2.40 @ 200°C)	$0.104+1.25\times10^{-4}$ T (0.13 @ 200°C)	3.14

Equations 5.3.1-5 are solved for temperature and, where appropriate, binder content by the Galerkin Finite Element Method that is incorporated in PDEase2 software (SPDE, Inc., Bass Lake, CA, marketed through

MacSyma, Inc., Arlington, MA). The software is run under the DOS system on a personal computer that contains a 80486 microprocessor operating at 66 MHz. The binder content at each time and position is expressed by the fractional conversion, , defined by,

$$\xi = \frac{w_0 - w}{w_0} = 1 - \frac{w}{w_0} \tag{5.3.6}$$

where w/w_0 is the ratio of polymer mass per unit volume of bed at time t to that initially and is found by integrating the rate expression.

Figure 5.24 shows calculated temperature and fractional conversion profiles with oven time at the center of a 0.6 cm thick iron powder green shape that is heated in nitrogen from 200°C to 300°C at 100°C/hr, then held at 300°C. As the part is heated, little binder decomposition occurs until the temperature reaches about 280°C. At that point, depolymerization of the binder occurs so rapidly that nearly 90% of the binder mass is converted to gas by the time the center temperature approaches 300°C. The remaining 10% of the binder mass is very slowly removed and about 8 hours total time (7 hours at 300°C) are required to remove 98% of the binder material and to reduce the binder content to 0.0006 g binder per g of iron. If the green part were heated in air so that the rate of depolymerization were doubled, 98% of the binder could be removed in only 4 hours. These calculated results seem reasonable in light of work by Badrinarayan, 1995, in which he examined a series of experimental "oxidation cycles" for debinding and oxidizing 1/4 in x 1 in x 3 in green test specimens made from iron powder. His thermally least intense protocol involved holding the green specimens at 200°C for 5 hours, ramping at 100°C/hr to 300°C, then holding at that temperature for four hours. He found no evidence of polymer binder remaining in the oxidized parts. In another study, he varied the time that the parts were held at 300°C from 4 to 16 hours to see what, if any, effect oven time would have on part strength and geometry. No effects were found, suggesting that a 4 hour hold at 300°C in air is sufficient to remove "all" binder, at least to within the limits of detectability.

If we accept the calculations as being reasonable and possibly conservative, we can explore strategies for improving binder removal and for reducing binder bake out times. As suggested by Figure 5.24, the most effective strategy is to raise the oven hold temperature. This strategy leads to increased kinetic rates of binder decomposition and dramatically reduces the time required to remove, arbitrarily, 99.8% of the binder from 17.5 hours at 300°C to about 2.5 hours at 350°C for a 0.6 cm thick part.

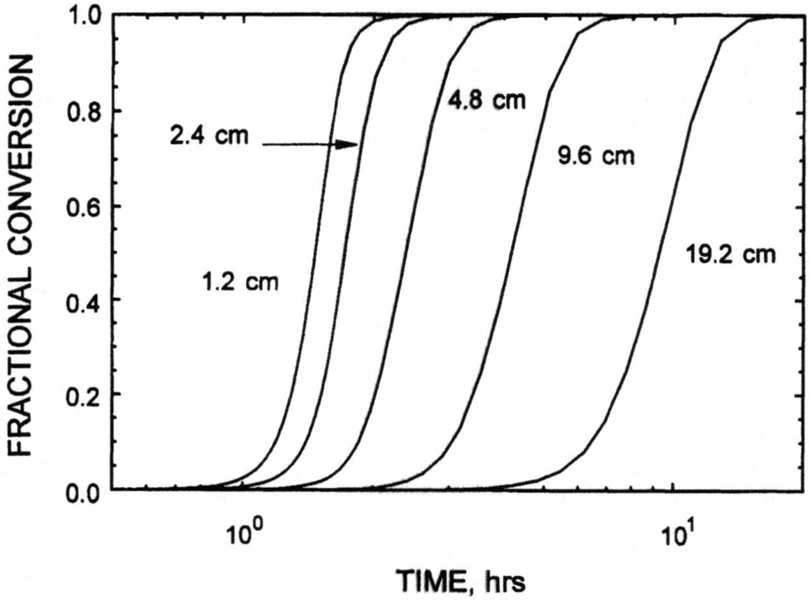

Figure 5.25 Calculated fractional conversion at centerline at 350°C as a function of part thickness.

Part and bed thicknesses strongly affect the required binder bake out time because both the conduction lengths and the total mass to be heated increase with increasing thicknesses. Figures 5.25 and 5.26 show the calculated, time dependent, fractional conversion and center temperatures, respectively, for a series of parts that range in thickness from 1.2 cm to 19.2 cm. The alumina packing remains at 2 cm on all sides for each part. As the part and bed thickness increases, the time required for center temperature to rise sufficiently to cause binder decomposition increases in a nonlinear manner. For example, the calculated center temperature of the 19.2 cm thick part has not even reached the 350°C oven hold temperature after 15 hours of heating. Consequently, the calculated fractional conversion of binder at part center does not reach $\xi = 0.998$ until the part has been heated for about 17.5 hours.

Obviously, account must be taken of part and bed size when establishing a binder burn out cycle. Carslaw and Jaeger (1959) show that the temperature change at a particular location within a slab during simple heating or cooling is a unique function of the non-dimensional heating or cooling time, given by $\alpha t/W^2$, where α is the thermal diffusivity of the solid, and W its half thickness. Since the thermal diffusivities of part and packing in this calculated example are similar and the heat of depolymerization is not an important issue at the low mass fraction of binder employed in the green part, the non-dimensional time idea should work to correlate binder burn out. That is, the time required for any particular fractional conversion should be linearly proportional to the square of the bed thickness.

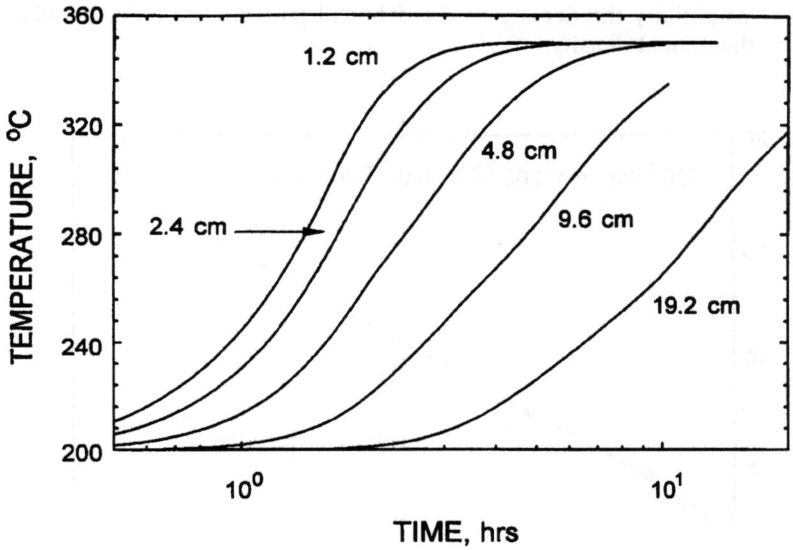

Figure 5.26 Calculated centerline temperature vs time as a function of part thickness for an oven temperature of 350°C.

Figure 5.27 shows that the $\alpha t/L^2$ correlation is approximately correct, although two line slopes are apparent. These correspond to that material, packing or part, that dominates the heat conduction problem. When the part thickness is small relative to that of the packing the correlation takes on a higher slope, corresponding to the higher α for the alumina packing, than when the part thickness is large relative to that of the packing.

5.3.2 Sintering

Once metallic or ceramic parts are debound, it is necessary to create autogenous interparticle bonding by exposing the powder mass to elevated temperature. For particulate assemblages, the fundamentals of sintering are well established (Jones, 1960, Lenel, 1980, Hausner and Mal, 1982, German, 1994). On an atomic level, sintering may be considered to be stress- or curvature-assisted mass transport. At first, this produces a microscopic solid-state weld at particle contacts which grows as sintering progresses. The high curvature of this weld "neck" provides the driving force for atomic diffusion in the absence of applied stress (i.e., isostatic pressing). Once a neck structure is created, parts have sufficient strength to be handled. The porosity is largely continuous with maximum pore size, making these parts most amenable to infiltration. This early stage of sintering is accomplished with minimal densification. Hence, a debind/modest sintering/infiltration process is desirable when minimum shrinkage and dimensional change are required. Actual part density at this stage is essentially the density of the debound part, on the order of 40% to 80% of theoretical density.

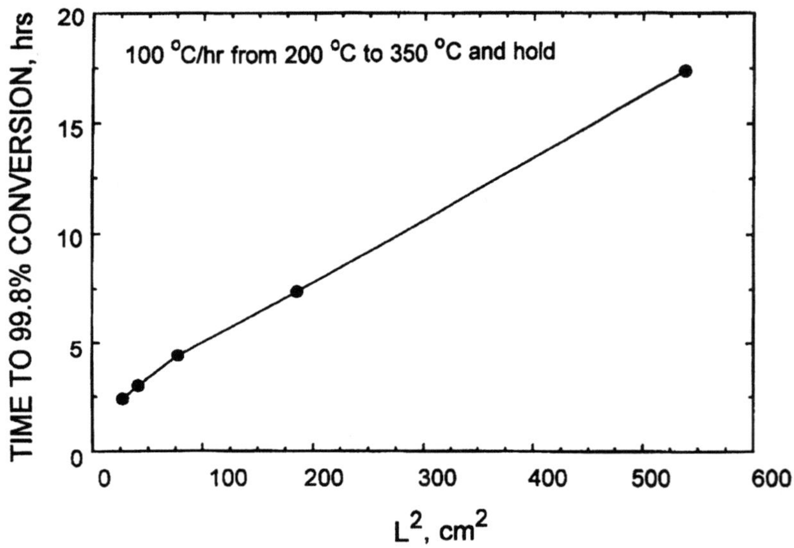

Figure 5.27 Correlation of centerline conversion with square of the bed thickness.

As sintering progresses, part dimensional changes occur. At just over 90% of theoretical density, the pore size decreases to a point where the "tunnels" of porosity become unstable and begin to collapse locally. Infiltration is not possible beyond this point, since the porosity becomes isolated. This has an impact on densification rate as well, as mass transport mechanisms involving evaporation/condensation and surface diffusion no longer operate. As densification progresses beyond this point, the isolated pores spheroidize and continue to shrink due to surface effects. These pores act as pinning points for prior particle boundaries. Another critical point is reached when the shrinking pores no longer effectively pin these boundaries, allowing them to move via grain growth kinetics. The densification rate is further decreased, since densification can only progress by volume diffusive processes.

The limitations associated with full density processing may be overcome by application of external tractions to the part. If the shape is simple, unidirectional pressing, either at room temperature or elevated temperature, may be employed. For complex geometries, some form of isostatic pressing is required. Two such processes are hot isostatic pressing or HIP, and a quasi-isostatic process called the Ceracon Process (Ferguson, 1984). With HIP, the part is sintered in a high pressure gaseous inert gas atmosphere. The porosity must be closed for HIP to be effective. Alternatively, the part may be encapsulated in metal foil or glass prior to HIP. The Ceracon ("CERAmic CONsolidation") Process is based on the quasi-isostatic stress redistribution that occurs when powder is unidirectionally pressed. The part is heated and embedded in a heated ceramic powder bed. Uniaxial pressure is applied to the ceramic powder which transfers the load to the part. The ceramic powder remains unsintered during part densification.

5.3.3 Infiltration

There are a variety of approaches which can be used to infiltrate parts. In general, it is desirable to infiltrate parts from one side so that there is a single flow front which moves through the part. Infiltration from multiple directions can lead to gas entrapment, cracking, and blistering. To insure that the infiltration is consistent from part to part, it is also desirable to have the capillary forces augmented by gravity or vacuum. If the infiltration is variable, shrinkage and part properties will also be variable. Parts can be infiltrated from the top or from the side. If the parts are infiltrated from the top, it is necessary to use two separate furnace cycles. In the first cycle, the parts are debound and sintered. The lower melting metal is then placed on top of the part, and it is infiltrated in the second cycle. If parts are infiltrated from the side, the lower melting metal is placed in close proximity to the green part, and the debinding, sintering, and infiltration can be performed in a single cycle.

Some sintering of the debound particles is generally required to create a rigid skeleton such that stresses generated during the infiltration process will not cause part distortion, but it is desirable to limit the sintering since an increase in particle-particle bonding results in greater part shrinkage. An excess of infiltrant can also cause part distortion; in general, the best parts are obtained when slightly less than the theoretical amount of infiltrant required to achieve full density is used (German, 1985).

Infiltration of a liquid into a solid particulate framework has been described by the relationship

$$r \propto \left(\frac{t\sigma}{\eta d} \right)$$

where r is the infiltration front penetration distance, t is the time, σ is the liquid metal surface energy, η is the liquid viscosity, d is the solid particle size and n is a constant approximately equal to 0.5 (Bunnell, 1995).

5.4 SUMMARY

As discussed in this chapter, there exist numerous transfer and indirect approaches for production of metallic and ceramic parts using solid freeform fabrication. Transfer technologies such as investment and sand casting result in the complete change of part material with retention of geometry. Indirect methods such as Laminated Object Manufacturing create ceramic laminates with fusible binder interlayers. Fused Deposition of Ceramics, Reactive Stereodeposition, Stereolithography, 3D Printing and Selective Laser Sintering produce a part composed of metallic or ceramic particles in a suitable binder. The binder is removed in a post processing or secondary treatment. Following binder removal, parts are sintered to high density or infiltrated with a metal or polymer. The commercial Rapidtool Process and metal-epoxy tooling are examples specific to Selective Laser Sintering.

In the next chapter, a more detailed study of the engineering science of one particular process, SLS, is given.

REFERENCES

Agarwala, M.K., R. van Weeren, R. Vaidyanathan, J. Burlew, R. Donaldson, P. Whalen, C. Ballard: "Structural Ceramics by Fused Deposition of Ceramics", *Solid Freeform Fabrication Proceedings*, Austin, Texas, 1995. The University of Texas, Austin, Texas, pp. 1-8.

Apelskog-Killander, L., "Rapid Mould: Epoxy-Infiltrated, Laser Sintered Inserts", *Rapid Prototyping J.*, 1996, 2, pp. 34-40.

ASM Handbook Committee, *Metals Handbook 10th Edition*, Volume 2 Properties and Selection: Nonferrous Alloys and Pure Metals, ASM International, 1990, pp. 45-49.

Badrinarayan, B. "Study of the Selective Laser Sintering of Metal-Polymer Powders," *Ph.D. Dissertation*, The University of Texas at Austin, Dec. 1995.

Badrinarayan, B., and Barlow, J. W., "Selective Laser Sintering of a Copper-PMMA System," *Solid Freeform Fabrication Proceedings*, The University of Texas: Austin, Texas, 1991, pp. 245-250.

Badrinarayan, B., and Barlow, J. W., "Manufacture of Injection Molds using SLS," *Solid Freeform Fabrication Proceedings*, The University of Texas: Austin, Texas, 1994, 371-378.

Barlow, J. W., *et al*, U. S. Patent Appl., 1994, "Mold Useful for Injection Molding of Plastics and Methods of Production and Uses Thereof."

Barlow, J. W., Beaman, J. J., and Badrinarayan, B., "A Rapid Mould Making System: Material Properties and Design Considerations," *Rapid Prototyping Journal*, 1996, 2, pp. 4-15.

Bernhardt, E. C., *Processing of Thermoplastic Materials*, Van Nostrand Reinhold Company: New York, 1959, pp. 359-378.

Biddle, K., *et al.*: "The Chemistry of Ethyl Silicate Binders in Refractory Technology", *Journal of Applied Chemistry and Biotechnology* 27, 1977, pp. 565-573.

Bird, R. B, Stewart, W. E., and Lightfoot, E. N., *Transport Phenomena*, J. Wiley: New York, 1960, p.748.

Bowman, L., "CAD to Product - the LOM Route", *Proceedings of the 2nd European Conference on Rapid Prototyping and Manufacturing*, University of Nottingham, 1993.

Bunnell, D. E., Das, S., Bourell, D. L., Beaman, J. J., and Marcus, H. L., "Fundamentals of Liquid Phase Sintering During Selective Laser Sintering," *Solid Freeform Fabrication*

Proceedings, The University of Texas: Austin, Texas, 1995, pp. 440-447.

Bunnell, D.E., "Fundamentals of Selective Laser Sintering of Metals, *PhD Dissertation* ," The University of Texas at Austin, Austin, Texas, 1995.

Carslaw, H. S., and Jaeger, J. C., *Conduction of Heat in Solids, 2nd Ed.*, Oxford University Press: New York, 1959.

Carter, W.T. and M.G. Jones, "Direct Laser Sintering of Metals", *Solid Freeform Fabrication Proceedings*, The University of Texas at Austin, Austin, TX, 1993, pp. 51-59.

Cima, M.J., E.M. Sachs, L.G. Cima, J. Yoo, S. Khanuja, S.W. Borland, B. Wu, R.A. Giordano: "Computer-Derived Microstructures by 3D Printing: Bio- and Structural Materials" , *Solid Freeform Fabrication Proceedings*, The University of Texas at Austin, Austin, Texas, 1994, pp. 181-190.

Cima, M.J., J. Yoo, S. Khanuja, M. Rynerson, D. Nammour, B. Giritlioglu, J. Grau, E.M. Sachs: "Structural Ceramic Components by 3D Printing", *Solid Freeform Fabrication Proceedings*, The University of Texas,Austin, Texas, 1995. pp. 479-488.

Comb, J.W. and W.R. Priedeman: "Control Parameters and Material Selection Criteria for Rapid Prototyping Systems", *Solid Freeform Fabrication Proceedings*, The University of Texas, Austin, Texas, 1993, pp. 86-91.

Comb, J.W., W.R. Priedeman, and P.W. Turley, "FDM Technology Process Improvements", *Solid Freeform Fabrication Proceedings*, The University of Texas,Austin, Texas, 1994, pp. 42-49.

Crockett, R.S., J. O'Kelly, P.D. Calvert, B.D. Fabes, K. Stuffle, P. Creegan, R. Hoffman: "Predicting and Controlling Resolution and Surface Finish of Ceramic Objects Produced by Stereodeposition Processes", *Solid Freeform Fabrication Proceedings*, The University of Texas,Austin, Texas,1995, pp. 17-24.

Deckard, L., and Claar, T. D., "Fabrication of Ceramic and Metal Matrix Composites from Selective Laser Sintered Ceramic Preforms," *Solid Freeform Fabrication Proceedings,* The University of Texas: Austin, Texas, 1993, 215-222.

DTM Corporation, *Guidelines for Using RapidCasting Polycarbonate Patterns in the Investment Casting Process*, DTM Corporation, Austin, TX, 1995.

Emblem, H.G., "Ethyl Silicate Bonded Refractories", *Materials Chemistry and Physics*, 8 (1983) pp. 265-277.

Ferguson, B.L. and O.D. Smith: "Ceracon Process", in *Metals Handbook 9th Edition*, Volume 7 Powder Metallurgy, ASM International, 1984, pp. 537-541.

German, R.M., *Powder Metallurgy Science*, Second Edition, Metal Powder Industries Federation, Princeton, NJ, 1994.

German, R.M., *Particle Packing Characteristics*, Metal Powder Industries Federation, Princeton, NJ, 1989.

German, R.M., *Liquid Phase Sintering*, Plenum Press, New York, NY, 1985.

Griffin, Curtis, JoDee Daufenbach, Scott McMillin: "Solid Freeform Fabrication of Functional Ceramic Components Using a Laminated Object Manufacturing Technique", *Solid Freeform Fabrication Proceedings*, The University of Texas,Austin, Texas, 1994, pp. 17-24.

Griffith, Michelle L. , John W. Halloran: "Ultraviolet Curing of Highly Loaded Ceramic Suspensions for Stereolithography of Ceramics", *Solid Freeform Fabrication Proceedings* , The University of Texas,Austin, Texas, 1994, pp. 396-403.

Griffith, Michelle L., Tien-Min Chu, Warren Wagner, John W. Halloran: " Ceramic Stereolithography For Investment Casting and Biomedical Applications ", *Solid Freeform Fabrication Proceedings,* The University of Texas,Austin, Texas,1995, pp. 31-38.

HA-16, Rohm and Haas, Philadelphia PA 19105.

Hausner, H.H. and M.K. Mal, *Handbook of Powder Metallurgy*, Second Edition, Chemical Publishing Company, New York, NY, 1982.

Janney M.A., "Method for Molding Ceramic Powders", U.S. Patent No. 4,894,194, Jan. 16, 1990.

Jones, W.D., *Fundamental Principles of Powder Metallurgy*, Edward Arnold Publishers, London, UK, 1960.

Jurand, R., ed., *Modern Plastics Encyclopedia*, McGraw-Hill: New York, 1989, pp 576-619.

Kashigawi, T., Hirata, T., and Brown, J. E., "Thermal and Oxidative Degradation of Poly(methyl methacrylate): Molecular Weight," *Macromolecules*, 18[2], (1985), pp.

131-138.

Kine, B. B. and Novak, R. W., "Acrylic and Methacrylic Ester Polymers," *Encyclopedia of Polymer Science and Engineering, 2nd Ed.*, Vol. 1, J. Wiley & Sons: New York, 1985, p.241.

Krantz, T., "Effect of density and composition on dimensional stability and strength of iron-copper alloys", *International Journal of Powder Metallurgy*, Volume 5, Number 3, 1969, pp. 35-43.

Lee, G.-H., Barlow, J. W., "Selective Laser Sintering of Calcium Phosphate Powders," *Solid Freeform Fabrication Proceedings*, The University of Texas,Austin, Texas,1994, 191-197.

Lenel, F.V., *Powder Metallurgy Principles and Applications*, Metal Powder Industries Federation, Princeton, NJ, 1980.

Marin, J. and Sauer, J. A. *Strength of Materials*, MacMillan Company: New York, 1954, pp 76-85, p. 508.

Masters, K., *Spray Drying Handbook*, J. Wiley & Sons: New York, 1985.

McAdams, W. H., *Heat Transmission, 3rd Ed.*, McGraw-Hill: New York, 1954, p. 224.

Menges, G. and P. Mohren, *How to Make Injection Molds*, Hanser Publishers, New York, 1986, p. 35.

Michaels, S., E.M. Sachs, M.J. Cima: "Metal Parts Generation by Three Dimensional Printing", *Solid Freeform Fabrication Proceedings*, The University of Texas,Austin, Texas, 1992, pp. 244-250.

Pang , T.H. and P.F. Jacobs, "Stereolithography Epoxy Resin Development: Accuracy and Dimensional Stability", *Solid Freeform Fabrication Proceedings*, The University of Texas,Austin, Texas, 1993, pp. 11-26.

Pitts, D. R., and Sisson, L. E., *Heat Transfer*, McGraw-Hill: New York, 1977, chapt. 11 & Appendix B.

Progelhof, R. C. and Throne, J. L., *Polymer Engineering Principles. Properties, Processes, Tests for Design*, Hanser/Gardner Publications: Cincinnati, Ohio, 1993, chapt. 5.

Rohsenow, W. M., Hartnett, J. P., and Ganic, E. N., Eds., *Handbook of Heat Transfer Fundamentals, 2nd Ed.*, McGraw-Hill: New York, 1985, pp. 3-113.

Sachs, E.M., S. Allen, M.J. Cima, E. Wylonis, H. Guo: "Production of Injection Molding Tooling with Conformal Cooling Channels using the Three Dimensional Printing Process", *Solid Freeform Fabrication Proceedings*, The University of Texas,Austin, Texas, 1995., pp. 448-467.

Sih, S. S. and Barlow, J. W., "Emissivity of Powder Beds," Marcus, H. L., *et al*, Eds., *Solid Freeform Fabrication Proceedings*, The University of Texas: Austin, Texas, 1995, 402-408.

Stuffle, K., A. Mulligan, J. Lombardi, P. Calvert: "Solid Freebody Forming of Ceramics from Polymerizable Slurry", *Materials Research Society Symposium Proceedings*, 286 (1993a) pp. 309-314.

Stuffle, Kevin, Anthony Mulligan, Paul Calvert, John Lombardi: "Solid Freebody Forming from Polymerizable Slurry", i *Solid Freeform Fabrication Proceedings*, The University of Texas,Austin, Texas, 1993.b, pp. 60-63.

Tadmor, Z. and Gogos, C. G., *Principles of Polymer Processing*, John Wiley & Sons: New York, 1979, chapt. 14.

Tobin, J. R., Badrinarayan, B., Barlow, J. W., Beaman, J. J., and Bourell, D. L., "Indirect Metal Composite Part Manufacture Using the SLS Process," *Solid Freeform Fabrication Proceedings*, The University of Texas,Austin, Texas,1993, pp. 303-307.

Vail, N. K., Barlow, J. W., and Marcus, H. L., "Silicon Carbide Preforms for Metal Infiltration by Selective Laser Sintering of Polymer Encapsulated Powders," *Solid Freeform Fabrication Proceedings*, The University of Texas,Austin, Texas, 1993, 204-214.

Vail, N. K., and Barlow, J. W., "Effect of Polymer Coatings as Intermediate Binders on Sintering of Ceramic Parts," *Solid Freeform Fabrication Proceedings*, The University of Texas: Austin, Texas, 1991, 195-204.

Vail, N. K., and Barlow, J. W., "A Method for Producing High Temperature Parts by Low Temperature Sintering," U. S. Patent No. 5,284,695, Feb. 8, 1994.

Vail, N. K., "Preparation and Characterization of Microencapsulated, Finely Divided Ceramic Materials for Selective Laser Sintering," *Ph.D. Dissertation*, The University of Texas at Austin, 1994.

Venus, A.D., "Improving Product Development for Automotive Cast Components", *Proceedings for the Dedicated Conference on Rapid Prototyping in the Automotive Industries*, Automotive Automation Limited, Croydon, England, 1995, pp. 307-319.

Walters, W., "Rapid Prototyping Using FDM: A Fast, Precise, Safe Technology", *Solid Freeform Fabrication Proceedings*, The University of Texas,Austin, Texas,1992, pp. 301-308.

Warner, C., "New Trends in Pattern Making Using LOM", *Proceedings for the Dedicated Conference on Rapid Prototyping in the Automotive Industries*, Automotive Automation Limited, Croydon, England, 1995, pp. 59-70.

Yoo, J., M.J. Cima, S. Khanuja, E.M. Sachs: "Structural Ceramic Components by 3D Printing", *Solid Freeform Fabrication Proceedings*, The University of Texas,Austin, Texas, 1993, pp. 40-50.

Young, A.C., O.O. Omatete, M.A. Janney, P.A. Menchofer: "Gel Casting - A New Ceramic Forming Process", *Journal of the American Ceramic Society* 74#3 (1991) pp. 612-618.

Young, A.G., "Linking Prototyping and Production", *IMS International Conference on Rapid Product Development*, Stuttgart, Germany, 1994.

Zong, G., Wu, Y., Tran, N., Lee, I., Bourell, D. L., and Marcus, H. L., "Direct Selective Laser Sintering of High Temperature Materials," *Solid Freeform Fabrication Proceedings,* The University of Texas: Austin, Texas, 1992, 72-85.

6

SLS PROCESS MODELING AND CONTROL

The SLS process is ultimately about successful control of heat transfer from an intense laser beam to a poorly conducting powder bed to produce geometrically precise localized heating and fusion of powder. Although some very interesting work on control of laser energy input to the powder through monitoring and feed back of the powder surface temperature has been recently reported (Benda, 1994), the SLS process, as commercially practiced, primarily relies on controlling all other aspects of the heat transfer problem. These include control of powder bed temperature, powder bed thermal properties, polymer sintering characteristics (where appropriate), and laser beam diameter, power, and scan rates to achieve parts with acceptable geometry and strength. These issues are briefly discussed below.

6.1 PROPERTIES OF POWDER BEDS

Because Selective Laser Sintering is a very dynamic process, its mathematical description necessarily involves the solution of unsteady state heat transfer equations (Carslaw and Jaeger, 1959) which contain three important material properties; the density, ρ, the specific heat, C_p. and the thermal conductivity, k. These parameters are usually combined as the thermal diffusivity, α,

$$\alpha = \frac{k}{\rho C_p} \tag{6.1.1}$$

Knowledge of these properties and their variation with temperature for the powder beds of interest is essential to the building of realistic process models.

6.1.1 Experimental Methods

The density of powder beds is a variable that depends not only on the density of the solid particles but also on their packing. The packing of particles depends, in turn, on the particle size distribution and on the particulars of the process used to pack the particles. Common packing processes that are used to characterize powder density include random placement of particles through "pour" or bulk density measurement where powder is poured to fill a known volume and then weighed or tap density measurement where the powder is "settled" or packed by tapping a graduated cylinder containing a known mass of powder and then measuring the volume (German, 1994). The SLS process generally produces bed densities that are closer to the random pour density than to tapped density values.

Some calculated estimates of powder bed density can be made in certain limiting cases. For example, a powder comprised of like-sized spheres, packed in a perfect simple cubic array, would have a density that is independent of particle size and given by $\rho = \pi\rho_S/6$, where ρ_S is the density of the solid particle. The relative density,

$$\rho_R = \frac{\rho}{\rho_S} \tag{6.1.2}$$

for this example would be 0.523 (52.3%) and its fractional porosity,

$$\varepsilon = \frac{V - V_S}{V} = 1 - \rho_R \tag{6.1.3}$$

where V is the volume of the system and V_S that of the solid, would be 0.477 (47.7%). Most real powder beds are not comprised of perfectly packed like-sized spheres but still have relative densities that lie between 40% and 60%. Therefore, approximately half of the powder bed volume is taken up by space that contains the surrounding gas.

The specific heat capacity of the powder bed can be estimated by,

$$C_P = \sum_n \omega_i C_{Pi} \tag{6.1.4}$$

Where ω_i and C_{Pi} are the weight fraction and specific heat of component i, respectively. Because the density of air is more than one thousand times smaller than that of the typical solid, the weight fraction of gas is too small for the specific heat of the gas component to have much influence on that of the bed, and the specific heat of the typical powder bed is just that of the solid particle. The specific heat of the solid or powder can be easily measured by a variety of methods, including Differential Scanning Calorimetry (Perkin-Elmer,1976). By this method, the specific heat of the unknown solid is measured with respect to a standard such as sapphire (Ginnings and Furukawa, 1953) in temperature increments of 10-20°C, at a fixed heating rate, in order to minimize various interferences. The specific heat capacity of the unknown sample is then computed by,

$$C_P = C_{PS}\frac{W_S}{W}\frac{D}{D_S} \tag{6.1.5}$$

where D signifies the instrument signal deviation from the base line for the unknown, W is the mass of the unknown, and subscript s designates the values when the sapphire standard is measured. These measurements are usually fit to a polynomial smoothing equation in temperature (Sih, 1996). Some typical measured specific heats for a variety of powder beds are shown in Figure 6.1.

Heat transfer through powder beds is limited to consideration of conduction and radiation modes. Generally, the interparticle spaces are too small to permit significant gas convection, and the transfer of heat through these spaces is primarily limited by thermal conduction through the gas. The thermal conductivity of most atmospheric gases is quite low and increases with increasing temperature. Air, for example, has a thermal conductivity, k_g, near 0.0065 W/m-°K at room temperature, and it increases linearly with increasing temperature to 0.049 W/m-°K at 325°C (McAdams, 1954). Gas thermal conductivity values are much lower than the typical thermal conductivity of solids, which range from near 0.2 W/m-°K for plastics to more than 400 W/m-°K for good metallic conductors. Consequently, the dominant resistance to heat transfer in the powder bed is that of the gas, and this resistance leads to similar powder bed conductivity values regardless of the solid conductivity of the particulate. For example, the conductivity of a typical polycarbonate powder is 0.065 W/m-°K, while that for a typical copper powder is only 0.24 W/m-°K at room temperature (Incropera and DeWitt, 1985), despite the fact that the conductivity for solid copper is approximately 2000 times that of solid polycarbonate (Nelson, 1993; Nelson, *et al*, 1993). Models that attempt to accurately predict the conductivity of powder beds must properly consider the resistance to heat transfer by the gas phase.

The thermal conductivity, k, of powder beds can be measured experimentally by either steady state or non-steady state methods. Steady state methods require measurement of the heat transmitted through the material between a temperature difference, $T_1 - T_2$, that is,

$$k = \frac{QL}{A(T_1 - T_2)} \tag{6.1.6}$$

where Q is the heat transfer rate through the material, L is bed thickness, and A is heat transfer area. Typical methods are available in *ASTM Standards* (1967) and various journal articles, for example Schroder (1963). Unsteady state methods generally require solution of the unsteady energy balance equation (Bird, Stewart, and Lightfoot, 1960),

$$\frac{\partial T}{\partial t} = \alpha \nabla^2 T \tag{6.1.7}$$

in order to evaluate the thermal diffusivity, α, and thereby determine k from equation (6.1.1) and knowledge of material density, ρ, and specific heat, C_p. While more steps are involved in evaluating k by unsteady state methods, they have been used extensively with SLS powders for the simple reason that the SLS process is also an unsteady-state process whose description requires knowledge of ρ and C_p.

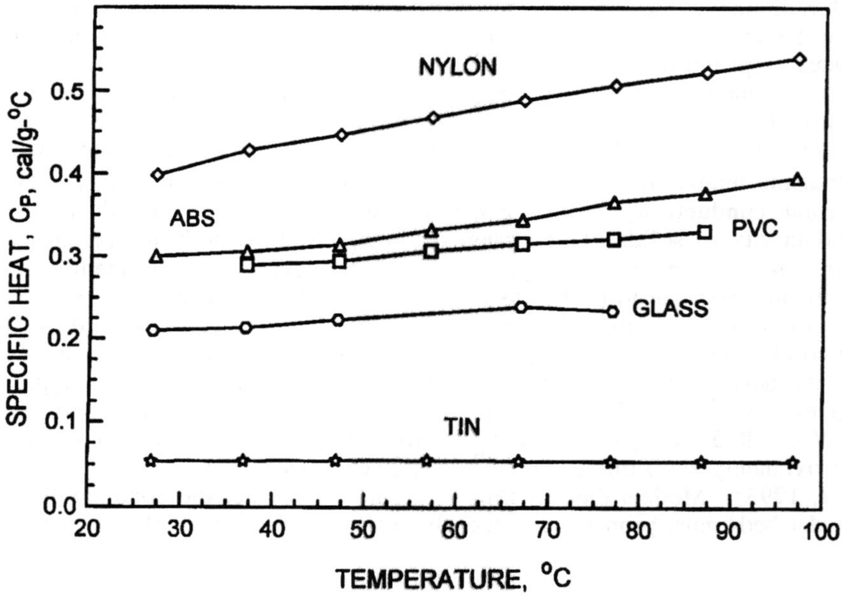

Figure 6.1 Specific heats of typical metallic, ceramic, and organic powders (from Xue, 1991).

Two unsteady-state measurement methods have been employed. In the first, powder is poured into a thin wall, aluminum, cigar humidor tube and a thermocouple is axially centered in the tube. The tube and its contents are placed in a water bath and allowed to equilibrate. It is then placed in a second water bath at a temperature typically displaced only 10°C with respect to that of the first bath, and the temperature at tube center is recorded as a function of time (Xue and Barlow, 1990, Xue, 1991). This small temperature change permits analysis based on constant physical properties. The process is then repeated by raising the temperature of the first bath by 20°C and placing the tube into the first bath after it has equilibrated in the second, and so forth. Typical data obtained by this method are shown in Figure 6.2. Thermal diffusivity obtained from this method are accurate to within ± 10% relative error (Xue, 1991).

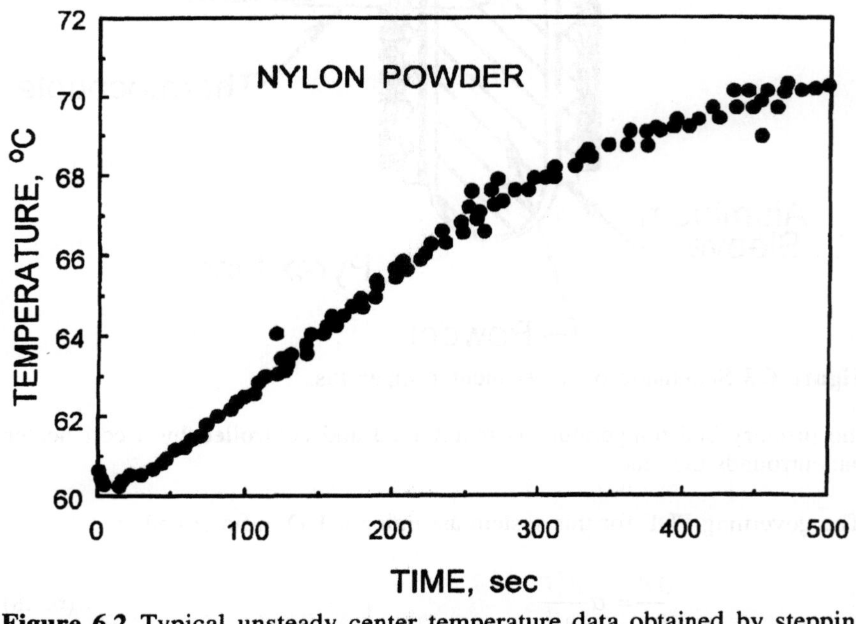

Figure 6.2 Typical unsteady center temperature data obtained by stepping the tube surface from 60 °C to 70°C (Xue, 1991).

Perhaps a better approach for obtaining bed thermal conductivity and bed surface emissivity data that is meaningful to the SLS process is to measure that emissivity using a laser beam with the same radiation frequency as is used by the SLS machine. A schematic of such a measurement system is shown in Figure 6.3 (Sih and Barlow, 1995, Sih, 1996). This system is a modified version of that developed by Ikeda, *et al.* (1985) and Taniguchi, *et al.* (1989) for measuring the properties of solids and powders. Here, the powder is contained in a thermally guarded tube that heats the bed to desired experimental temperature. The powder is then exposed to an expanded CO_2 laser beam at its top surface. A thermocouple is mounted 1 cm below the surface to record the temperature. The temperature rise is again limited to less than 20°C, relative to the initial bed temperature, to permit thermal analysis that is based on constant thermal properties.

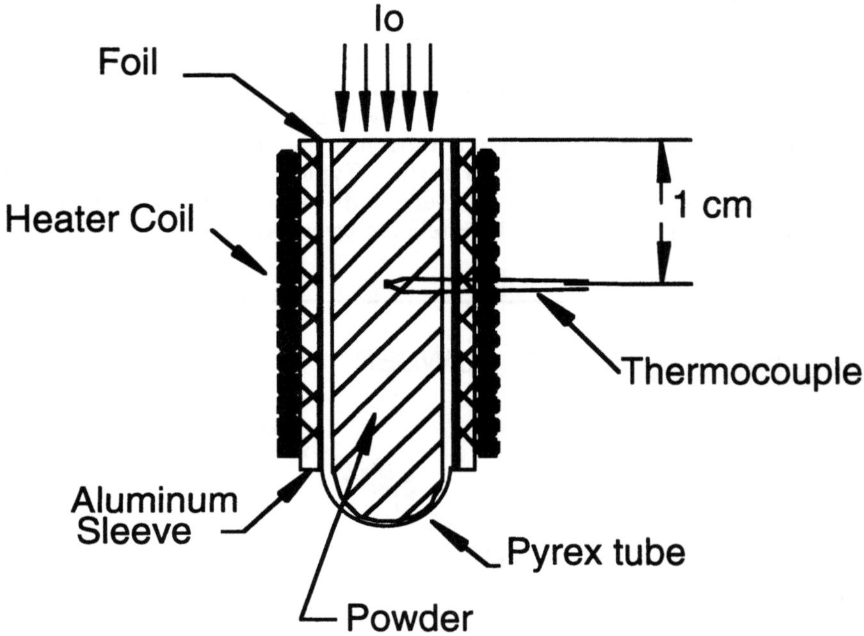

Figure 6.3 Schematic of experimental apparatus.

The primary bed temperature is maintained and controlled by a coil heater that surrounds the tube.

The governing PDE for this system assuming a 1-D, infinite bed is

$$\frac{\partial T}{\partial t} = \alpha \frac{\partial^2 T}{\partial z^2} \tag{6.1.8}$$

with initial and boundary conditions,

$$\begin{aligned} T(z,0) &= T_0 \\ T(\infty,t) &= T_0 \\ -k\left(\frac{\partial T}{\partial z}\right)_{z=0} &= q = \varepsilon_{RS} I_0 \end{aligned} \tag{6.1.9}$$

where T_0 is the initial bed temperature, z the distance from the surface of the bed to a point in the interior, ε_{RS} the emissivity of the powder bed surface, and I_0 the laser irradiance received by the surface of the powder bed. The solution to this PDE is given by Carslaw and Jaeger (1959) as,

$$T-T_0 = \frac{2\varepsilon_{RS} I_0}{k}\sqrt{\alpha t}\ ierfc\left(\frac{z}{2\sqrt{\alpha t}}\right) \tag{6.1.10}$$

where *ierfc* is the integral complementary error function.

The thermal diffusivity of the powder bed is determined by measuring the temperature rise at position L in the bed at two times, t_1 and $2t_1$, see Figure 6.4. The Temperature Ratio, R,

$$R = \frac{T(L,2t_1)-T_0}{T(L,t_1)-T_0} \tag{6.1.11}$$

is then formed and compared to the model, equation (6.1.10),

$$R = \frac{\sqrt{2}\ ierfc\left(\frac{1}{2\sqrt{2}\sqrt{F}}\right)}{ierfc\left(\frac{1}{2\sqrt{F}}\right)} \tag{6.1.12}$$

where F is the Fourier Number,

$$F = \frac{\alpha t_1}{L^2}\ . \tag{6.1.13}$$

R is clearly a function only of the Fourier Number, F. Consequently, F can be determined from measured R and equation (6.1.12). The solution to equation (6.1.12) can be easily built into a trial and error computer program for determining F, then α, and finally k. Relative error, as judged by scatter about the mean, appears to be ± 10%, see Figures 6.6 through 6.9.

Unless noted otherwise, all experimental thermal conductivity data reported in this chapter were obtained by either of the two methods discussed above. Generally, data that are limited in temperature to less than 100°C were obtained by the first method. Additional data are available in Sih (1996).

Surface absorbance, α_R, assumed to be equal to surface emissivity, ε_R, is also measured with the apparatus in Figure 6.3 by noting differences in the thermal rise behavior of the powder bed between exposure, at the same laser power, of the natural surface and the same surface covered with a thin layer of carbon black powder, assumed to have $\varepsilon_{RS} = \alpha_{RS} = 1.0$. The thermal rise rate is measured in the linear region of the thermal rise curve in both measurements, see Figure 6.4, and the bed heat capacity is assumed constant. Consequently, the governing equation is

$$\varepsilon_R = \varepsilon_{RS} \frac{(dT/dt)_R}{(dT/dt)_{RS}} \tag{6.1.14}$$

Errors in this method are primarily those associated with determining the rise rates and are estimated to be ± 10% relative error. A comparison between measurements and theory may be seen in Figures 6.11 and 6.12.

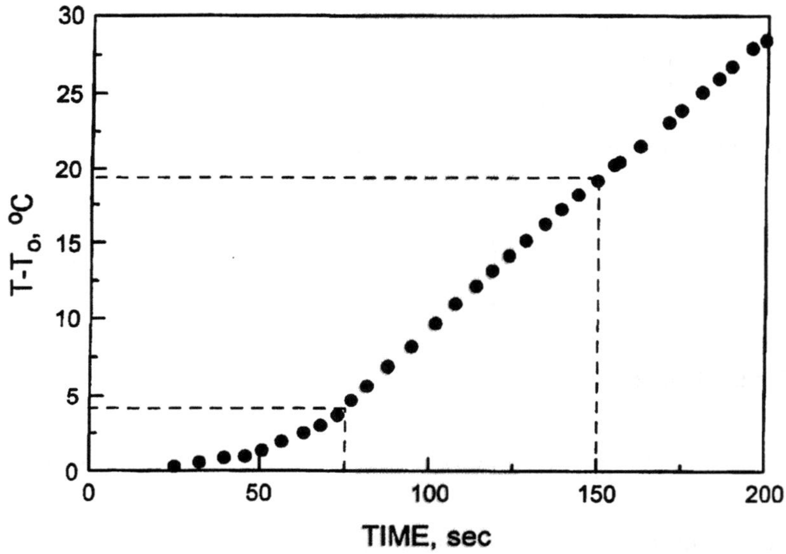

Figure 6.4 Typical temperature rise *vs* time data, showing selection of points for determining temperature ratio, *R*.

6.1.2 Thermal Property Models

Several thermal conductivity models have been investigated, developed, and discussed for use in modeling the Selective Laser Sintering process (Sih, 1996, Xue and Barlow, 1991, Sih and Barlow, 1994). For sake of brevity, only two models, the YK (Yagi and Kunii, 1957) and the ZSD (Zehner and Schlunder, 1970, Damkohler, 1937) will be mentioned here. Both models attempt to properly weigh the relative contributions to powder conductivity by conduction through the solid, through the gas, and by interparticle radiation, and both can predict the thermal conductivity of powder beds with fair accuracy. They differ in the assumptions that are used in estimating these contributions.

The high temperature form of the YK model is given by,

$$\frac{k}{k_s} = \frac{(1-\varepsilon)}{\frac{k_g}{k_s} + \frac{1}{\frac{1}{\psi} + \frac{D_p h_{rs}}{k_g}}} + \frac{\varepsilon D_p h_{rv}}{k_g}$$

$$\psi = 0.02 \times 10^{2(\varepsilon - 0.3)} \qquad (6.1.15)$$

where k_g is the thermal conductivity of the interstitial gas, k_s is the thermal conductivity of the solid particle, ε is bed void fraction, D_p is particle diameter, h_{rs} is radiation heat transfer coefficient between particles, and h_{rv} is radiation heat transfer coefficient between neighboring void spaces. These coefficients are completely defined in the 1957 article. Although empirical, the YK equations have been successfully used for predicting the conductivity of packed beds in the chemical process industry.

The ZSD model is based on the use of the analogy between heat and mass transfer to relate the conductivity of free fluid, k_g, and solid fluid composite conductivity, k_C, to bed porosity. The model uses the work by Currie (1960) that showed, experimentally, that the apparent diffusivity of a gas through a packed bed of solid particles, D, is reduced, relative to its diffusivity in an unobstructed gas mixture, D_g, by the relationship,

$$\frac{D}{D_g} = 1 - \sqrt{1-\varepsilon} \qquad (6.1.16)$$

where ε is the fractional porosity of the bed. An analogy between heat and mass transfer suggests that the thermal conductivity of a powder bed could be written as,

$$\frac{k}{k_g} = (1 - \sqrt{1-\varepsilon}) + \sqrt{1-\varepsilon}\,\frac{k_C}{k_g} \qquad (6.1.17)$$

where k_C is the thermal conductivity of the particulate and entrained fluid. The mass transfer analog to $k_C = 0$ is $D = 0$ in equation (6.1.21); that is, no diffusion through the solid particles. However, k_C is not zero because both the entrained fluid and the particulate have finite thermal conductivity.

Figure 6.5 shows a cell of the powder bed that provide the basis for calculating k_C. The composite phase of entrained fluid has dimensions, $0 \leq z \leq 2$ and $-1 \leq r \leq 1$.

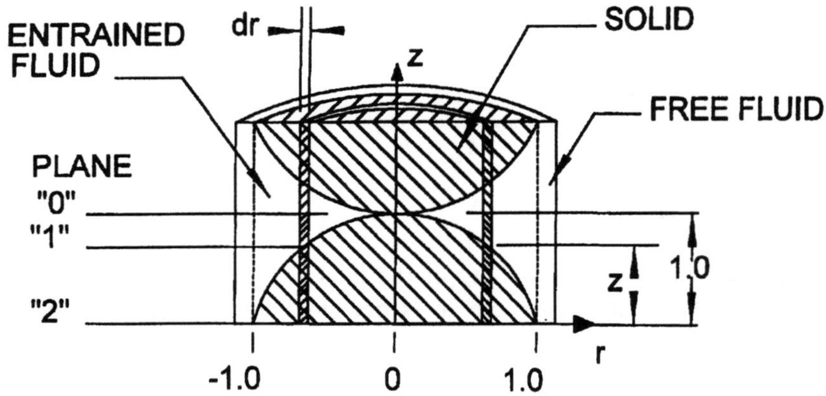

Figure 6.5 Cell for calculating k_C *(*Zehner and Schlunder, 1970).

At steady state, the temperatures at planes 0, 1, and 2, of the cell are assumed to be independent of the free fluid or composite phases found in each plane; that is, no r directed heat transfer is considered. With reference to Figure 6.5, a series resistance model is adopted for conduction through the composite phase,

$$q = \frac{T_0 - T_1}{\frac{1-z}{k_g} + \frac{z}{k_S}} \tag{6.1.18}$$

where q is the flux passing through an element of area, $2\pi r dr$. Integrating over r, the total heat transfer Q is

$$Q = \int_0^1 q 2\pi r dr = \int_0^1 \frac{T_0 - T_2}{\frac{1-z}{k_g} + \frac{z}{k_S}} 2\pi r dr = k_C \frac{T_0 - T_2}{1} \pi (1)^2 \tag{6.1.19}$$

Rearranging equation (6.1.19) yields,

$$\frac{k_C}{k_g} = \int_0^1 \frac{2}{\left(\frac{k_g}{k_S}-1\right)z + 1} r\,dr \tag{6.1.20}$$

The relationship between r and z needs to be specified before equation (6.1.20) can be evaluated. This depends, of course, on the assumed shape of the solid particles. Zehner and Schlunder (1970) define a general surface, in cylindrical coordinates, by

$$r^2 + \frac{z^2}{(B-(B-1)z)^2} = 1 \tag{6.1.21}$$

where B is a parameter that controls the orientation and shape of the particle surface. The most common shape of particulate, namely spherical, is obtained by setting $B = 1$.

Equation (6.1.21) can be rearranged to express r and dr as a function of z. This function can be substituted into equation (6.1.20) and integrated to give

$$\frac{k_C}{k_S} = \frac{2}{\left(1-\frac{Bk_g}{k_S}\right)}\left[\frac{\left(1-\frac{k_g}{k_S}\right)B}{\left(1-\frac{Bk_g}{k_S}\right)^2} ln\left(\frac{k_S}{Bk_g}\right) - \frac{B+1}{2} - \frac{B-1}{\left(1-\frac{Bk_g}{k_S}\right)}\right] \tag{6.1.22}$$

The porosity of the bed is a function of the parameter, B. Zehner and Schlunder (1970) have shown that,

$$\varepsilon = 1-\left[\frac{B}{(B-1)^3}(B^2-4B+2ln(B)+3)\right]^2 \tag{6.1.23}$$

by integrating throughout the cell to determine the volumes of the solid and fluid phases. These are then substituted into equation (6.1.3) to obtain equation (6.1.23). An approximate inverse function of equation (6.1.23), given by

$$B = 1.25\left[\frac{1-\varepsilon}{\varepsilon}\right]^{\frac{10}{9}} \tag{6.1.24}$$

permits an estimate of B to be made from knowledge of the bed porosity, ε. According to equation (6.1.23), a bed of spherical particles (B=1.) should have a fractional porosity near 0.55. This is close to what is usually observed.

Sih and Barlow (1994, 1995) recently proposed a high temperature extension to the Zehner and Schlunder (1970) model by adding a Damkohler (1937) inspired term to account for radiation between bed particles,

$$k_R = 4F_V \sigma D_P T^3 \tag{6.1.25}$$

where σ is the Stefan-Boltzmann constant, F_V is a view factor term, D_p is the mean particle diameter, and T is the local absolute temperature of the bed. A view factor that is found to work well is given by Wakao and Kato (1969) as,

$$F_V = \frac{2}{(2/\varepsilon_R - 0.264)} \tag{6.1.26}$$

where ε_R is the emissivity of the powder bed. Procedures for describing powder bed emissivity are described below in equations (6.1.29)-(6.1.36). The thermal conductivity by radiation in the free fluid phase, see Figure 6.5, is also estimated to be proportional to a characteristic size, the pore size, approximately εD_p. Consequently, equation (6.1.17), modified by consideration of radiation heat transfer in a series resistance model, takes the form,

$$\frac{k}{k_g} = (1-\sqrt{1-\varepsilon}\left[1+\frac{\varepsilon k_R}{k_g}\right]+\sqrt{1-\varepsilon}\left[\frac{k_C}{k_g}+\frac{k_R}{k_g}\right] \tag{6.1.27}$$

Equation (6.1.27) is called the ZSD equation.

Figures 6.6, 6.7, and 6.8, compare predictions by both the ZSD and YK models with experimental values for powder beds containing alumina (Landolt-Bornstein, 1980), hydroxyapatite (Sih, 1996), and ABS plastic (Sih, 1996) powders, respectively. While both models give reasonable predictions, the ZSD model seems to be more accurate than the YK model, not only in these instances but generally (Sih, 1996).

The prediction of the thermal conductivity of powder beds that contain polymer coated particles requires estimation of k_s from the composition and morphology of the agglomerated particulate. The morphology is difficult to describe, however as discussed above only a crude estimate of k_s need be made because of the overwhelming dominance of the gas phase resistance to heat transfer. Accordingly, (Badrinarayan and Barlow, 1990) the problem has been idealized by considering one-dimensional conduction through a sphere which has a core of inorganic material, thermal conductivity k_1 and radius R_1, and which is uniformly coated by the polymer with conductivity k_2 and thickness, t_c. Following and correcting Swift (1966), the thermal conductivity of the composite sphere is then given by,

$$\frac{1}{k_S}=\left(\frac{A}{k_1^2}-\frac{2}{k_1}\right)ln\left[\frac{A-k_1}{A-\frac{k_1 t_c}{R_1}}\right]+\frac{t_c}{R_1 k_2}+\left(\frac{1-\frac{t_c}{R_1}}{k_1}\right) \qquad (6.1.28)$$

where,

$$A=2\left(\frac{t_c}{R_1}\right)k_2+2\left(1-\frac{t_c}{R_1}\right)k_1$$

$$\frac{t_c}{R_1}=\frac{1-(1-\phi_c)^{1/3}}{(1-\phi_c)^{1/3}}$$

and ϕ_c is the volume fraction of coating. Figure 6.9 shows a comparison between thermal conductivity measurements (Sih and Barlow, 1994) and prediction by equations (6.1.33) and (6.1.34). Agreement is seen to be quite good.

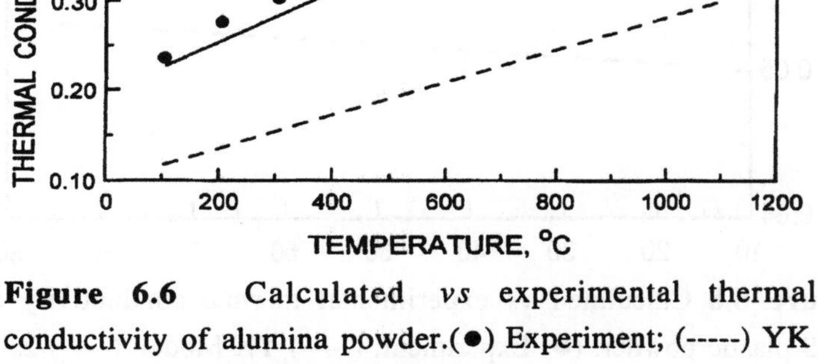

Figure 6.6 Calculated *vs* experimental thermal conductivity of alumina powder.(●) Experiment; (-----) YK Model; (——) ZSD Model (Sih, 1996)

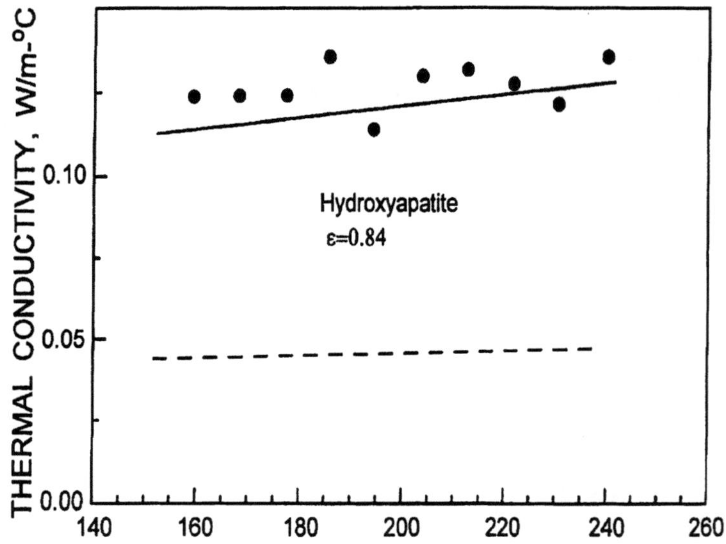

Figure 6.7 Calculated *vs* experimental thermal conductivity of hydroxyapatite powder.(●) Experiment; (-----) YK Model; (——) ZSD (Sih, 1996)

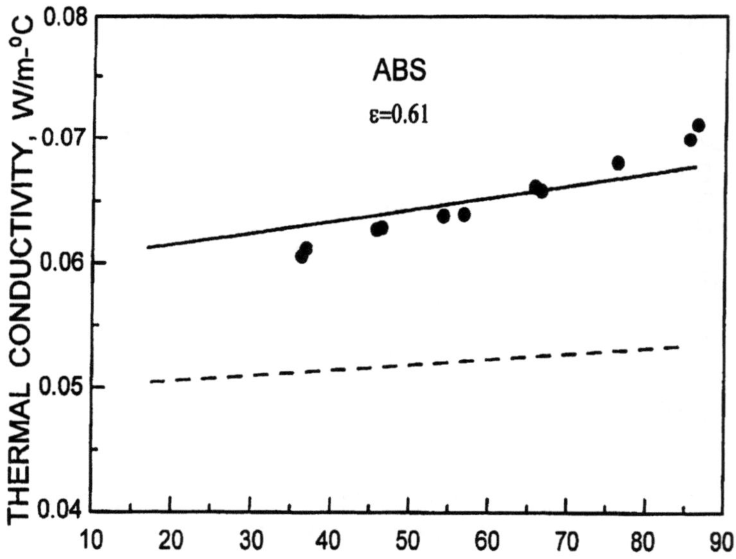

Figure 6.8 Calculated *vs* experimental thermal conductivity of ABS plastic powder. (●) Experiment; (-----) YK Model; (——) ZSD (Sih, 1996)

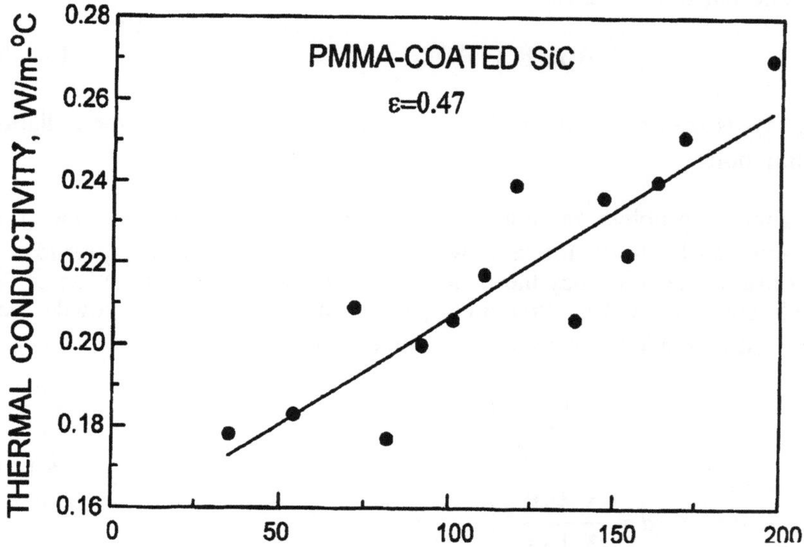

Figure 6.9 Comparison of experimental (●) and calculated (----) thermal conductivity of PMMA-coated SiC powder (Sih, 1996).

The emissivity of powder beds is needed for all calculations that involve heat transfer by thermal radiation, see for example equation (6.1-26). We have developed a simple estimation of the surface emissivity, ε_R, that can be useful when experimental techniques are not available (Sih and Barlow, 1995). We begin with the usual assumption that the irradiated body is "gray" with no transmission of received radiation, so that $\alpha_R = \varepsilon_R$. The emission of radiation from the heated powder bed is further assumed to be caused by emission from the heated particles and from the cavities that are open to the surface, all at the same temperature. Although the walls of the cavities are made by the same particles, the emissivity of the cavity is a function of its geometry, particularly the fraction, f, of total cavity surface that is cut away by the emitting hole (Bird, *et al*, 1960, DeWitt and Nutter, 1988),

$$\varepsilon_H = \frac{\varepsilon_S}{\varepsilon_S + f(1-\varepsilon_S)} \tag{6.1.29}$$

where ε_S is the emissivity of the solid particle and ε_H is that of the hole. Equation (6.1.29) suggests that the hole emissivity is always higher than that of the material comprising the cavity and may even approach 1.0, the emissivity of a black body radiator, as f becomes small. We approximate the powder emissivity, ε_R, as the area fraction weighted average of the hole and solid emissivity values,

$$\varepsilon_R = A_H \varepsilon_H + (1 - A_H)\varepsilon_S \tag{6.1.30}$$

where A_H is the area fraction of the surface that is occupied by the radiation emitting holes.

The general problem remains to describe the parameters A_H and f in equations (6.1.29-30) for a powder bed. German (1989) provides a relationship that describes the mean pore diameter, d, as a function of the particle size, D_p, and the fractional packing density, ρ_R, for a powder bed comprised of randomly packed, single-sized spheres,

$$d = \frac{2}{3}\frac{D_P(1-\rho_R)}{\rho_R} \quad \text{or}$$

$$d = \frac{2}{3}\frac{D_P\varepsilon}{1-\varepsilon} \tag{6.1.31}$$

where ε is the fractional porosity of the bed, see equation (6.1.3), above. German also estimates that there are approximately 2.92 pores per sphere in a randomly packed bed. The surface of the randomly packed bed should therefore have $(2.92)^{2/3} = 2.043$ pores per sphere in the plane. The area fraction, A_H, occupied by the pores is then,

$$A_H = \frac{2.043\frac{\pi d^2}{4}}{2.043\frac{\pi d^2}{4} + \frac{\pi D_P{}^2}{4}} . \tag{6.1.32}$$

Substitution of equation (6.1.31) into equation (6.1.32) yields,

$$A_H = \frac{0.908\varepsilon^2}{1.908\varepsilon^2 - 2\varepsilon + 1} \tag{6.1.33}$$

We estimate f by considering each pore to be a cylinder of length, L, so that

$$f = \frac{\frac{\pi d^2}{4}}{\frac{2\pi d^2}{4} + \pi d L} = \frac{1}{2 + \frac{4L}{d}} \tag{6.1.34}$$

The length, L, can be determined from the definition of porosity, ε,

$$\varepsilon = \frac{2.92\frac{\pi d^2}{4}L}{2.92\frac{\pi d^2}{4}L + \frac{1}{6}\pi D_P^3} \tag{6.1.35}$$

Using equation (6.1.35) with equation (6.1.31) and 6.1.34) finally leads to

$$\varepsilon_H = \frac{\varepsilon_S\left[2 + 3.082\left(\frac{1-\varepsilon}{\varepsilon}\right)^2\right]}{\varepsilon_S\left[1 + 3.082\left(\frac{1-\varepsilon}{\varepsilon}\right)^2\right] + 1} \,. \tag{6.1.36}$$

Equation (6.1.36) suggests that the emissivity of the hole is independent of particle diameter (for like-sized spheres) and a function only of solid emissivity and the bed porosity. As shown in Figure 6.10, the calculated emissivity of the hole decreases slightly as the bed porosity increases and also increases as the emissivity of the solid increases. The first trend occurs because f increases with increasing porosity causing the hole to become a less perfect radiator. The second trend simply notes the importance of the solid emissivity in determining the emissivity of the powder bed.

Comparison between experiment and prediction by equations 6.1.30, 6.1.33, and 6.1.36 is shown in Figures 6.11 and 6.12 for alumina and silicon carbide powders, respectively. Generally, agreement lies within the experimental error band for all cases studied. The model depends strongly on the emissivity of the solid, and generally this is a function of temperature that must be selected with some care from other sources. The references to solid emissivities employed in these examples are taken from Rohsenow, *et al* (1985).

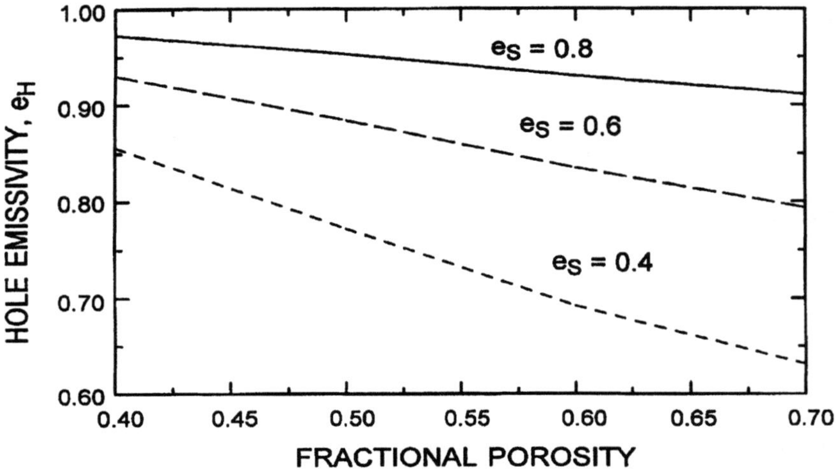

Figure 6.10 Calculated hole emissivity as a function of bed porosity and solid emissivity (Sih and Barlow, 1995).

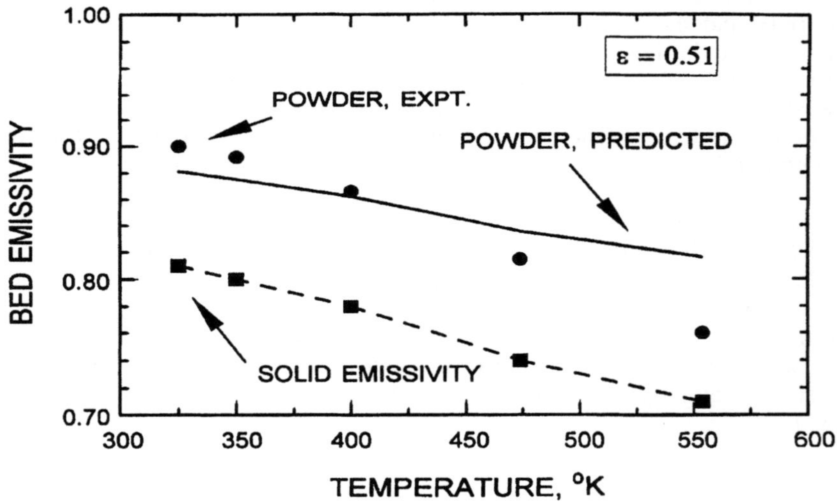

Figure 6.11 Comparison of predicted and measured emissivity for powder beds that contain 15 μm alumina. Bed porosity is 0.51 (Sih and Barlow, 1995)

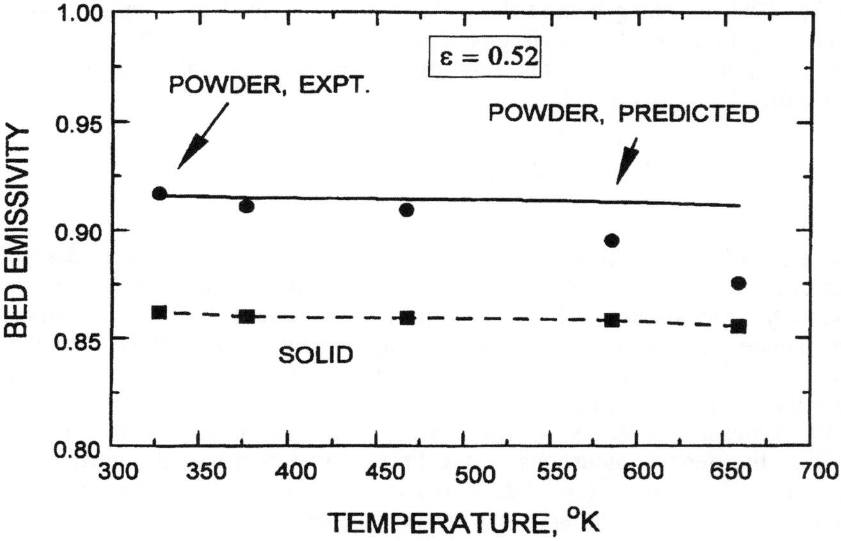

Figure 6.12 Comparison of predicted and measured emissivity for powder beds that contain 7μm silicon carbide. Bed porosity is 0.52 (Sih and Barlow, 1995)

6.2 SINTERING KINETICS

During the SLS process, the laser beam transmits a pulse of thermal energy to the powder surface that is typically 20 msec in duration. A small portion of this energy is absorbed and conducted through the powder layer heating it to the sintering temperature. Due to rapid heat loss by radiation from the surface of the heated layer, significant particle fusion must occur within seconds in order to produce parts with sufficient strength. From a practical process engineering prospective, what matters is the rate of sintering of a particular powder under the laser beam. This is difficult to measure under actual SLS processing conditions, although some valuable process information about a new powder system can be gained by using the SLS process to generate a series of "single layer" tests at different laser power and scan rates. The qualitative, relative strengths of the single layers can help the operator to find a suitable set of operating parameters.

6.2.1 Amorphous polymers

The next level of process understanding, a mathematical model of the SLS process, requires some knowledge of the sintering kinetics of the powder. The direct measurement of sintering kinetics under the laser beam has proved difficult, however it has proved possible to measure sintering kinetics of amorphous glasses in a separate experiment, model the results, and extrapolate to laser processing conditions (Nelson, 1993, Nelson, *et al*, 1993). The reasoning behind this approach is described by Frenkel (1945). According to Frenkel, the viscous sintering of amorphous polymer glasses can be approximated by the rate of coalescence of spherical adjacent particles,

$$\left(\frac{x}{R}\right)^2 = \frac{3}{2}\left(\frac{\sigma t}{R\eta}\right) \tag{6.2.1}$$

where x is half the thickness of the neck that is formed between adjacent particles of radius R, at time t, σ is the surface tension, and η is the viscosity, see Figure 6.13. Equation (6.2.1) arises from an assumed mechanism that equates the energy dissipated by viscous flow with that gained by the decrease of surface area during densification. With some variations in model forms, other investigators (MacKenzie and Shuttleworth, 1949, Scherer, 1977, 1984, Kuczynski, *et al*, 1970, Ristic, 1987) have verified the general assertions of the Frenkel equation that the normalized rate of coalescence, $\partial(x/R)/\partial t$, increases with increasing surface tension, decreasing initial particle size, and decreasing melt viscosity.

Surface tension values are much the same, 20-30 dyne/cm at 180 °C, for a wide variety of polymers (Wu, 1971, Hobbs, *et al*, 1988), and they decrease slowly with increasing temperature. As discussed in Chapter 4, polymer melt viscosity generally follows an Arrhenius relationship with temperature, given by,

$$\eta = A exp(\Delta E / RT) \tag{6.2.2}$$

where ΔE is the activation energy for viscous flow, A a prefactor in appropriate units, R the gas constant, and T absolute temperature. The activation energies for most polymers are typically in the range 3-30 kcal/mol (Nelson and Barlow, 1990) and their melt viscosities typically decrease by a factor of two for a 25 °C increase in temperature (Tadmor and Gogos, 1979). This behavior suggests that rapid sintering rates are best achieved by raising the powder bed temperature as high as possible, consistent with the simultaneous need to prevent uncontrolled sintering of the entire powder bed. Typically, this is accomplished by setting the powder bed temperature near the polymer glass transition temperature T_g, if amorphous, or near the polymer melting temperature T_m, if semi-crystalline, then using the relatively low heat input from the laser beam to locally heat and rapidly sinter the powder.

According to equation 6.2.1, the key powder parameters that control sintering are the particle size and the melt viscosity. While smaller particles will sinter more rapidly than larger ones, the SLS process appears to limit particle size to a minimum of about 2 μm. Certainly sub-micron particles have proved difficult to process in most cases as a result of excessive triboelectric charging of the particles by the SLS powder spreading mechanism. Particles are limited in maximum size to that of a part layer, approximately 125 μm, although surface quality of the part may further limit the maximum value. Generally, particles and polymer-coated composite particles in the size range 10-75 μm show the best over-all spreading, fusion, and part surface characteristics. Given these practical constraints, adjusting the particle size to adjust sintering rate has not proved useful.

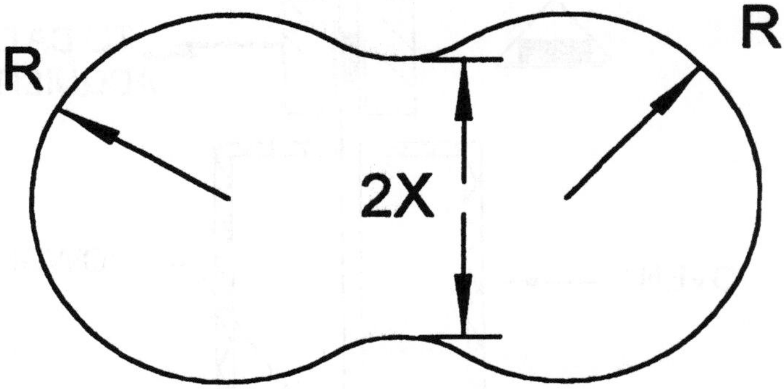

Figure 6.13 Schematic of viscous sintering.

Equation (6.2.1) is a conceptually useful representation of the relationships between important parameters that control sintering rates in amorphous systems, however it may not be sufficiently global to describe the sintering kinetics of a bed of particles. For example, Scherer (1977,1986) developed a model that approximates the porous powder bed as a network of cylinders. This approach provides an explicit relation between porosity, ε, and strain rate, $\partial e/\partial t$, via,

$$\frac{\partial e}{\partial t} = -0.721\left(\frac{\sigma}{\eta R}\right)\frac{\left(2-3C\frac{R}{\ell}\right)}{\left[\left(\frac{R}{\ell}\right)\left(1-C\frac{R}{\ell}\right)^{2}\right]^{1/3}} \qquad (6.2.3)$$

$$\varepsilon = 1-(1-\varepsilon_0)exp(-3e)$$

where C = 1.200 and R/ℓ is the ratio of cylinder radius to cylinder length. Sun, *et al*, 1990, combined the Scherer model with the YK equation for bed conductivity, equation (6.1.15), and a one-dimensional energy balance, equations (6.1.8) and (6.1.9), to estimate laser sintering in a bed of ABS powder. Their approach did a fairly good job of fitting and estimating the effects of laser power, bed temperature, and laser scan rate on sintering depths, probably because it correctly recognized the strong influences of temperature on viscosity, through equation (6.2.2), and of porosity on conductivity, through the YK model. It suffered, however, from an arbitrary and experimentally unverifiable estimation of porosity.

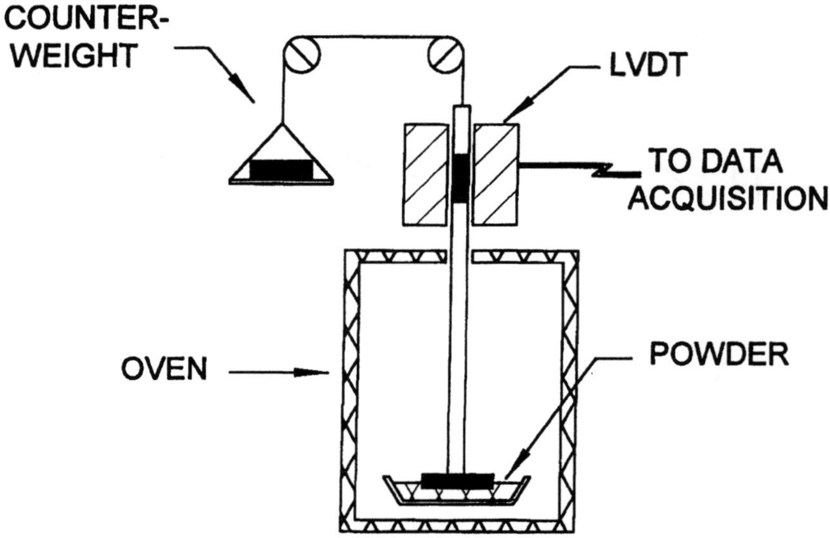

Figure 6.14 Schematic apparatus for measuring sintering kinetics of polymeric powders.

This success led to the development of an experimental protocol for measuring the kinetics of sintering of thermoplastic powders (Nelson, *et al*, 1993). Figure 6.14 shows a schematic of the apparatus that was developed for this purpose. The apparatus consists of a small oven that is isothermal to within ± 1°C. A Schaevitz Engineering, Type 500 HR, Linear Variable Differential Transformer (LVDT) is fixed to the top of the oven to measure the change in position of the iron core that is attached to the ceramic rod that passes through the oven wall and connects to a 5 cm dia. brass plate. The signal from the LVDT is conditioned and amplified by a Schaevitz Engineering ATA-101 transducer amplifier, and logged by a personal computer that is equipped with a Metrabyte Corporation DASH-16, data acquisition system. The rod and plate assembly sits on the surface of the powder and is counter-balanced, externally, to adjust and control the normal stress. This is necessary to prevent enhancing the sintering rate by the weight of the rod and plate. Each material may require a different minimum stress to avoid this problem. Polycarbonate, PC, for example, shows no rate enhancement at normal stresses below 0.4 g/cm^2.

To ensure measurement of isothermal sintering rates, operating temperatures and thickness of powder beds are chosen so that the rate of sintering is much less than the rate of heat transfer. By calculation, the center of a 0.6 cm thick polymer powder bed heats to 90% of equilibrium within 6-8 min. For PC powder, sintering temperatures are selected to be within 10-50 °C above T_g, where sintering takes 90-120 min to complete.

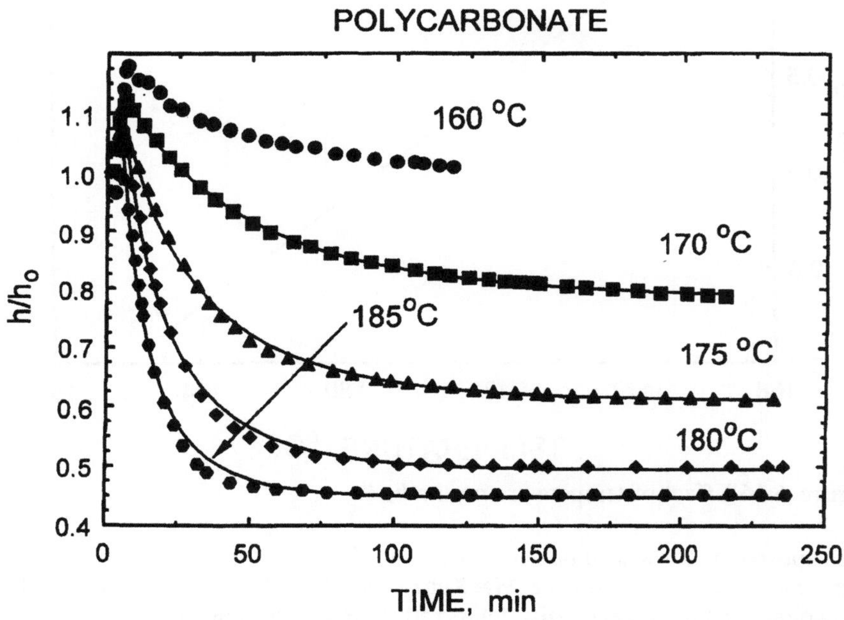

Figure 6.15 Oven sintering curves for polycarbonate at various temperatures (Nelson, *et al*, 1993)

Figure 6.15 shows typical measurements of PC powder sintering. The curves are normalized for initial powder bed depth, h_o, and only the data obtained after the first 10 min or so of heating is considered for kinetic analysis. The initial expansion of the bed is apparently caused by a change in particle shape from oblate spheroid to more spherical as the polymer is able to relax above the glass transition temperature, T_g. Preheating the powder for 10 min at 170 °C, followed by cooling it to room temperature, and loading it into the oven in the usual way is sufficient to eliminate this effect (Nelson, *et al*, 1993). The other interesting feature of the data is that h/h_o approaches lower long time plateau values as sintering temperature is raised.

This behavior is probably a consequence of the particle size distribution in the powder; that is, the larger particles are slower to sinter and establish the plateau. Raising the bed temperature allows more sintering of the larger particles to occur and reduces the h_∞/h_0 plateau.

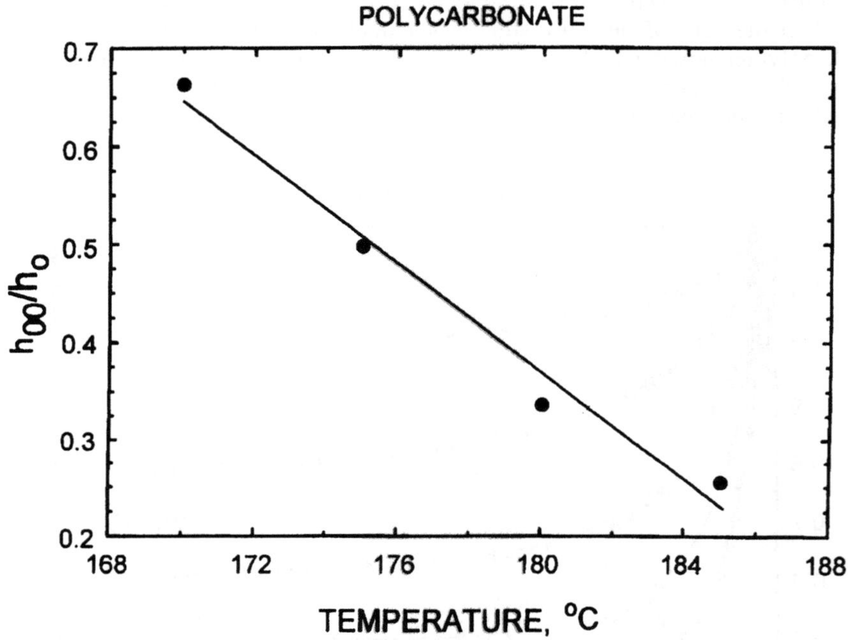

Figure 6.16 Effect of temperature on the h_∞/h_0

One approach toward modeling the sintering behavior is to write an empirical model, based on the MacKenzie and Shuttleworth (1949) model, to account for the presence of a temperature dependent plateau,

$$-\frac{de}{dt} = k'(\varepsilon - \varepsilon_\infty) \tag{6.2.4}$$

where ε and ε_∞ are the porosity and plateau porosity, respectively, and k' is a pseudo-rate constant that should be related to the ratio σ/η of the polymer. Equation (6.2.4) can be written in terms of the bed height, h, by noting that no material flows across the perimeter boundary of the disk platen during the sintering test described above. Consequently, $V = Ah$, where A is the platen area. If we then, as an approximation, associate h_∞ with zero porosity (at each temperature), equation (6.2.4) becomes,

$$-\frac{dh}{dt} = k'(h/h_\infty)(h-h_\infty) \tag{6.2.5}$$

which can be integrated to give,

$$\frac{h}{h_0} = \frac{\frac{h_\infty}{h_0}\left(\frac{1}{1-\frac{h_\infty}{h_0}}\right)}{\left(\frac{1}{1-\frac{h_\infty}{h_0}}\right)-exp(-k't)} \cong \frac{A_1}{A_2-exp(-k't)} \tag{6.2.6}$$

The solid lines in Figure 6.15 are the result of fitting equation (6.2.6) to the experimental data with a standard Marquardt-Levenburg least squares minimization method, following Press, *et al*, 1988. Generally, a good fit to the data is possible with this model, especially when the three parameter form is used, see equation (6.2.6), however A_1 and A_2 are not truly independent parameters. Interestingly, the fitted parameter, A_2, can be used to determine h_∞/h_0 from which A_1 can be calculated. The calculated value agrees with the curve-fitted value to within -20% relative error. Figure 6.16 shows the observed relationship between h_∞/h_0 and temperature.

The parameter k' follows the Arrhenius behavior that is expected from its assumed dependency on $1/\eta$, equation (6.2.2). Figure 6.17 shows that a good straight line with negative slope results from plotting $\ln(k')$ *vs* $1/T$. The slope of this line should correspond to $-\Delta E_A/R$, and this expectation leads to a value of $-\Delta E_A = 28.4$ kcal/mol,. This activation energy compares fairly well with the value $-\Delta E_A = 21.5$ kcal/mol that was obtained by Rahim (1979) from capillary flow measurements on polycarbonate between 280 and 320°C.

The apparent rate constant, k', for polycarbonate sintering is consequently given by,

$$k' = 1.627 \times 10^{12} exp\left(-\frac{2.84 \times 10^4}{RT}\right) \text{min}^{-1} \tag{6.2.7}$$

The k' reported here is similar in magnitude at processing temperatures, but is somewhat less temperature sensitive, than that derived previously by Nelson, *et al*, 1993, from the same data base. Its use will provide slightly broader thermal and density fronts than were estimated previously.

The SLS Model, described below, uses equation (6.2.7) with equation (6.2.4) to predict differential changes in thermal and density fronts. In this estimation, ε_∞, is taken to be zero. This approximation at the high temperatures used in SLS processing is suggested by the decrease in the h_∞/h_0 ratio, Figure 6.16.

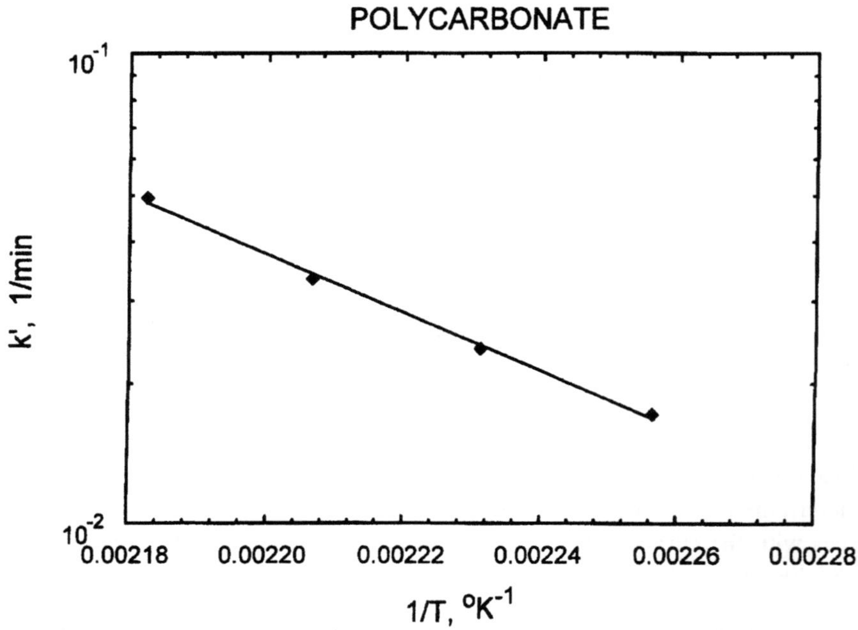

Figure 6.17 Construction for evaluating Arrhenius rate parameters.

6.2.2 Mixtures Containing Polymer Binders

Two methods have been used to produce inorganic powder compositions that can be shaped by the SLS process. The simpler method is to physically mix the inorganic powder with binder powder. This method is not commercially important due to problems associated with powder segregation during shipping and with poor binder efficiency compared with the preferred spray-coating method (Vail and Barlow, 1991, 1992). It is, however, an easy method with which to explore the influences of filler content on sintering rates, and these explorations on mixed powders can provide some insights into the behavior of the spray coated materials.

Figure 6.18 shows a typical sintering curve (Badrinarayan, 1995) that was obtained for a equivolume mixture of 5-10 μm copper powder and an experimental PMMA binder (Vail, *et a*l, 1994). At 50 volume percent polymer binder, the composition appears to oven sinter in a manner that is similar to that for a pure polymer, see Figure 6.15. This similarity is by design, see Chapter 4. As the fraction of polymer in the mixed system is varied, however, some important and interesting differences are seen. For example, Figure 6.19 shows that the rate constant, k', decreases with decreasing binder content. This observation is amplified in Figure 6.20, where the initial rate of consolidation of the powder bed, *-d(h/ho)/dt*, is plotted vs the binder content at several sintering temperatures. Since only the polymer can sinter at the temperatures employed, it is reasonable to expect that the observed sintering rate for the mixture, $R_M = -[d(h/h_0)/dt]_{M,t=0}$, should be proportional to the volume fraction of binder, ϕ_B, and its initial sintering rate, $R_B = -[d(h/h_o)/dt]_{B,t=0}$

$$R_M = R_B \phi_B \tag{6.2.8}$$

The lines in Figure 6.20 are computed from equation (6.2.8). Good agreement between experiment and this prediction is seen for binder content greater than 25 vol%. Below this binder level, very little bed height consolidation occurs and measurements of sintering kinetics are very small and uncertain. The reason for this behavior is probably related to porosity of the mixed powder system insofar as the binder particles at low binder content can sit in the interstices of the inorganic powder. The binder may well sinter, however at low binder contents that sintering will not cause bed consolidation to occur.

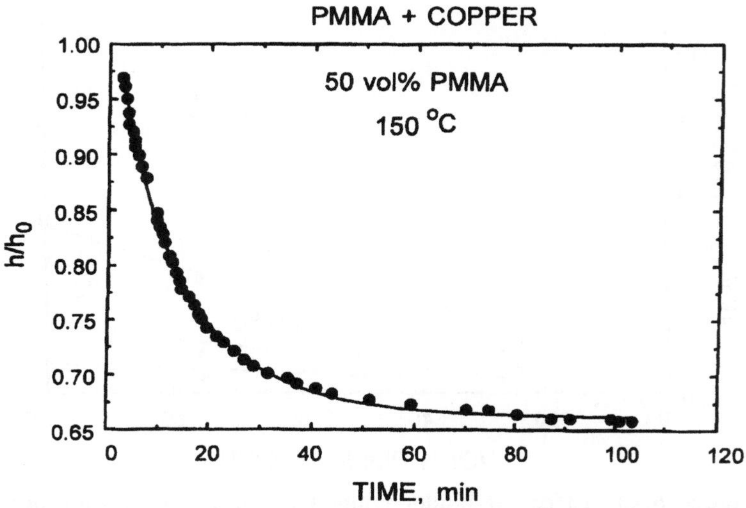

Figure 6.18 Typical oven sintering curve for an equivolume mixture of PMMA binder and copper powders at 150°C (Badrinarayan, 1995).

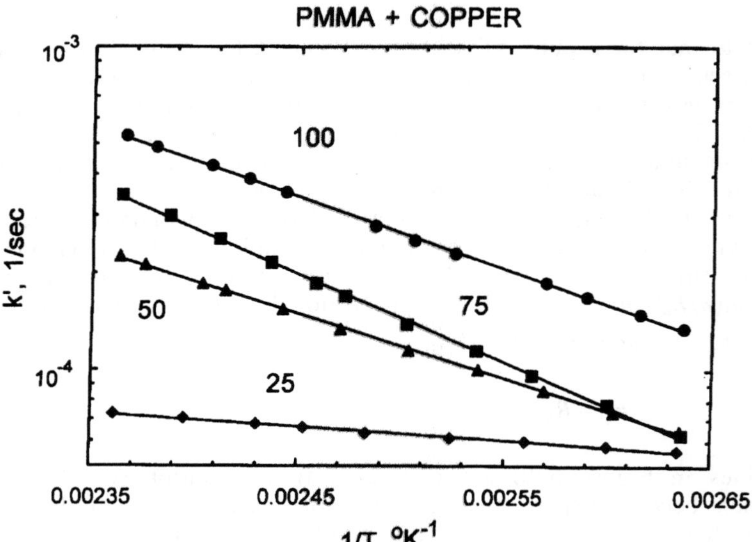

Figure 6.19 Summary of Arrhenius constructions from oven sintering of mixed powders of PMMA binder and copper. Line labels indicate the vol% of binder in the mix (Badrinarayan, 1995)

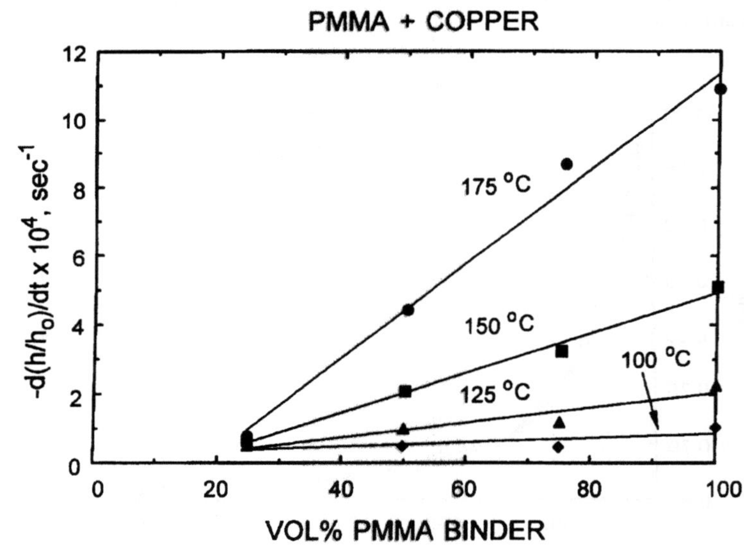

Figure 6.20 Effect of binder content on initial oven sintering rates (Badrinarayan, 1995).

The observation that sintering is proportional to binder content has two important ramifications. First, less sintering implies less dimensional shrinkage and greater dimensional accuracy. Indeed, parts that are made from either mixed or spray coated powders can be more easily be made accurately than parts that are prepared from pure polymer powders. Second, in mixed powders there is a binder limit below which sintering of the binder will not cause the production of an acceptably strong part. This point perhaps less obvious, although all would agree that the statement is true in the limit of no binder!

A closer examination of Figure 6.19 reveals an interesting feature of the oven sintering of mixed powders; the temperature dependency of the sintering rate also decreases with decreasing binder content. Figure 6.21 expands this observation. Whether or not there is actually a peak in temperature sensitivity at 75 vol% binder is certainly debatable, however it is interesting that a "transition" in temperature sensitivity appears to occur from a polymer powder dominated system to an inorganic powder dominated system near the equivolume composition. This result suggests that sintering rates become less controlled by binder viscosity as the inorganic powder content is increased, not just at a limit, as suggested by Figure 6.20, but through much of the compositional range below 50 vol% binder. The reason for this behavior is not known, but is probably related to the increasing dominance of interactions between the binder and inorganic surfaces at high loading of inorganic powder. After all, the probability that a polymer particle will contact and sinter with another polymer particle is only (1/4)(1/4)= 0.0625 when the volume fraction of polymer particles in the mix is 0.25.

Oven sintering tests (Nelson, et al, 1995a) on PMMA-coated SiC powder containing 20 vol% polymer show only 3-5 % reduction in normalized bed height after 20 min at 150 °C (45 °C above T_g), a result that is qualitatively consistent with the behavior of mixed systems and with the idea that consolidation occurs primarily by polymer sintering. Assuming a bed of polymer and filler each at 50% relative density, a 10% reduction in height is the maximum reduction that could occur due to total consolidation of the polymer in a powder bed that initially contains 20 parts by volume of solid polymer, 80 parts by volume inorganic filler. This has obvious practical ramifications in the area of dimensional stability; polymer/filler systems show much less shrinkage and distortion than pure polymer systems.

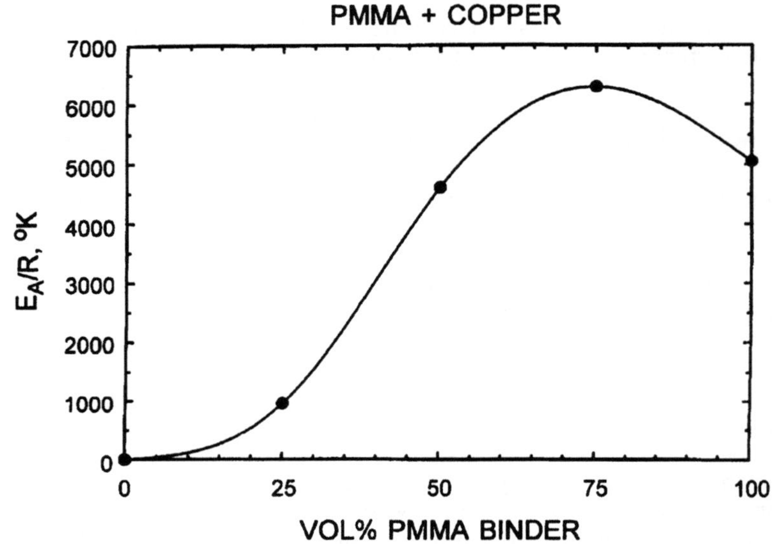

Figure 6.21 Influence of binder content on the activation energy for sintering in a mixed system (Badrinarayan, 1995)

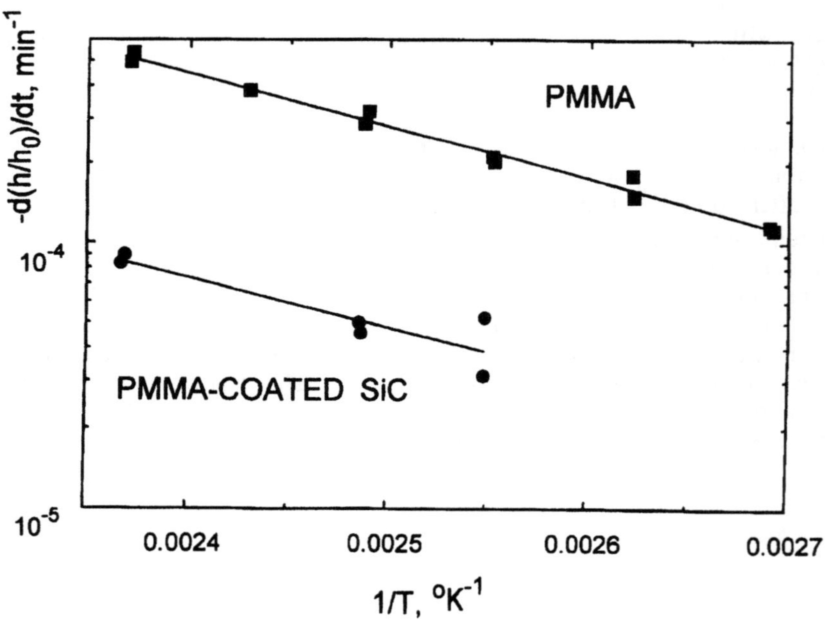

Figure 6.22 Effect of temperature on oven sintering rates of PMMA binder and PMMA binder-coated SiC powders (Nelson, et al, 1995a).

Initial sintering rates (Nelson, et al, 1995a) of the PMMA binder and of the PMMA binder-coated SiC powders are shown in Figure 6.22. As in the case of mixed powders, Figure 6.20, the initial sintering rate is greatly reduced, relative to that of the pure binder, by the presence of the inorganic filler, primarily for the reason cited immediately above. In contrast to the mixed system, however, the coated system maintains about the same temperature sensitivity as the binder. That is $\Delta E_A/R$ ranges between 4000 and 4700 °K for both materials, even though the binder content is only 20 vol%. This suggests that viscous (or diffusive) sintering still dominates the sintering kinetics of the coated materials, in direct contrast to the behavior shown by mixed materials, Figure 6.21. This should be expected because the contacts between polymer-coated particles should diffuse and sinter in much the same way as contacts between pure polymer particles.

6.3 PROCESS MODELS AND SIMULATIONS

6.3.1 SLS Process Models

One-dimensional thermal models of the SLS process have been developed that can estimate, with fair accuracy, the effect of laser scanning parameters on sintering depths and binder degradation in powders that contain amorphous polymers (Nelson, *et al*, 1993, Nelson, et al, 1995a, Vail, *et al*, 1996). At first glance, one might expect that a full three-dimensional model of heat transfer in the powder bed would be required to accurately model the process, however Festa, *et al*, (1988), have shown that a one-dimensional approximation is sufficient when the parameter, N, defined by,

$$N = \frac{VR}{\alpha} \tag{6.3.1}$$

where V is the beam speed, R the beam radius, and α, the thermal diffusivity of the material, is much greater than 3.9. Due to the poor thermal diffusivity of powders, N is typically greater than 1000, and temperature profiles predicted by the one-dimensional approximation differ by less than 5% from those predicted by the three-dimensional calculation (Sun, 1991).[1]

The unsteady state, one dimensional, energy balance is given by,

$$\rho C_P \frac{\partial T}{\partial t} = \frac{\partial}{\partial z}\left(k \frac{\partial T}{\partial z}\right) + G(z,t) \tag{6.3.2}$$

[1]For recent related work in this area see Dalgarno, *et al*, (1996), Ryder, *et al*,1996, and Williams, *et al*, 1996

where the boundary conditions are,

$$@z = 0$$

$$-k\frac{\partial T}{\partial z} = \delta(t)\alpha_R <I> - h(T-T_\infty)$$

$$\delta(t) = 1 \quad @t_i < t_i + \tau_p \text{ and } t < \tau$$

$$\delta(t) = 0 \quad @t_i + \tau_i < t < t_{i+1} \text{ or } t > \tau$$

$$i = 1 \ldots np-1$$

$$@\, z = \infty,\ t>0$$

$$\frac{\partial T}{\partial z} = 0$$

and ρ is the density, C_p the heat capacity, $T(t,z)$ the temperature, and k the thermal conductivity of the bed. The linear distance from the bed surface is given by z and t denotes time. The parameter $G(t,z)$ represents the heat sink due to depolymerization of the polymer or to melting of a semicrystalline polymer and is negative where these thermal events are being considered. Otherwise, for amorphous polymers, $G(t,z) = 0$. The absorptivity of the powder to laser radiation is represented by $\alpha_R = \varepsilon_R$, $<I>$ is the average intensity of the beam, τ is the duration during which laser energy is supplied, τ_i the duration of the laser pulse due to overlap of adjacent raster scan vectors, t_i is the time between pulses, np the number of pulses due to overlap, h is the combined radiation and convection heat transfer coefficient, and T_∞ the environmental temperature.

Figure 6.23 shows that during the laser sintering process each single point on the surface of the powder bed can receive multiple energy pulses with varying intensity, provided the beam radius, R, is greater than the scan spacing, $SCSP$, the usual mode of operation. Figure 6.23 also suggests that the laser beam intensity profile, $I(r,\omega)$, is Gaussian,

$$I(r,\omega) = I_0 exp\left(-\frac{2r^2}{\omega^2}\right) = \frac{2P}{\pi\omega^2} exp\left(-\frac{2r^2}{\omega^2}\right) \qquad (6.3.3)$$

where P is the laser power, r the radial distance from the center of the beam, and ω is the beam radius that is commonly measured by laser beam profilometers. ω is defined to be that value where I/I_o = exp(-2)= 0.135 (Duley, 1983).

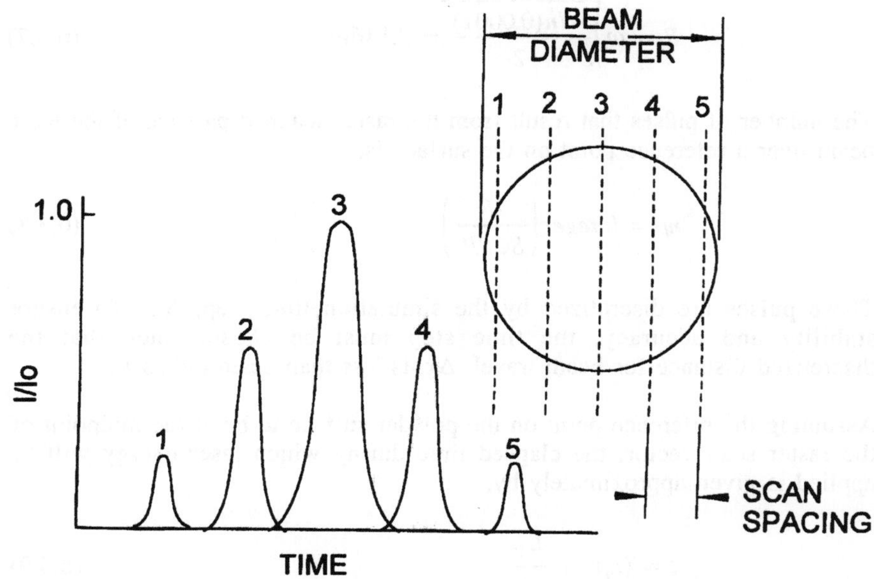

Figure 6.23 Relationships that lead to multiple laser pulses at a point on the surface.

This choice permits calculation of $I_o = 2P/\pi\omega^2$ from measurement of total laser power. The average laser intensity, $<I>$, along a scan line is given by,

$$< I > = \frac{\int_0^{S_x} I_0 exp\left(-\frac{2r^2}{\omega^2}\right)dx}{\int_0^{S_x} dx} = \frac{I_0\omega}{2S_x}\sqrt{\frac{\pi}{2}}\frac{erf(\sqrt{2}\,S_x/\omega)}{exp(2\Delta^2/\omega^2)} \tag{6.3.4}$$

where Δ is the distance from the particular scan line to beam center,

$$\Delta = abs(n_C - n_S)SCSP \tag{6.3.5}$$

and n_C and n_S are the numbers assigned to the scan line at beam center and the particular scan line, respectively, see Figure 6.23. The length of the scan line can be calculated by,

$$S_x^2 = R^2 - \Delta^2 \tag{6.3.6}$$

where R is the beam radius. The beam radius, R, used in our calculations is determined by the distance where the laser intensity is reduced to 0.0001 of the maximum,

$$R = \omega\sqrt{\frac{-ln(0.0001)}{2}} = 2.146\omega \tag{6.3.7}$$

The number of pulses that result from the raster scanned passage of the laser beam over a reference point on the surface is,

$$np = Integer\left(\frac{2R}{SCSP}\right) \tag{6.3.8}$$

These pulses are discretized by the simulation time step, Δt. To ensure stability and accuracy, the time step must be chosen such that the discretized distance for beam travel, Δx, is less than beam radius R.

Assuming the reference point on the powder surface to be at the midpoint of the raster scan vector, the elapsed time during which laser energy will be applied is given approximately by,

$$\tau = (np-1)\frac{L_V}{V} \tag{6.3.9}$$

where L_V is the length of the raster scan vector. Between each successive pulse to the reference point, a time delay will occur due to travel of the beam, see Figure 6.23. During these delay periods, $\delta = 0$ and the surface cools by radiation and convective losses, see equation (6.3.2), at a rate that is governed by the combined heat transfer coefficient, $h = h_c + h_r$, for natural convection and radiation. The coefficient h_c is estimated from the relationships by Fishenden and Saunders (McAdams, 1954) for transfer from heated plates to air,

$$\frac{h_C L}{k_f} = a(Gr\,Pr)^b$$

$$a = 0.54,\, b = 1/4 \quad 10^5 < (Gr\,Pr) < 2 \times 10^7 \tag{6.3.10}$$

$$a = 0.14,\, b = 1/3 \quad 2 \times 10^7 < (Gr\,Pr) < 3 \times 10^{10}$$

where Gr and Pr are the Grashof and Prandtl numbers. The radiation contribution, h_r, is estimated by

$$h_r = \frac{\alpha_R \sigma T^4}{(T - T_\infty)} \tag{6.3.11}$$

where σ is the Stephan-Boltzman constant and absorbtivity, α_R, equals the emissivity, ε_R, by the usual gray body approximation, see Section 6.1.

The thermal model, equations (6.3-2), is solved with an Crank-Nicholson implicit finite difference scheme (Ames, 1977) that is unconditionally stable with discretization errors of the order of $(\Delta t^2 + \Delta z^2)$. A time increment of 5×10^{-5} sec was chosen based on applied energy density stability studies. Node spacings in the z or depth direction were typically taken as 0.004 cm (1.5×10^{-4} in). These values are fine enough to capture the rapid heating rates and changes in bed ρ, C, k, and composition that occur. Bed depths near 0.06 cm (0.15 in) were found to provide a reasonable approximation of the infinite boundary condition in equation (6.3.2), above. Typical simulations were carried to about 30 sec to approximate the time usually required to form a single layer.

The pulses of laser energy input are calculated using equations (6.3.2)-(6.3.11) and either experimental emissivity data or equation (6.1.30). The material properties are continuously recalculated at each grid point and time increment to reflect changes in bed porosity, ε, thermal conductivity, k, and binder content, w. The kinetic model for sintering, equation (6.2.4) is used to describe the densification by sintering, where experimentally determined values for the Arrhenius factors are used to estimate k' at processing temperatures. The appropriate thermal conductivity model is evaluated at each point determined from the porosity ε.

Compositional changes due to binder depolymerization are similarly handled. Here, the kinetics of polymer decomposition are added to or substituted, depending on the circumstances, for the kinetics of porosity collapse. In addition, the energy associated with depolymerization must be properly counted in the heat sink term, $G(z,t)$ in equation (6.3.2), because the endothermic heat of depolymerization, estimated as the negative of the heat of polymerization ΔH_R, 13.8 kcal/mol for PMMA (Sawade, 1969), is potentially large. Its consideration (Vail, *et al*, 1996) requires the simultaneous solution of equation (6.3.2) and

$$G(z,t) = -\frac{d(w/w_0)}{dt}\frac{\Delta H_R}{M}\phi_{PO}\rho \qquad (6.3.12)$$

where ρ is the bed density, M the molecular weight of the repeat unit, ϕ_{PO} the initial weight fraction of polymer in the bed, and where $-d(w/w_o)/dt$ is the rate of disappearance of polymer according to the discussion in Chapter 4. For the typically 20 vol% ($\phi_{PO} = 0.09$) binder loading used in the study of polymer-coated SiC particles (Vail, *et al*, 1996), $G(z,t)$ is effectively zero during laser heating. That is, nearly all of the absorbed energy is used to heat the substrate. Consequently, equation (6.3.2) can be solved directly using $G(z,t)$ =0 without concern that the heat of depolymerization is distorting the unsteady state temperature profiles. The polymer decomposition can then be independently calculated from equations (4.24)-(4.27) and the time and temperature history at each node. For the sake of completeness, however, $G(z,t)$ was retained in the model results discussed below. The simulation limits the polymer decomposition rates to those values that correspond to temperatures below 500 °C. This limitation is required to prevent the model from over predicting the degradation of the binder. Some justification for this restriction can be found in the literature. For example, Duley and Mueller (1992) in their discussion of the laser welding of polyethylene argue that the surface temperature of the polymer should be limited to 387 °C, the decomposition temperature of the polymer. Combustion studies (Liu and Kannatey-Asibu, 1993) similarly argue that the surface temperature of the polymer under the flame is limited by the boiling temperatures of the degraded materials that are being produced. TGA studies on the binders in this study indicate that they rapidly degrade at temperatures near 450 °C (see Chapter 4), slightly below the imposed limit.

6.3.2 SLS Model Verification

Nelson, *et al,* 1993, compared calculated and observed sintering depths in polycarbonate powder that was deliberately processed at conditions where layer-layer bonding was incomplete. This allowed them to determine the depth of fusion, experimentally, by taking thin slices through the part cross section in a direction perpendicular to the layers. These slices could then be mounted on a microscope stage and back-lighted. Those regions of the slice that were sintered or fused transmitted light while the less sintered material scattered light and appeared dark to the viewer, see Figure 6.24. The sintering depth could then be determined from section photographs. Alternately, they measured the densities of the parts, ρ, and assumed that the fused regions of the parts were completely dense with polycarbonate density, ρ = 1.2 g/cm^3, and that the less fused regions were at the bed density, ρ_B = 0.45 g/cm^3. From a simple mass balance, they could estimate the fractional degree of sintering, F_S,

$$F_S = \frac{\rho - \rho_B}{\rho_S - \rho_B} \qquad (6.3.13)$$

These measurements are compared with one another and presented as a function of SLS process parameter, *SCSP*, in Figure 6.28, below. They are found to agree within about 0.0025cm.

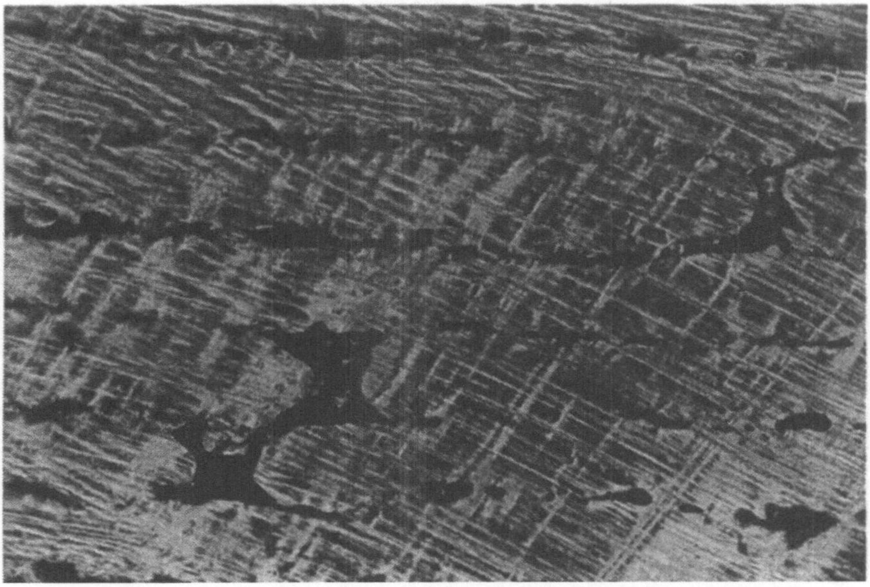

Figure 6.24 Photograph of SLS part made from PC powder at P = 16 W, V = 55.1 cm/sec, ω = 622 μm, and $SCSP$ = 0.058 cm (Nelson, 1993).

From Figures 6.24 and 6.25, one can conclude that the parts comprised of alternating high and low density layers at the SLS processing conditions employed. Larger beam diameter, ω, appears to lead to somewhat larger penetration depth, than smaller beam diameter, as does decreasing the scan spacing, $SCSP$, to provide more energy pulses per surface point. These issues are discussed in greater detail below.

Figure 6.26 shows typical calculated density profiles by the SLS model, described above. The transition from unsintered to fully sintered material is calculated to be quite sharp, typically 0.005 cm, and the transition moves into the bed with decreasing velocity after the laser beam moves on and the temperature begins to cool. These calculated results are qualitatively consistent with experimental evidence, Figure 6.24. The calculated temperature profiles, Figure 6.27, initially show the large gradients associated with low thermal diffusivity and high laser energy per unit area inputs. The material sinters, ε decreases, and k increases, so that the thermal gradients become smaller with increasing time, and relatively more energy is transported through sintered material than could be were sintering not possible. This collapse of porosity with a commensurate increase in bed thermal conductivity is responsible for the good layer to layer adhesion that can be achieved relatively modest energy input, relative to filled systems.

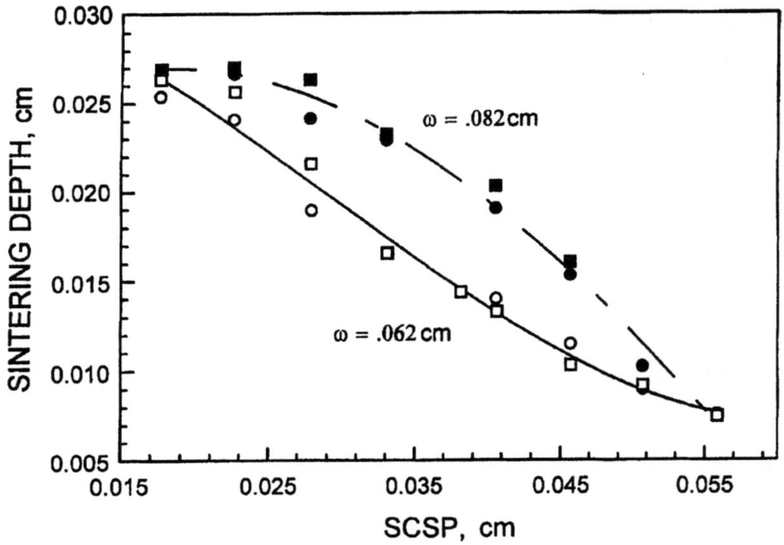

Figure 6.25 Effects of scan spacing, *SCSP*, and laser beam diameter on observed sintering depth in PC powder. ■, ❑, from optical; ●, ○, from density (Nelson, *et al*, 1993)

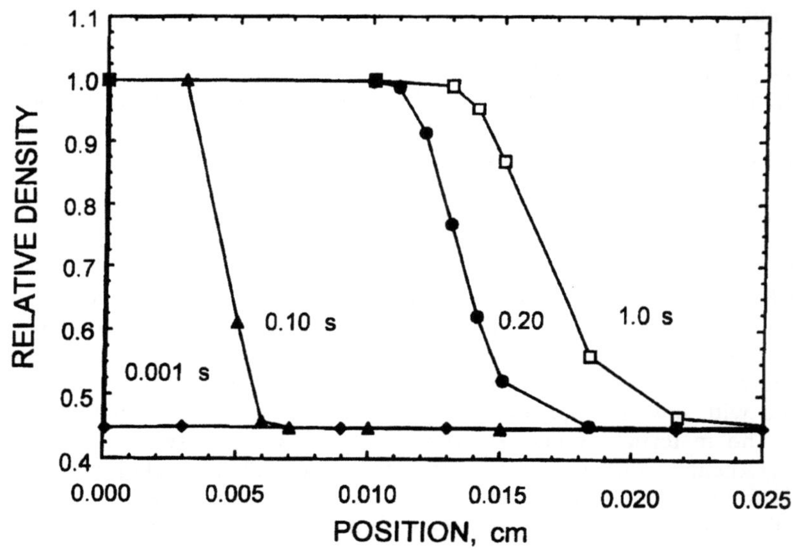

Figure 6.26 Calculated density profiles in PC powder at various times (Nelson, 1993).

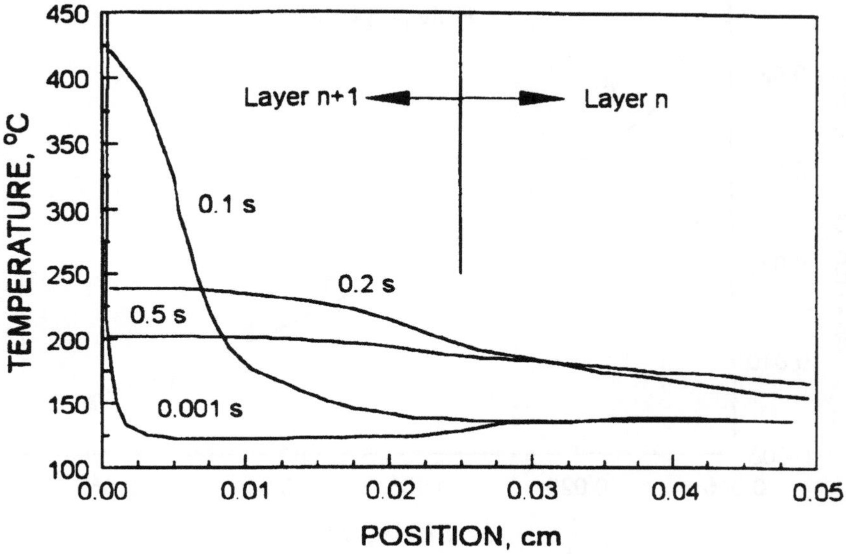

Figure 6.27 Calculated temperature profiles (Nelson, 1993).

Figure 6.28 compares observed with calculated sintering depths in PC powder as a function of scan spacing, *SCSP*. Generally the model is able to predict sintering depth to within an accuracy of ±0.001 in, a value that is near the limits of measurement. We conclude that the model does a fairly good job of predicting the effect of scan spacing on sintering depth.

The sensitivity of calculated sintering depths to SLS parameters other than scan spacing were also explored by Nelson, *et al*, 1993, in two case studies in which the parameters were varied one at a time. One of these case studies, SCSP= 0.033 cm, is presented in Table 6.1 below. Each parameter was varied by ± 10% from the base case. The table shows the calculated absolute deviations from the base case sintering depth of 0.015 cm and a fixed scan vector length of 2.54 cm. Generally, sintering depth is seen to be proportional to laser power, although the proportionality is far from one to one, proportional to beam radius, bed temperature, and inversely proportional to beam speed and air velocity, and scan spacing (Figure 6.28). The relative influence of air velocity is quite small, indicating that most of the cooling at the bed surface occurs by radiation, not convection. Likewise, the influences of beam radius and bed temperature on sintering depth appear small relative to scan spacing, *SCSP*, (Figure 6.28), laser power, *P*, and beam speed, *V*.

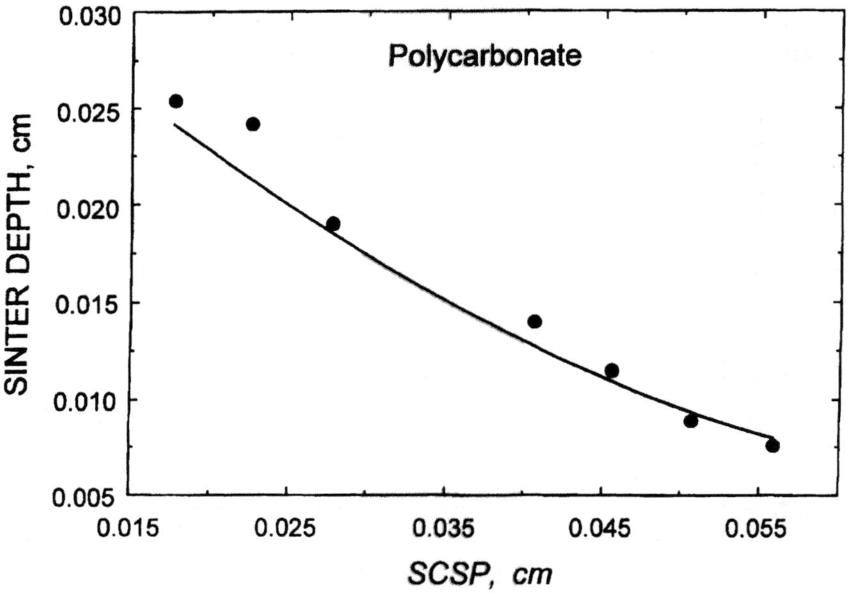

Figure 6.28 Comparison between calculated and observed sintering depths for ω = 0.062 cm, P = 16 W, V = 55.1 cm/sec. ● optical measurements; — calculated.

Table 6.1 Sensitivity of Sintering Depth to SLS Parameters at *SCSP* = 0.033 cm

Parameter	Base Value	Positive Dev. of Parameter	Sinter Depth Deviation	Negative Dev. of Parameter	Sinter Depth Deviation
Sinter depth	0.015 cm				
Laser Power, P	15 W	+2 W	+ 0.00132 cm	- 2 W	-0.00124 cm
Beam radius, ω	0.0622 cm	+0.0063 cm	+0.000804 cm	- 0.0063 cm	-0.000488 cm
Beam speed, *V*	55.1 cm/sec	+ 5.5 cm/sec	-0.00179 cm	- 5.5 cm/sec	+ 0.00207 cm
air vel.	305 cm/min	+305 cm/min	-0.000513 cm	- 305 cm/min	+0.000025 cm
Bed Temp.	125 °C	+ 10 °C	0.000925 cm	- 10 °C	-0.000060 cm

Observations such as those in Table 6.1, prompted Nelson, *et al*, 1993, to conclude that a combined parameter, since called the *Energy Density* (Nelson, et al, 1995a),

$$\textit{Energy Density} = \frac{P}{\pi R^2}\frac{2R}{V}\frac{2R}{SCSP} \propto \frac{P}{V(SCSP)} \qquad (6.3.14)$$

could be a useful correlating parameter for sintering depth and other energy driven phenomena. The first term in equation (6.3.14) is proportional to the laser flux adsorbed under the laser beam, $\alpha_R \langle I \rangle - h(T-T_\infty)$. The second term approximates the time (as a maximum) that a circle with radius R is under the moving laser beam. The third term approximates the number of pulses received by any point on the surface, see Figure 6.23. Equation (6.3.14) is identical in form, although not in derivation, to the Exposure, E_{av}, derived in Jacobs (1992) to explain the effect of beam speed on cure of photopolymerizable materials in the StereoLithography process.

The *Energy Density* parameter has been successfully used to correlate SLS parameters with sintering behavior. For example, Nelson, 1993, shows a qualitative correlation between optical sintering and *Energy Density*, see Figure 6.29. As *Energy Density* is increased from 1.5 to 2.2 cal/cm^2 sintering is seen to qualitatively increase.

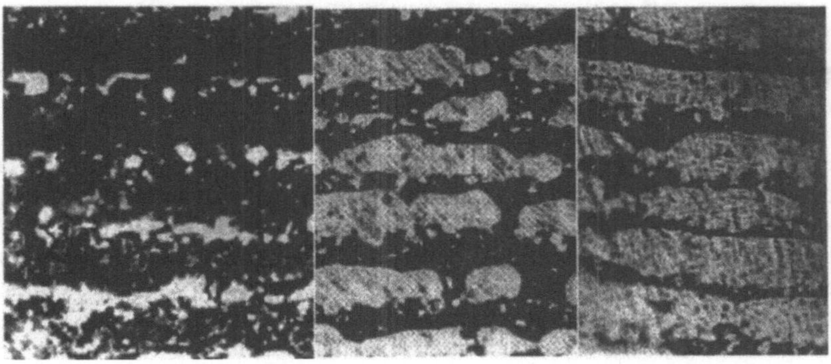

A B C

Figure 6.29 Effect of *Energy Density* on sintering of polycarbonate. *Energy Density*= (A) 1.50 ; (B) 1.85 ; (C) 2.19 cal/cm^2

Correlation of sintering with *Energy Density* in composite systems that contain large amounts of inorganic particulate is somewhat more difficult, partly because there is no obvious change in light transmission associated with sintering in these materials, and partly because the density changes are also small and more difficult to measure. Some attempts have nonetheless been made. Nelson, et al, 1995a, were able to see some changes in layer definition of fracture surfaces in polymer-coated silicon carbide powders. As shown in Figure 6.30, a clearly layered structure was apparent in this system at *Energy Density* near 0.6 cal/cm^2 .

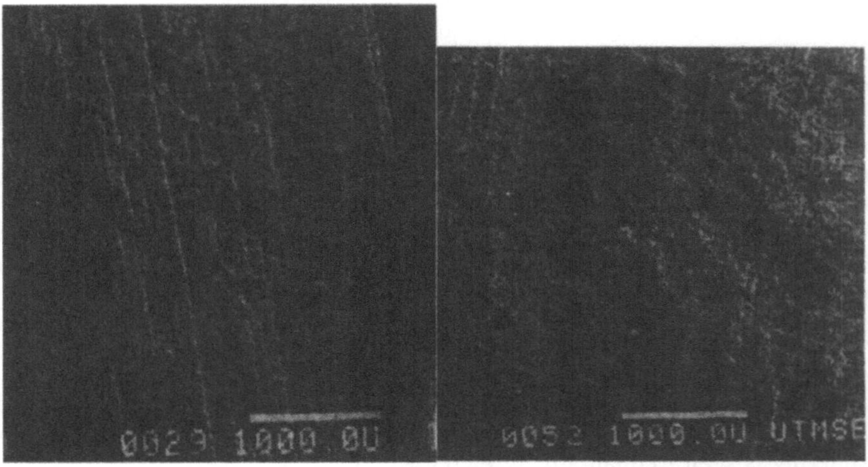

A B

Figure 6.30 Effect of *Energy Density* on sintering of polymer-coated SiC. *Energy Density*= (A) 0.6 ; (B) 1.3 cal/cm^2 (Nelson, et al, 1995a)

Increasing the *Energy Density* to 1.3 cal/cm^2 resulted in fracture surfaces with very little discernable layer formation. Badrinarayan, 1995, attempted to correlate changes in density of test bars made from 40 vol% PMMA/ 60 vol% copper mixed powders with changes in *Energy Density*. In addition, he used the Nelson, *et al*, 1993 model to predict these changes.His experimental results, Figure 6.31, show considerable scattering about the theoretical line, however the observed trend is reasonable.

6.3.3 Polymer Degradation

Measurable amounts of polymer degradation occur during the laser processing of shapes from polymer-coated ceramic powders by Selective Laser Sintering (SLS). Because the binder is important to achieving strong green parts that can be handled with minimal breakage during post-processing operations, it is essential to minimize the extent of binder losses occurring. As the first step toward understanding the mechanisms of binder degradation, a thermal model of the physical system has been developed and applied to beds of polymer-bound silicon carbide and of polymer-bound glass (Vail, *et al*, 1996, Vail and Barlow, 1994). Two different polymer binders were studied. Space limitations permit only the simulation of PMMA-bound silicon carbide powder to be discussed here. The model uses the same equations discussed earlier, to describe one-dimensional laser energy input, heat conduction, and decomposition. The simulation parameters and thermal properties of the various components in the powder beds that were simulated are shown in Table 6.2 and 6.3, respectively. The sources of the thermal property data are given in Vail, *et al*, 1996. The thermal properties of the powder beds were computed from the properties given in these tables by the methods described in Section 6.1.

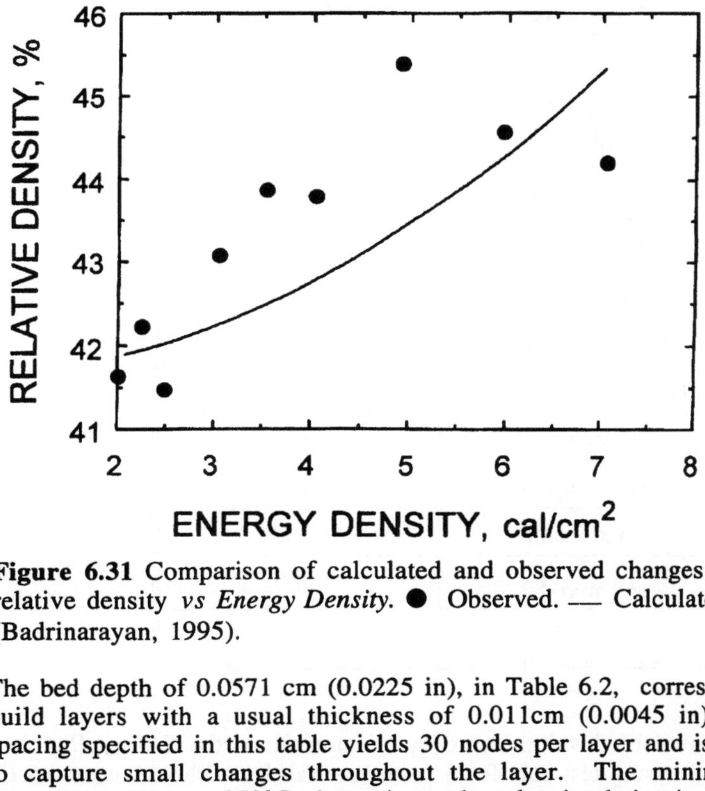

Figure 6.31 Comparison of calculated and observed changes in relative density *vs Energy Density*. ● Observed. — Calculated (Badrinarayan, 1995).

The bed depth of 0.0571 cm (0.0225 in), in Table 6.2, corresponds to five build layers with a usual thickness of 0.011cm (0.0045 in). The node spacing specified in this table yields 30 nodes per layer and is fine enough to capture small changes throughout the layer. The minimum surface temperature, $T_{S,min}$, 250°C, determines when the simulation is stopped. It is well below the temperature where polymer degradation occurs and corresponds to about two seconds of simulated time.

Figure 6.32 compares predicted and observed PMMA degradation over the experimental range of *Energy Density*. Here, the experimental polymer content is determined gravimetrically by noting the change in mass when a sample is heated to 300°C and held there for several hours. The agreement of predictions with experiments is quite good, within ± 5% relative error, and the model appears to capture the non-linear increase in polymer decomposition with increasing *Energy Density*. It is notable that some degradation occurs even at low *Energy Density*, however the degradation rapidly increases when the *Energy Density* is increased beyond about 1.0 cal/cm^2.

Table 6.2 Simulation Parameters for Coated Powders

Parameter	PMMA/SiC
D_p, m	12.4
Coated Vol. Fraction Polymer	0.237
Coated Wt. Fraction Polymer	0.236
Total Wt. Fraction Polymer	0.197
Bed Depth, cm	0.0571
Node Spacing, cm	3.81×10^{-4}
ω, m	230
Air Temp., C	104
Init. Bed Temp.,°C	92
$T_{S,max}$,°C	500
$T_{S,min}$,°C	250

Figure 6.33 shows the development of green strength as the *Energy Density* is increased in forming test bars for 3-point bending from PMMA-coated SiC powder. Initially, bend strength appears to increase linearly with increases in *Energy Density*, however the rate of strength increase declines with further increases in *Energy Density* beyond about 1.0 cal/cm^2. Comparison of Figures 6.32 and 6.33 suggests that the non-linear rate of strength increase could be caused by degradation of the polymer binder. Binder degradation could also be responsible for the reduction in relative density observed in the

copper/PMMA system at *Energy Density* values greater than 5 cal/cm^2.

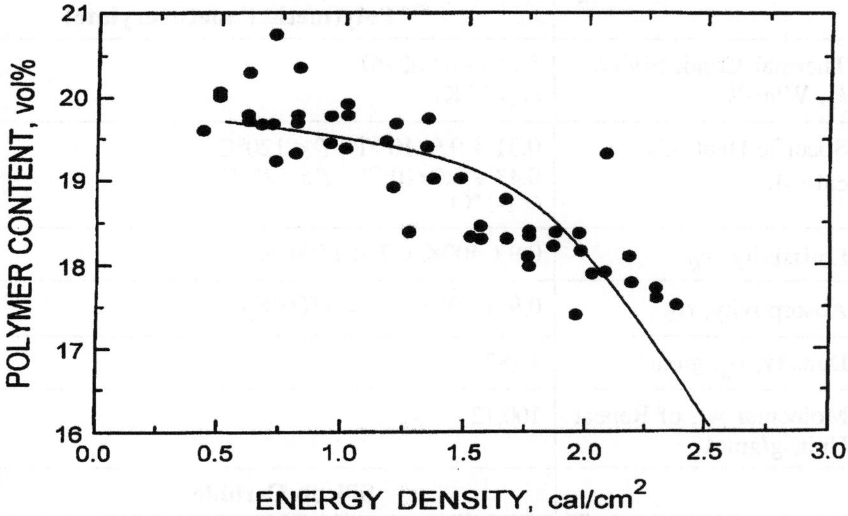

Figure 6.32 Comparison of predicted and observed polymer degradation as a function of *Energy Density* for PMMA/SiC powder (Vail, *et al*, 1996).

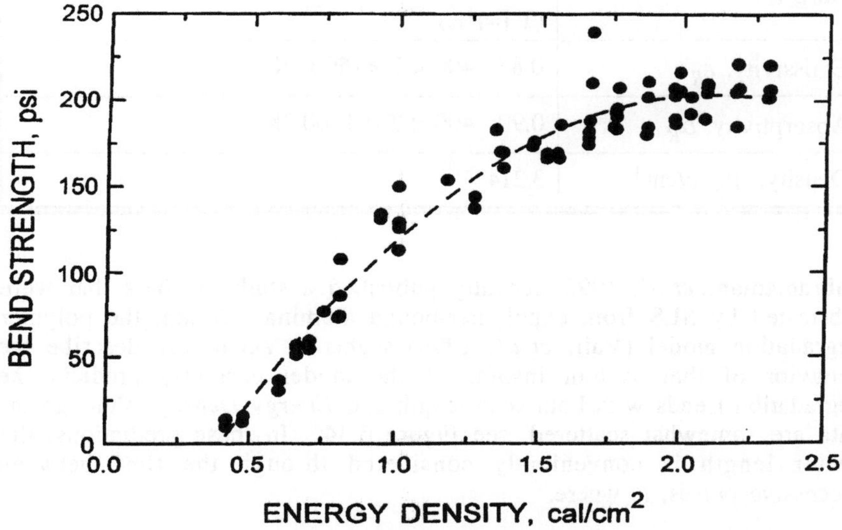

Figure 6.33 Effect of *Energy Density* on green strengths of bars made from PMMA/SiC powder (Vail, *et al*, 1996).

Table 6.3 Thermal Property Data Used in Simulations

Property	Value
	Poly(methyl methacrylate)
Thermal Conductivity, k_s, W/m-K	5.663 - 0.00276T (T[=] °K)
Specific Heat, C_P, cal/g-K	0.31 + 9.9x10^{-5}T ; T < 120°C 0.43 + 5.7x10^{-5}T ; T > 120°C (T [=] °C)
Emissivity, ε_R	0.9 (400K < T < 1500°K)
Absorptivity, α_R	0.9 (400K < T < 1500°K)
Density, ρ_S, g/cm^3	1.187
Molecular wt. of Repeat Unit, g/gmol	100.12
	Silicon Carbide
Thermal Conductivity, k_s, W/m-K	$\log(k_s) = 7.2018 - 1.8069 \log(T)$ T [=]°K
Specific Heat, C_P, cal/g-K	0.0405 + 5.48x10^{-4}T - 4.03x10$^{-7}T^2$ + 1.054x10$^{-10}T^3$ (T [=]°K)
Emissivity, ε_R	0.85; 400 < T <1500 °K
Absorptivity, α_R	0.90; 400 < T < 1500 °K
Density, ρ_S, g/cm^3	3.214

Subramanian, *et al*, 1995, recently published a study on bars that were fabricated by SLS from copolymer-bound alumina. Again, the polymer degradation model (Vail, *et al*, 1996) seems to accurately describe the behavior of that system insofar as the model correctly predicts the degradation trends with both scan length and *Energy Density*, although the data are somewhat scattered, see Figure 6.34. In these predictions, the vector length is conveniently considered through the time between successive pulses, t_i, where,

$$t_i = \frac{L}{V} \tag{6.3.15}$$

where L is the vector length, see equation (6.3.2). This approach seems to capture the kinetic aspects of the problem. That is, the ***Energy Density*** parameter alone is useful for correlating the effects related to energy input at fixed vector length, however the degree of sintering or the degree of polymer degradation depends on the net energy per unit area, the difference between that supplied and that lost by radiation, convection, etc. The loss terms are time dependent; longer vector length leads to longer time between pulses and to cooler surfaces and lower sintering or polymer degradation. An experimental example of this behavior is seen in Figure 6.35

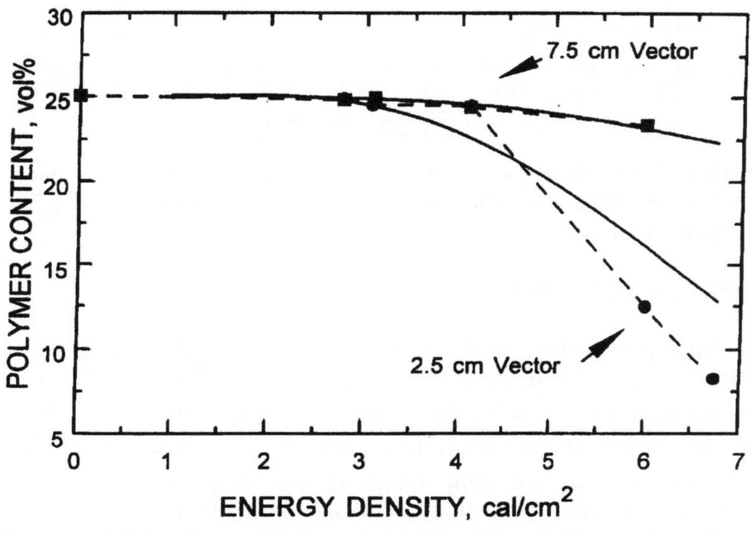

Figure 6.34 Comparison of predicted and observed polymer degradation for copolymer-coated alumina powder, scanned with different vector lengths. Solid lines are predicted. (Vail, *et al*, 1996).

6.4 COMMERCIAL SLS PROCESS BEHAVIOR/PROPERTIES

Current commercial SLS materials can be divided into three broad classifications: amorphous polymers, semicrystalline polymers, and binder systems. The general characteristics of these three classes have already been described in Chapter 4. In Sections 6.4.1 and 6.4.2, the SLS processing behavior of these polymers and the part properties of commercial materials will be discussed. SLS binder systems will be mentioned briefly in Section

6.4.3. These materials were extensively reviewed in Chapters 4 and 5.

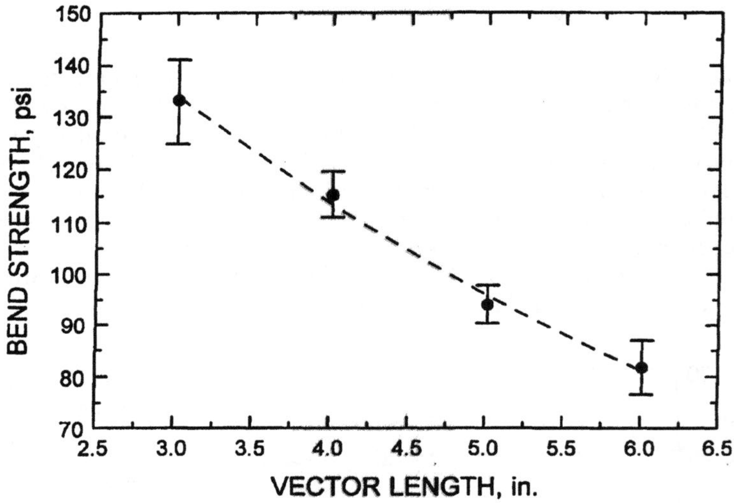

Figure 6.35 Effect of vector length on bend strength of bars made from PMMA-coated SiC at constant *Energy Density* = 1.75 cal/cm^2 (Vail, *et al*, 1996)

6.4.1 Amorphous Polymers

As discussed previously, amorphous polymers exhibit a glass transition temperature, T_g. These materials, which are solid below T_g, become leathery, rubbery and then liquid as they are heated above T_g. Powders begin to consolidate as they are heated above T_g. The rate of consolidation gradually increases as the temperature is increased above the transition point. In general, rapid consolidation is only possible at temperatures well above T_g.

There is no volume change associated with the glass transition and, hence, process shrinkage is limited to thermal contraction. In the SLS process, a portion of this shrinkage occurs as each new layer is built, while the remaining shrinkage occurs after the entire part has been built and is cooled from the part bed temperature to the breakout temperature. Layerwise shrinkage can cause wall distortion. As each new layer is built, it is bonded to solid material which has already shrunk. As a result, the shrinkage of the new layer is constrained and the vertical walls of a part can become "tapered" as is shown in Figure 6.36.

This effect can be eliminated by constructing what are called "bases" directly below all downward facing surfaces. A base is a thin, lightly sintered piece of material which typically covers the entire build surface. The first layer of each new part surface is bonded to a base. The base constrains shrinkage on the first part layer and prevents wall distortion. The bases are very brittle and can be easily broken off the part during the breakout process. Layerwise shrinkage can also cause parts to warp. Such warpage can be minimized by using bases and keeping the part bed temperature near the T_g where stress relaxation is activated. If the part bed temperature is increased beyond the T_g, the support powder will begin to fuse during the build, making part breakout difficult or impossible.

Figure 6.36 The effect of constrained layerwise shrinkage on vertical walls. The shrinkage of each new layer is constrained by the solid part which has already been constructed. This boundary effect can lead to tapered vertical walls.

It is difficult to produce fully dense SLS parts from most amorphous polymers. As described above, the part bed temperature for these materials must be kept near T_g. Part layers must be heated with the laser to well above this temperature in order to achieve a consolidation rate sufficient to allow dense parts to be produced within the time scale of the process. However, when part layers are heated to such temperatures, process control becomes difficult. The resulting parts typically exhibit material degradation due to local overheating and extensive growth due to heat diffusion beyond the part boundaries. As a result, parts made from most amorphous powders are only partially consolidated.

Typically, the total SLS process shrinkage for these porous parts is only 0.5 - 1.5%. In general, the production of SLS parts with good dimensional accuracy is facilitated by these relatively low shrink values.

It is possible to build useful SLS parts from a number of amorphous polymers. Commercially available amorphous polymers include a polycarbonate-based material and an acrylic-based material which is sold under the name "TrueForm PM". The most significant difference between the two materials is that the mean volume average particle size of the acrylic-based powder is 30 microns and the shape of the particles is spherical, while the mean volume average particle size of the polycarbonate-based powder is 100 microns and the shape of the particles is irregular. SEM micrographs of the two powders are shown in Figure 6.37.

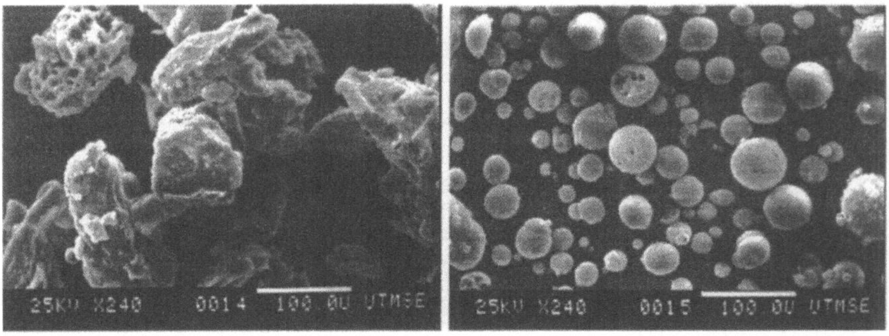

Figure 6.37 SEM micrographs of SLS polycarbonate (left) and TrueForm PM polymer (right) powders. The scale bar is equal to 100 microns.

The smaller particle size of the acrylic-based material allows parts with greatly improved feature detail and edge definition to be produced. The difference in edge definition which can be obtained with the two materials is illustrated in Figure 6.38.

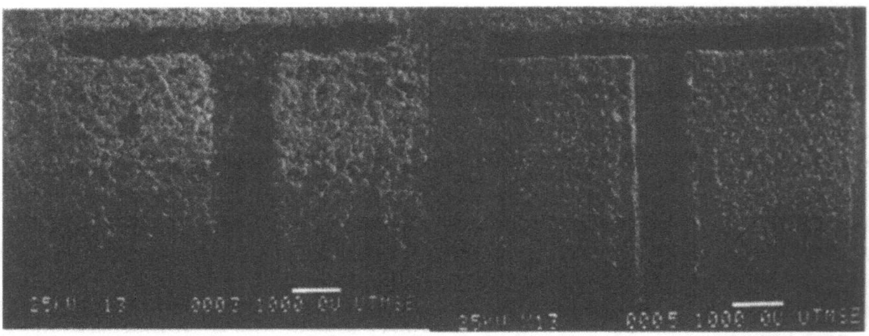

Figure 6.38 SEM micrographs showing edge definition in SLS polycarbonate (left) and TrueForm PM (right) parts. The scale bar is equal to 1 mm.

The properties of parts made from SLS polycarbonate and TrueForm PM polymer are listed in Table 6.4. The mechanical data were generated using ASTM tests.

Table 6.4 Properties of SLS Parts Made From Polycarbonate and TrueForm

Property	Unit	SLS Polycarbonate	TrueForm
Tensile Strength	MPa	23	10
Tensile Modulus	MPa	1200	1100
Tensile Failure Strain	%	5	1
Un-notched Izod	J/m	107	21.9
Notched Izod	J/m	53	8.1
DTUL (0.45 MPa)	°C	138	62
DTUL (1.8 MPa)	°C	136	59
Surface Roughness, Ra (As-Processed)	microns	28	5.5
Surface Roughness,Ra (Finished)	microns	7	0.6

The contrast in surface roughness for the two materials - both as-produced and after finishing - is particularly striking. The improved surface finish which can be obtained with the TrueForm PM polymer powder is due to the smaller size of the particles as well as their shape. Particle shape is an important characteristic for amorphous polymers because only limited shape rearrangement takes place during the densification process. Hence, powders which contain irregularly shaped particles produce rougher surfaces than powders which contain spherical particles even when their mean particle sizes are nearly equivalent. In general, it is also true that porous parts such as those made with SLS polycarbonate and TrueForm PM polymer are much easier to sand and polish than fully dense, ductile parts such as those made with SLS nylon materials.

The good dimensional accuracy, feature resolution and surface finish which can be obtained with these parts makes them particularly well suited for pattern-based applications such as the manufacture of silicone rubber and cast epoxy molds. In general, these applications do not require the part strength and durability which can be obtained with semicrystalline polymers. It is possible, however, to infiltrate porous polymer parts with epoxies and other thermoset resins to greatly increase their mechanical properties. The porosity of the parts also allows them to be used as investment casting patterns. Porous plastic patterns, unlike most fully dense plastic patterns, do not expand and cause shell cracking during the pattern removal process. An example of an investment casting pattern made from TrueForm PM polymer is shown in Figure 6.39.

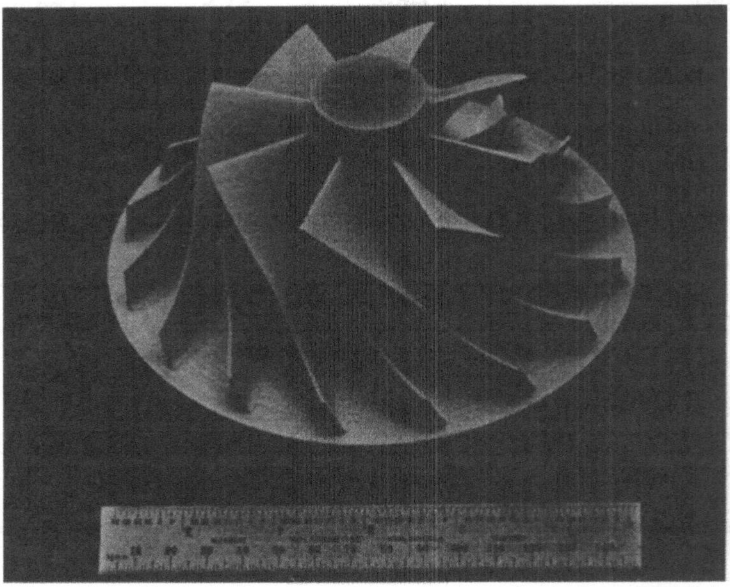

Figure 6.39 An investment casting pattern made from TrueForm PM polymer.

6.4.2 Semicrystalline Polymers

Semicrystalline polymers exhibit a melt transition, T_m, and, at a lower temperature, a T_g. These materials are solid below the melt transition and become liquid at T_m. Because most polymers contain a distribution of molecular weights and crystal sizes, the melting process actually occurs over a temperature range rather than at a discrete temperature. In general, rapid powder consolidation is possible at the temperature where melting is complete.

As semicrystalline polymers are cooled from the melt, polymer crystals nucleate and grow. The rate at which this occurs depends on the individual polymer and the crystallization temperature. Some semicrystalline polymers are able to crystallize much faster than others as a result of their chain structure. The minimum rates for a particular polymer are found at crystallization temperatures near T_m or T_g. The maximum rate is found between these two temperatures (Tadmor and Gogos, 1979).

There is a significant volume change associated with melting and crystallization. In the SLS process, the shrinkage associated with crystallization can occur as each new layer is built. This layerwise shrinkage can cause wall distortion and part warpage, just as it can for amorphous polymers. However, the shrinkage stresses for high modulus semicrystalline polymers are generally so large that it is not possible to keep the parts flat by using bases or even constraint structures. If, however, the rate of crystallization is slow relative to the vertical build rate, it is possible to maintain the part in a supercooled liquid state during the layerwise build process. In this state, the part is incapable of supporting stresses. It is possible to control this rate by selecting a polymer which crystallizes slowly and by keeping the part bed temperature - which is effectively the crystallization temperature - near the onset of melting. In this case, the shrinkage associated with crystallization occurs after the part has been built and begins to cool in the support powder. As will be discussed in Section 6.5.1, parts can still develop warpage during this portion of the process if the thermal environment is not properly managed.

It is much easier to produce fully dense SLS parts from semicrystalline polymers than from amorphous polymers. As described above, the part bed temperature for semicrystalline polymers must be kept near the onset of melting. As a result, it is possible to melt and rapidly consolidate part layers by heating them to just above this temperature with the laser. The laser energy requirement is dominated by the polymer heat of melting. Fully dense parts can be produced if the melt viscosity is low enough to allow complete consolidation to occur within the time scale of the process.

The total SLS process shrinkage for semicrystalline polymers is typically 3 - 5%. It is possible to build dimensionally accurate parts from such materials if the shrinkage is predictable but, in general, the most accurate parts are built from materials which have lower shrinkages.

It is possible to build useful SLS parts from a number of semicrystalline polymers. The commercially available material is based on a polyamide (nylon) chemistry (Forderhase *et al*, 1994, Forderhase *et al*, 1995). The T_m for this polymer is approximately 186°C. Powders with mean volume average particle sizes in the 50 -120 micron range are available. It is instructive to compare some of the properties of SLS nylon parts to those of compression molded parts made from the same material. These data are given in Table 6.5.

The properties of the compression molded parts are representative of fully dense, isotropic parts. The densities of the SLS and compression molded parts are nearly the same, indicating that the SLS parts are essentially void free. Further support for this conclusion comes from micrographs of polished part cross sections. Only occasional voids are found in the SLS parts. The mechanical properties of the SLS and compression molded parts are also very similar. The somewhat lower tensile strength, tensile failure strain, and notched Izod values for the SLS parts probably reflects the sensitivity of these properties to even a small number of voids. Nonetheless, these data demonstrate that it is possible to produce plastic SLS parts which have properties which are nearly indistinguishable from those of pressure consolidated parts.

Table 6.5 Properties of Nylon Parts Made by SLS and by Compression Molding

Property	Unit	SLS Parts	Compression Molded Parts
Tensile Strength	MPa	36	44
Tensile Modulus	MPa	1400	1400
Tensile Failure Strain	%	24	30
Notched Izod	J/m	70	108
Density	g/cm^3	0.95 - 1.0	1.04

The properties of these nylon parts can be modified by making composite parts from blends of nylon powder and discontinuous reinforcements. Nylon-based composite parts have been made from both glass bead and glass fiber reinforcements (Forderhase *et al*, 1995). In general, the changes in part properties associated with the addition of these reinforcements to the polymer follow expected trends. For example, in Figure 6.40, tensile moduli for glass bead composites are plotted as a function of the volume fraction reinforcement.

The moduli increase monotonically with increasing volume fraction reinforcement. At 0.3 volume fraction reinforcement, the modulus of the composite parts is approximately twice that of the unreinforced parts (1400 MPa). It should be possible to predict the moduli of such composites from the properties of the components. It is necessary to assume that the composites are void free and that there is complete coupling between the phases. A common model is (Halpin, 1992):

$$\frac{E_c}{E_m} = \frac{(1 + \zeta\beta V_f)}{(1 - \beta V_f)}$$

$$\beta = \frac{\left(\frac{E_f}{E_m} - 1\right)}{\left(\frac{E_f}{E_m} + \zeta\right)} \tag{6.4.1}$$

where E_c is the modulus of the composite, E_m is the modulus of the polymer, and E_f is the modulus of the glass beads. The factors ζ and β are parameters which depend on the boundary condition between the matrix and reinforcement as well as the moduli of the matrix and reinforcement. In the case of the bead reinforced composites, ζ is assumed to be 2. The moduli calculated from equation 6.4.1 are also plotted in Figure 6.40. The predicted values are in reasonable agreement with the experimental values up to a reinforcement volume fraction of approximately 0.3. This demonstrates that the SLS composite parts are well consolidated. At higher reinforcement loadings, however, the experimental moduli no longer increase in a predictable manner. Micrographs of part cross sections show that these parts have a large number of voids. This porosity exists because at higher reinforcement loadings there is not enough polymer to fill the interstitial void space between the packed beads. Hence, 0.3 volume fraction reinforcement represents a practical upper limit for these composites.

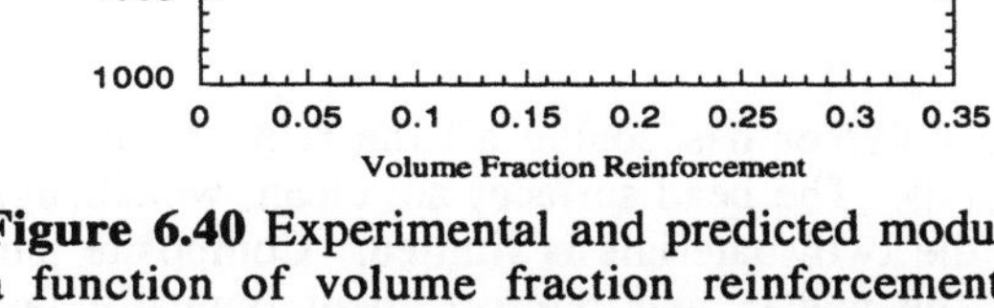

Figure 6.40 Experimental and predicted moduli as a function of volume fraction reinforcement for SLS nylon-based glass bead composites. The bead diameter is 35 microns. The predicted values were calculated using equation 6.4.1.

The tensile strength and failure strain of bead reinforced parts are a function of bead diameter. In Figure 6.41, tensile strength and failure strain data are plotted as a function of bead diameter. The data are for composites which contain 0.30 volume fraction reinforcement.

The tensile strength and failure strain both increase with decreasing bead diameter. This trend is expected, since smaller particles provide the most homogeneous stress distribution in parts (Nielson, 1974). The tensile strength of the composite parts made with 35 micron diameter beads is about 25% higher that for unreinforced parts (36 MPa). Since glass is a brittle material, the composite part failure strains are less than that of the unreinforced parts (24%), although they are relatively high for this type of composite.

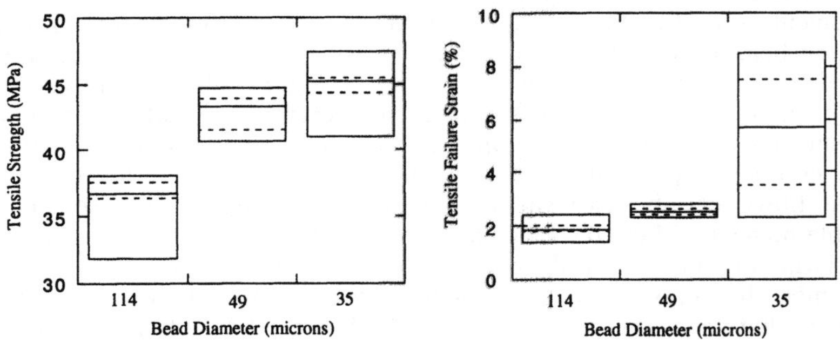

Figure 6.41 Tensile strength and failure strain as a function of bead diameter for SLS nylon-based glass bead reinforced composites. The volume fraction reinforcement is 0.3. The bottom and top of each box represents 5% and 95% of the data. The middle line represents the median value (50%) of the data, while the lower and upper dashed lines represent 25% and 75% of the data, respectively.

A typical tensile fracture surface for a glass bead reinforced part is shown in Figure 6.42.

It appears that the polymer has coated all the reinforcement particles and that there are no voids. The bead surfaces are clean, which suggests that the adhesion between the two materials is limited. Composite parts with poor adhesion between the matrix and reinforcement typically fail by gradual damage accumulation rather than catastrophic crack initiation and propagation. This probably explains the relatively high failure strains observed for the composite parts.

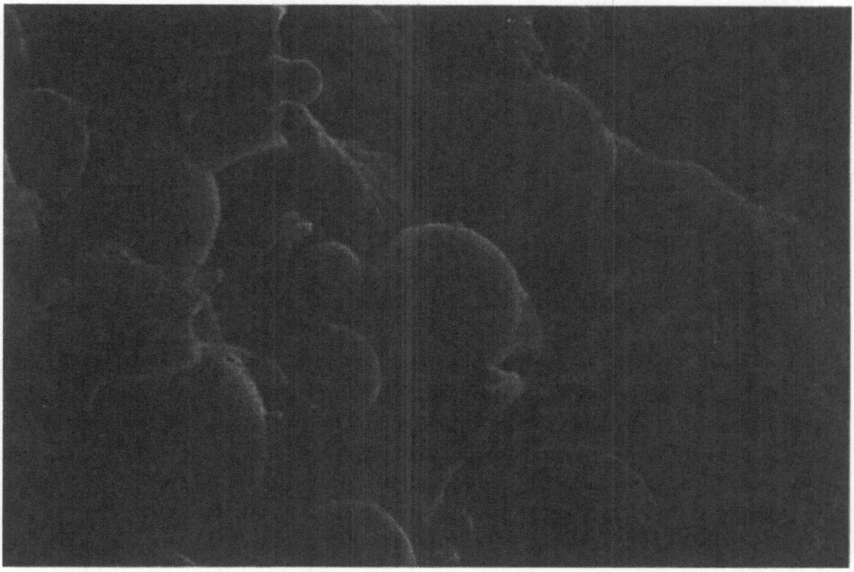

Figure 6.42 A fracture surface for a SLS nylon-based glass bead reinforced composite. The volume fraction reinforcement is 0.3 and the bead diameter is 35 microns. The dominant failure mode appears to be polymer/reinforcement debonding.

The mechanical property trends for fiber reinforced parts are similar to those for bead reinforced parts. The properties of fiber reinforced parts show considerable anisotropy, however. The tensile modulus and strength in the x direction (horizontal) are both significantly higher than in the y direction (vertical) as shown in Figure 6.43. This effect is consistent with fiber alignment in the direction of powder delivery by a roller. The data in Figure 6.43 were generated by testing tensile coupons built in both the x and y directions.

These materials would also be expected to shrink differently in the x and y directions. To build accurate SLS parts from such a material, it would be necessary to apply separate scale factors to the x and y axes. Current software already allows this to be done.

This composite approach is quite versatile. A variety of part properties can be "tuned" by selecting the appropriate reinforcement. The most significant limitation is associated with the flow behavior of the powder/reinforcement blend. The blend must be free flowing in order to form layers in the SLS process. Very small (< 10 microns) particulates and high (> 5) aspect ratio fibers inhibit flow and make it difficult to form layers.

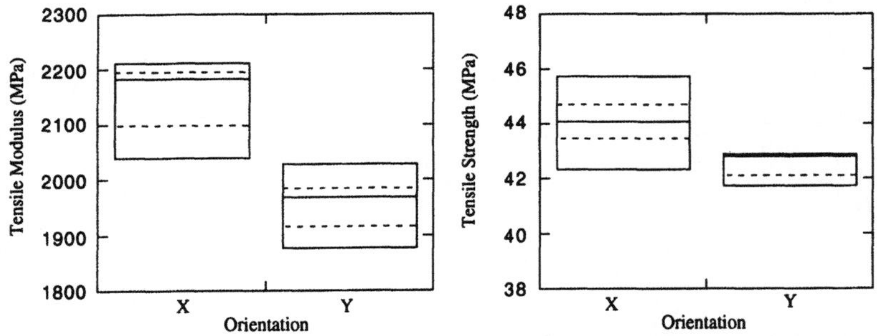

Figure 6.43 Tensile modulus and strength as a function of orientation for SLS nylon-based glass fiber reinforced composites. The volume fraction reinforcement is 0.15, the fiber diameter is 16 microns, and the aspect ratio is 4.4. The data are plotted in the same manner as Figure 6.41. The x test direction corresponds to the direction of roller travel in the SLS machine; the y direction is transverse to this. The data indicate that there is some preferred fiber orientation in the direction of roller travel.

Nylon and glass bead reinforced nylon materials are commercially available. The composite material has a volume fraction reinforcement of 0.3 and is sold by DTM Corp. under the name "ProtoForm". The properties of parts made from nylon and ProtoForm composite are listed in Table 6.6. The mechanical data were generated using ASTM tests.

The tensile modulus of composite material is approximately twice that of the unreinforced material, while the flexural modulus of the reinforced material is approximately five times greater than that of the unreinforced material. The tensile strength for the ProtoForm composite is approximately 30% higher than that of unreinforced nylon, while the heat deflection temperature of the ProtoForm composite is 90°C higher than that of unreinforced nylon. As expected, the tensile failure strain and notched Izod values for the composite are lower than those of the unreinforced nylon due to the presence of the brittle glass phase. The two nylon-based systems are complimentary; the unreinforced material offers outstanding ductility with good strength and heat resistance, while the composite material offers outstanding modulus with even higher strength and heat resistance. The reinforced parts can also generally be finished more easily than the unreinforced parts.

The excellent mechanical properties of the nylon-based parts makes them particularly well suited for use as functional prototypes. They can also be used as durable patterns for manufacturing silicone rubber and cast epoxy molds, although parts with higher resolution and smoother surfaces can be produced from TrueForm PM polymer. An example of a functional prototype part made from nylon is shown in Figure 6.44.

Table 6.6. Properties of SLS Parts Made from SLS Nylon and ProtoForm Composite.

Property	Unit	SLS Nylon	ProtoForm Composite
Tensile Strength	MPa	36	49
Tensile Modulus	MPa	1400	2800
Tensile Failure Strain	%	24	6
Flexural Modulus	MPa	870	4300
Unnotched Izod	J/m	1370	440
Notched Izod	J/m	70	68
DTUL (0.45 MPa)	°C	163	188
DTUL (1.8 MPa)	°C	44	134
Surface Roughness, Ra As-Processed	microns	12	15
Surface Roughness, Ra Finished	microns	5	4

6.4.3 Binder Systems

These materials were discussed extensively in Chapter 4 and 5 and earlier in this chapter. Briefly, the binder systems are composed of a powdered high temperature material - typically a metal or ceramic - and a polymer binder. The polymer binder can be mixed in with or coated on the powdered material. In contrast to the composite materials discussed in Section 6.4.2, these composite parts are porous and the total polymer content is generally less than 20 volume %. In the SLS process, laser energy is used to melt the polymer binder and form an intermediate part which is then post processed in a high temperature furnace process to remove the binder and sinter the high temperature particles. The intermediate SLS parts have relatively low strengths (< 3 MPa) compared to those of the amorphous and semicrystalline parts discussed in Sections 6.4.1 and 6.4.2. In addition, the SLS portion of the process shrinkage can be very low (< 0.2%) for these materials. As a result, it is possible to make intermediate SLS parts with excellent dimensional accuracy.

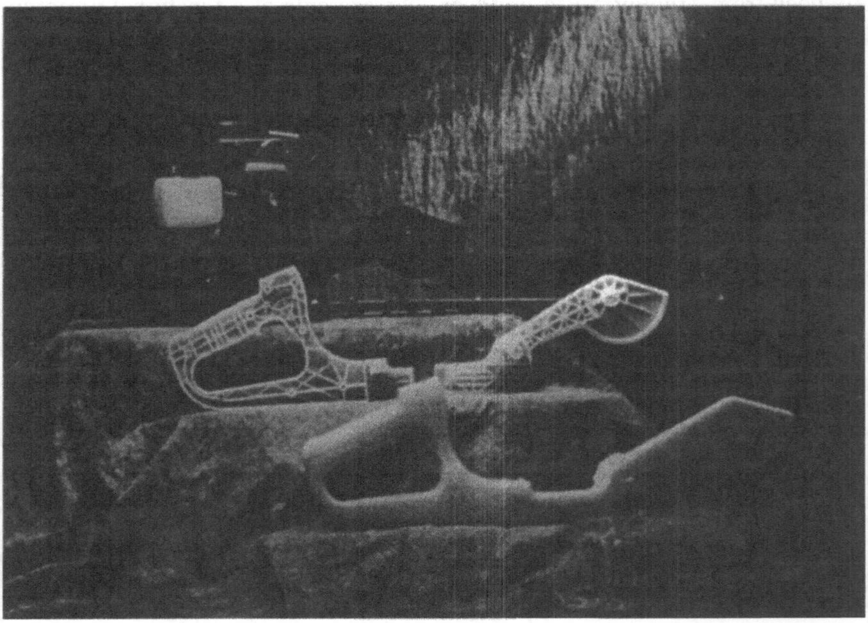

Figure 6.44 A functional prototype part made from SLS nylon. The part was built in pieces and assembled.

6.5 SLS PROCESS CONTROL CONSIDERATIONS

In this section, key process elements which influence part accuracy - thermal control, beam position control, and scale/offset compensation - will be discussed.

6.5.1 Thermal Control

The part bed and feed surface temperatures are actively controlled on current commercial systems. As was discussed in Sections 6.4.1 and 6.4.2, the part bed surface temperature must be typically kept just below the softening or melting temperature of a material to minimize layerwise shrinkage stresses and prevent part warpage. As was also discussed, however, if the part bed temperature is increased much above this temperature, the support powder can fuse, making part breakout difficult or impossible. For some materials, such as most amorphous polymers, the temperature window to produce acceptable parts can be 5 - 10°C, while for others, such as some semicrystalline polymers, it can be as small as 1°C. For materials with a small temperature window, excellent thermal control of the entire build surface is required.

The peak temperature of scanned regions is not actively controlled on current commercial systems. It is possible to indirectly control this temperature by increasing or decreasing the *Energy Density*. For a fixed *Energy Density*, however, there can be considerable variation in the peak temperatures found in different scanned regions within a particular layer as well as from layer to layer. A layer can contain a single or multiple cross sections. These cross sections can also vary in size and shape. When there is only a single, small cross section in the scan path, the time between adjacent scans is relatively short. When there is a larger cross section, or multiple cross sections in the scan path, however, the time between adjacent scans tends to be longer. As a result, small, isolated cross sections are typically heated to higher temperatures than other cross sections. These higher temperature regions can exhibit excessive growth and degradation as well as greater shrinkage as discussed in Section 6.3. A variety of techniques can be used to limit part temperature variability. One method is to build a set of bars outside the part boundaries, perpendicular to the scan direction, to insure that the time between adjacent scans remains above some minimum value. The bars are discarded when the parts are complete. Obviously, the disadvantage of this technique is that the parts build at a slower rate. More elegant approaches, such as temperature feedback control, have been explored (Benda, 1994).

Temperature control in the build volume can also be important. As was discussed in Section 6.4.2, parts made from semicrystalline polymers generally solidify and shrink as they are cooled, typically over a period of hours, from the build temperature to the breakout temperature. If such parts are cooled inhomogeneously, significant thermal stresses can develop. If these stresses are unbalanced, the parts can warp. This type of distortion is often referred to as "post-build" warpage, to distinguish it from the "in-build" warpage which occurs at the build surface.

If parts do exhibit post-build warpage, the options for modifying the thermal environment in the build volume are limited. In some cases, parts cool inhomogeneously because they are near the piston or side walls. Typically, the piston and side walls are cooler than the parts and support powder. Temperature gradients in these regions can be minimized by actively controlling the temperatures of the piston and side walls. The only practical way to modify temperature profiles away from these surfaces is to flow gas through the porous powder. This practice is commonly referred to as "downdraft". The effect of the gas flow is to accelerate the redistribution of heat in the build volume and minimize the potential for temperature gradients.

The warpage of nylon parts has been minimized by controlling the piston temperature and by using downdraft (Forderhase, 1994, 1995). The reduction in warpage associated with piston temperature control is illustrated in Figure 6.45. In this case, warpage was characterized by calculating the diameter of a circle which would pass through the bottom surface of the part. The larger the curvature diameter, the flatter the part. The parts were 89 mm long coupons, built approximately 25 mm from the piston surface.

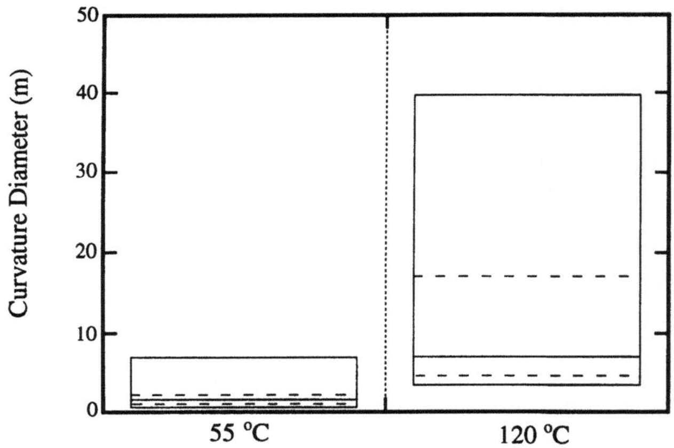

Figure 6.45 Part curvature diameter as a function of piston temperature. The larger the curvature diameter, the flatter the parts. The parts were 89 mm long coupons, built approximately 25 mm from the piston surface. The bottom and top of each box represents 5% and 95% of the data. The middle line represents the median value (50%) of the data, while the lower and upper dashed lines represent 25% and 75% of the data, respectively.

These data demonstrate that increasing the piston temperature from 55°C to 120°C dramatically reduces the warpage of parts built near the piston surface. The coupons built with the 120°C piston temperature have less than 0.25 mm warpage across their length.

The effect of both piston temperature control and downdraft on part warpage is illustrated in Figure 6.46 In this case, part warpage was characterized by measuring plane form, which is the deviation of the bottom surface of a part from a theoretical flat plane. The part is an enclosure geometry, 185 mm long and 135 mm wide, built approximately 40 mm from the piston surface.

These data show that the effect of piston temperature is not significant for parts built at this distance from the piston surface. This is not a surprising result, since the powder is an excellent insulator. At the same time, the impact of downdraft on these parts is quite significant. The plane form is reduced from 0.40 mm to 0.20 mm as the downdraft flowrate is increased from 0 to 24 standard liters per minute.

6.5.2 Beam Position Control

Beam positioning is accomplished with mirror galvanometers on current commercial systems. A vector is drawn by moving the mirrors at a constant velocity with the laser turned on. In order to draw accurate vectors, the signals to start and stop the galvanometer mirror movement must be properly coordinated with the signals to turn the laser on and off (Nelson, et al, 1995b). The timing of these events is important because there is a finite time required to accelerate and decelerate mirror galvanometers as is shown in Figure 6.47. These data were generated by scanning a position sensitive detector.

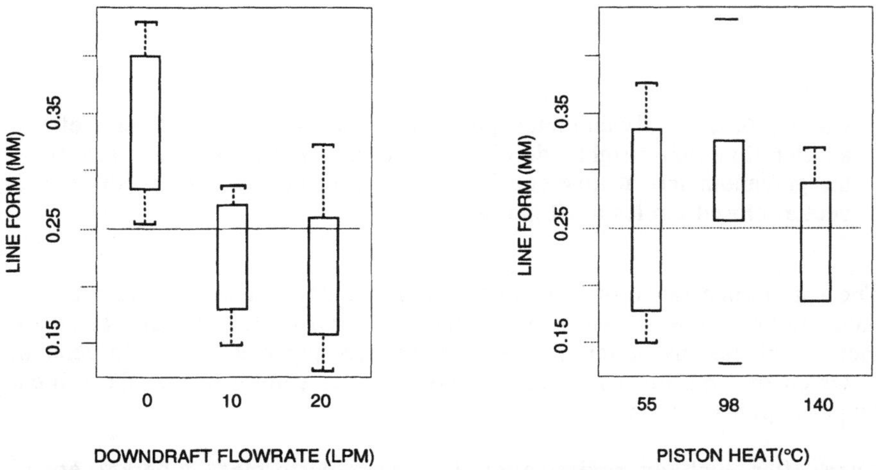

Figure 6.46 Plane form as a function of downdraft flowrate and piston heat temperature. The smaller the plane form value, the flatter the parts. The part was an enclosure geometry, 185 mm long and 135 mm wide, built approximately 40 mm from the piston surface. The box represents 50% of the data, while the upper and lower brackets represent 5% and 95% of the data, respectively. The horizontal line corresponds to 0.25 mm total warpage.

In order to minimize potential dimensional errors associated with the time required to accelerate and decelerate the mirror galvanometers, a number of parameters can be tuned. The two most important are the Laser On (LO) and the Laser Off (LF) commands (GSI, 1988). The LO command specifies a delay time between when the ramp signal is sent to the mirror galvanometers and when the laser is turned on, while the LF command specifies a delay time between when the stop signal is sent to the mirror galvanometers and when the laser is turned off.

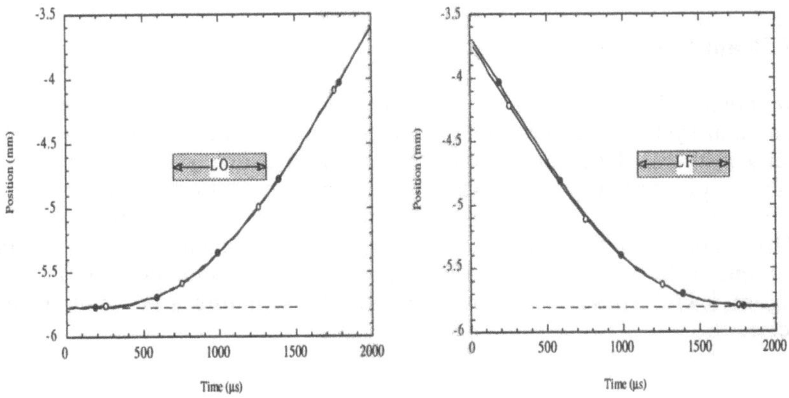

Figure 6.47 Galvanometer position as a function of time (left: acceleration curve; right: deceleration curve). A start signal is sent to the galvanometers at time equal zero. Legend: open circles = 100 mm vector, closed circles = 10 mm vector

The appropriate range of LO and LF values can be determined from model data similar to that shown in Figure 6.47. Optimized LO and LF values are then established by building parts with different parameter sets. In general, a design of experiments (DOE) methodology is employed and the critical output is part accuracy.

Even after such an optimization has been performed, inherent errors associated with the mirror galvanometers can affect part accuracy. These errors are a function of a number of factors including galvanometer design, the mass of the mirrors, and the scan speed. As an example, dynamic error data for two GSI galvanometers - G325 and M3 - are shown in Figure 6.48. These errors were characterized using a simple test where two series of marks (305 mm long) are burned on a sheet of Mylar. The first series is drawn at a slow scan speed (0.86 m/s) while the second is drawn at a faster scan speed (2.36 m/s). The dynamic error is assumed to be the difference in length between the two sets of marks.

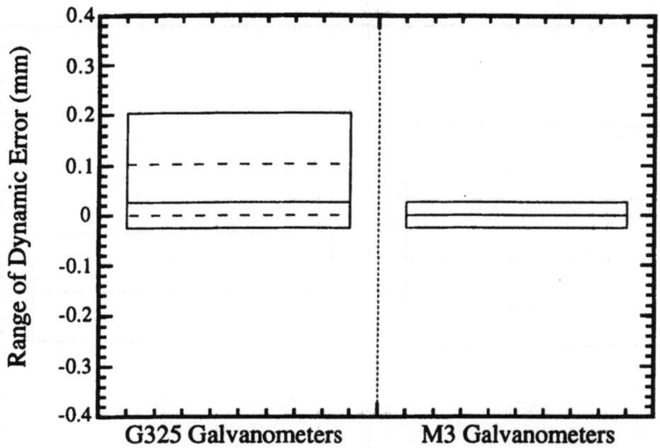

Figure 6.48 An example of dynamic errors characterized for GSI G325 and M3 galvanometers. The errors were measured along the x and y axes across a 305 mm distance. The data are plotted in the same manner as those in Figure 6.45.

The dynamic error range (90%) for the G325 is approximately 0.25 mm, while it is 0.05 mm for the M3. In addition, the absolute positional accuracy for the G325 is 0.076 mm, while the positional accuracy M3 is 0.006 mm. The impact of these errors on part accuracy has been evaluated by building polymer coated metal parts with both sets of galvanometers. Since the polymer coated metal material exhibits essentially no shrinkage in the SLS process, scan errors are assumed to account for a significant fraction of the total error. Dimensional accuracy data for a model part are plotted in Figure 6.49.

The range of error for the parts built with the first set of galvanometers is about twice that for the parts built with the second set. These data demonstrate that beam positioning errors can be minimized with optimized mirror galvanometer design and tuning.

6.5.3 Shrinkage/Offset Compensation

As was discussed in Section 6.4, most materials shrink during the SLS process. In addition, the finite diameter (0.4 mm) of the laser beam and growth can cause material to fuse outside the desired part boundaries. In order to build accurate parts, it is necessary to scale .STL files (see Chapter 3) to compensate for shrinkage and offset .STL file surfaces in order to compensate for the beam diameter and growth. There is software available to accomplish this if the correct scale and offset factors are known. These parameters can be determined by trial and error, or by building and measuring a model part (Nelson, et al, 1995b). An example of such a model part is shown in Figure 6.50.

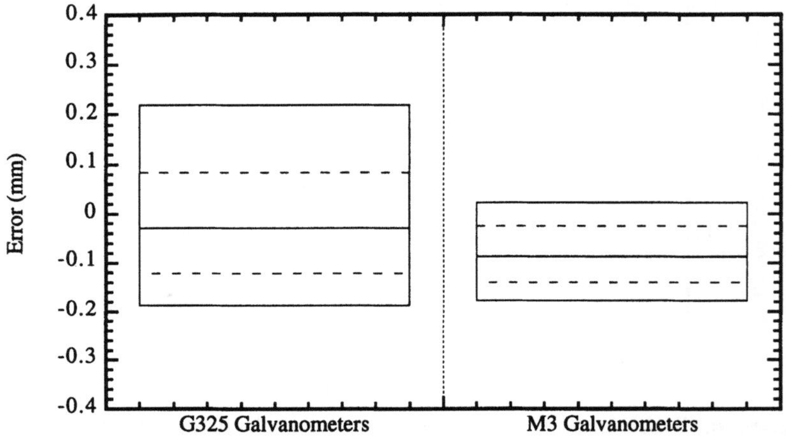

Figure 6.49 Dimensional accuracy data for a part built using GSI G325 and M3 galvanometers. The material is polymer coated metal. The part has measured dimensions from 5.1 mm to 165 mm. The data are plotted in the same manner as those in Figure 6.45.

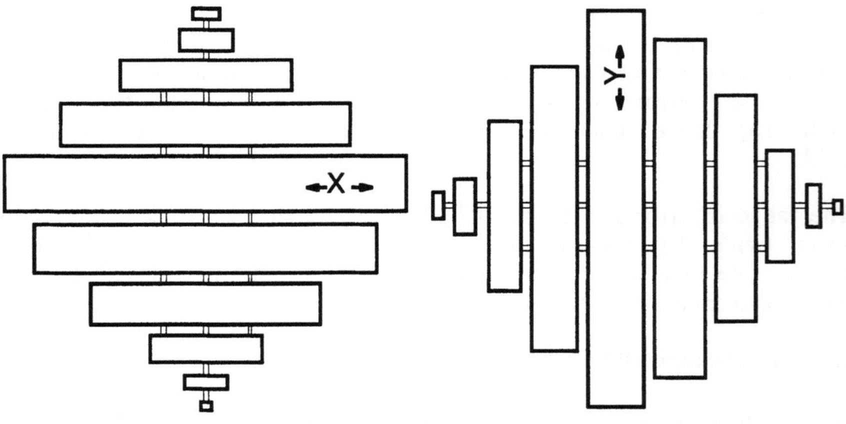

Figure 6.50 A model part used to determine scale and offset values for polymer-based parts. This part has nominal dimensions from 5 to 178 mm (0.2 to 7.0 inches). Typically, one part is built is built parallel to the direction of roller travel (*x*) and another perpendicular to this (*y*).

This part is built without any scale or offset compensation and then measured to determine the error (relative to nominal) for each dimension. A plot of error versus nominal dimension is then made as is shown in Figure 6.51.

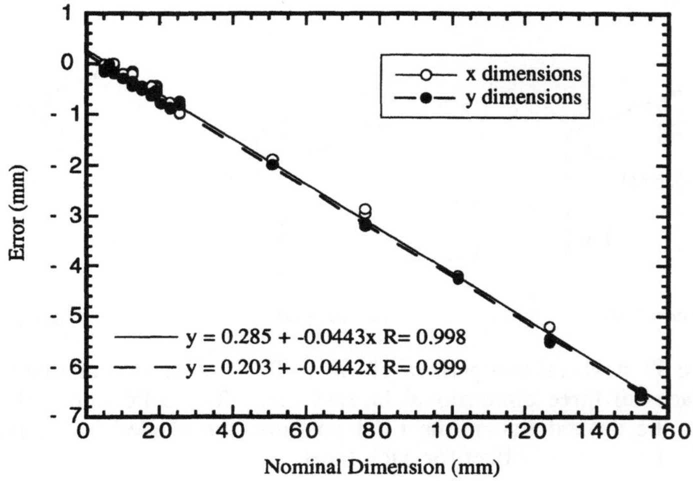

Figure 6.51 A plot of error versus nominal dimension for the part shown in Figure 6.50. The material is SLS nylon. The slope of the best linear fit is equal to the shrinkage, while the intercept is equal to twice the offset.

The slope and intercept of a linear data fit are then determined. The slope is equal to the shrinkage value and the intercept is equal to twice the offset value. The scale value required to compensate for the shrinkage is equal to 1/(1-shrink).

There are a number of potential issues associated with determining scale and offset factors as well as compensating the .STL files. For the scale factor, the most significant problem is that shrinkage does not always occur in a uniform manner. As was described in Section 6.4.1, the shrinkage of a new layer can be constrained by the existing part substrate. Support powder trapped within an enclosed area can also constrain shrinkage. In addition, as was described in Section 6.5.1, there can be considerable spatial variation in the part surface temperature within a layer. The high temperature regions will generally shrink more than the lower temperature regions. As a result, the best scale value for a particular geometry may be different than the one determined from a model part. Obviously, this is a more significant issue for high shrink materials.

There can also be some variability in the offset correction. This variability is caused by the additional growth which can occur in higher temperature part regions. In addition, it is not simple to move the surfaces of three dimensional faceted files uniformly. The current methods for offset correction include dihedral offset, normal offset, and constant offset as shown in Figure 6.52

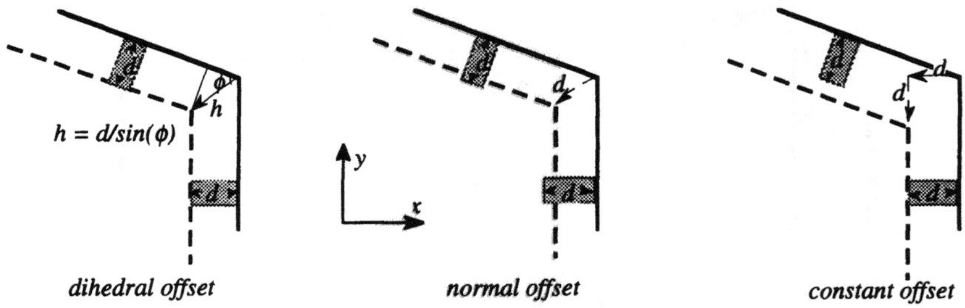

Figure 6.52 A visual comparison of three offset methods available to offset the surfaces of three dimensional faceted files. The solid line is the initial position, the dashed line is the final position, the shaded boxes represent the desired amount of offset for each facet.

The dihedral offset method uses the dihedral angle between connecting facets to calculate the new offset position of the vertex. For the normal offset method, a vector normal is calculated at each vertex based on the normals of the adjacent facets. The vertex is then moved along this vector normal by an amount equal to the offset value to the new offset position. In the case of the constant offset method, the vertex is moved first in the x direction by an amount equal to the offset value and then in the y direction by an amount equal to the offset value. The dihedral offset method is generally more accurate than the normal offset and constant offset methods, although all have advantages for particular geometries. Ultimately, the issues associated with offsetting three dimensional faceted files can be eliminated by offsetting two dimensional slices instead.

6.6 SLS PART ACCURACY - BENCHMARK DATA

Accuracy is a much discussed topic in SFF process literature. Despite this interest, the available database is somewhat limited. A large amount of dimensional tolerance data for a model stereolithography part have been reported (Jacobs, 1992, Pang, 1994). In addition, dimensional tolerance data for different rapid prototyping technologies have been compared in a number of studies (Schmidt, 1992, Van Putte, 1992, Schmidt, 1994, Juster, 1994, Baldwin, 1995). In this section, dimensional tolerance data for commercial SLS materials will be presented. These data were generated to demonstrate current state-of-the-art as well as to illustrate the impact of some of the material and process considerations discussed in Sections 6.4 and 6.5.

While the focus of this section is on dimensional tolerances, it should be noted that other accuracy related considerations can be equally important. For some applications, the visual quality of parts is critical. Here, feature definition, resolution, and surface finish are most important. The impact of particle size and shape on these characteristics has already been discussed in Section 6.4.1. For other applications, the ease of part fit and assembly is critical. Here, the consistency and repeatability of part dimensions are likely to be more important than absolute tolerances.

6.6.1 Model Part

A large fraction of the prototype parts made today are models of plastic injection molded parts. Typically, such parts have relatively thin walls (2 -3 mm) and a variety of features including ribs, posts, and holes. A model part was designed to be representative of this type of geometry. Drawings of this part with the measured dimensions are shown in Figure 6.53 The part has a total of 64 measured dimensions ranging in size from 1.27 mm to 152.4 mm. This part was used to benchmark the dimensional tolerances of all the plastic-based materials. In the case of the polymer coated metal material, the benchmarking was performed with the core and cavity sections for the model part. Drawings of the core and cavity sections with dimensions are shown in Figure 6.54.

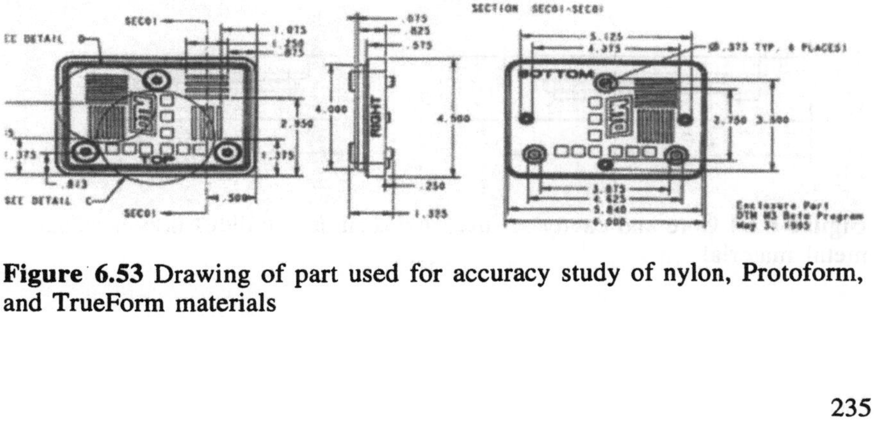

Figure 6.53 Drawing of part used for accuracy study of nylon, Protoform, and TrueForm materials

6.6.2 Experimental Protocol. Parts were built with a total of five different materials: polycarbonate, acrylic-based TrueForm PM, nylon, glass reinforced nylon or ProtoForm composite, the polymer-coated metal or RapidSteel material. These materials were discussed in detail in Sections 6.3, 6.4 and Chapter 5. In the case of the polymer-coated metal, dimensional tolerances were characterized both before and after the furnace infiltration step. In the case of the polycarbonate and nylon-based materials, parts were built on machines equipped with both G325 and M3 galvanometers. The characteristics of these galvanometers were discussed in Section 6.5.2.

In each case, two parts were built from each material using either default scale and offset values, or values calculated from a model geometry such as the one described in Section 6.5.3. These parts were then measured and corrected scale values were determined from the errors in the overall dimensions. An additional two parts were then built with the corrected scale values. The latter data set is reported below. Distortion was not an issue; all of the parts had less than 0.25 mm warpage across the longest part dimension.

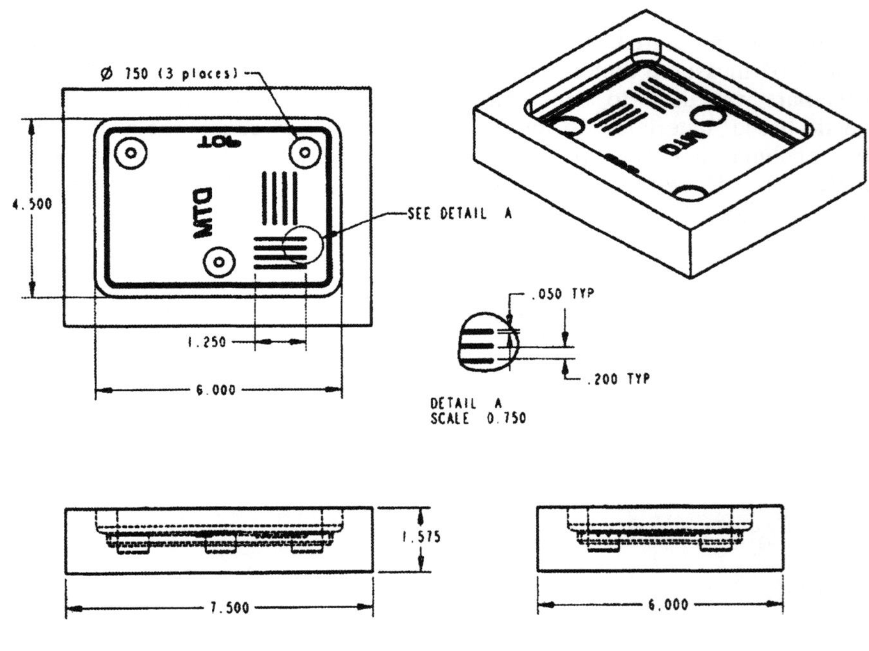

Figure 6.54 Core and cavity set used for accuracy study of polymer coated metal material.

6.6.3 Accuracy Results

Data for the semicrystalline nylon-based materials are shown in Figure 6.55 These data demonstrate that the range of error is reduced by approximately 30 - 40% when the parts are built with a machine equipped with M3 galvanometers instead of G325 galvanometers. This trend is consistent with that demonstrated for polymer coated metals (Figure 6.49). The range of error for nylon and ProtoForm composite is similar.

Dimensional error data for the amorphous systems - polycarbonate and TrueForm PM polymer - are shown in Figure 6.56 Again, the range of error for the polycarbonate parts built with the M3 galvanometers is about 30 - 40% less than that for the parts built with the G325 galvanometers. In addition, the range of error for the TrueForm PM polymer parts is about 50% less than that for polycarbonate parts built with equivalent galvanometers.

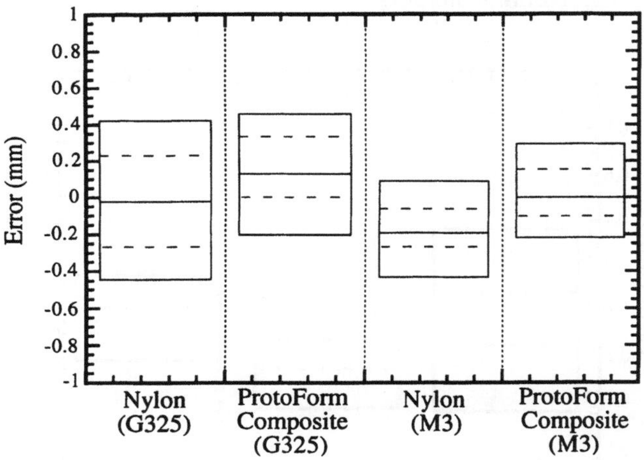

Figure 6.55 Dimensional error data for nylon-based materials. The first set of nylon and ProtoForm composite parts were built on a machine equipped with G325 galvanometers, while the second set were built on a machine equipped with M3 galvanometers. The bottom and top of each box represents 5% and 95% of the data. The middle line represents the median value (50%) of the data, while the lower and upper dashed lines represent 25% and 75% of the data, respectively.

Dimensional error data for both polymer coated green parts and copper infiltrated parts are shown in Figure 6.57. The error range (90%) for the green parts is ± 0.1 mm - better than for any other current SLS material. After copper infiltration, the error range for the same parts is ± 0.25 mm.

There is a strong correlation between the dimensional accuracy of parts and SLS shrinkage. This is demonstrated in Figure 6.58, which is a plot of RMS error versus SLS shrinkage. As can be seen, the RMS error for the lowest shrink material (polymer coated metals) is about 3-4 times lower than the RMS error for the highest shrink material (nylon). Polycarbonate is the only material that does not follow the general trend of increasing RMS error with increasing material shrinkage.

The fact that the most accurate parts can be made with the low shrink materials is not surprising. In fact, the dependence is more dramatic than shown here if first time build results, rather than "corrected" build results are compared. For the high shrink materials, the range of error for the first time builds is about 50% higher than it is for the corrected builds, while for the low shrink materials, the range of error is essentially the same for first time and corrected builds. These results indicate that the most significant improvements in dimensional accuracy can be achieved by either reducing intrinsic material shrinkage to <1% or by reducing the variability which is inherent with high shrink materials.

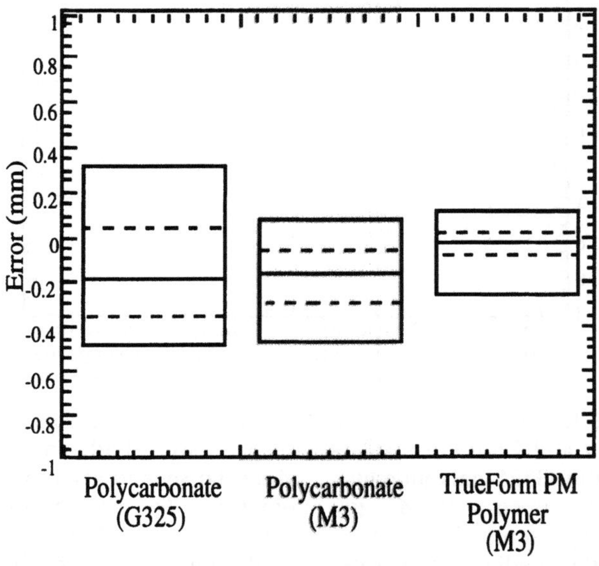

Figure 6.56 Dimensional error data for polycarbonate and TrueForm PM polymer materials are shown. The first set of polycarbonate parts were built on a machine equipped with G325 galvanometers, while the second set of polycarbonate parts and the TrueForm PM polymer parts were both built with a machine equipped with M3 galvanometers. The data are plotted in the same manner as those in Figure 6.55

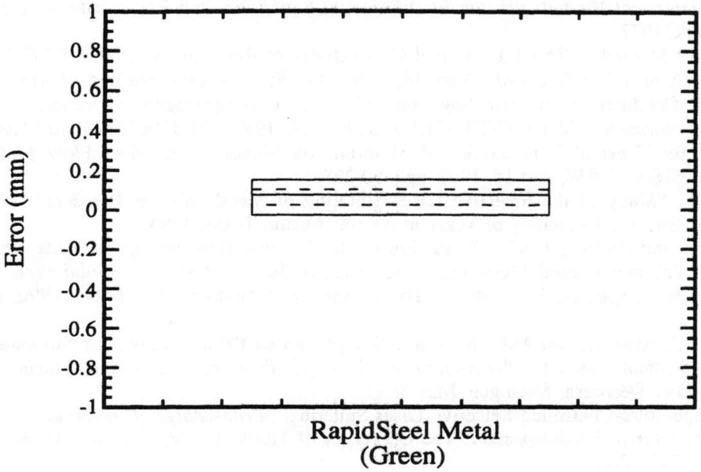

Figure 6.57 Dimensional error data for RapidSteel green parts and copper infiltrated parts. The green parts were built with a machine equipped with M3 galvanometers. The data are plotted in the same manner as those in Figure 6.55.

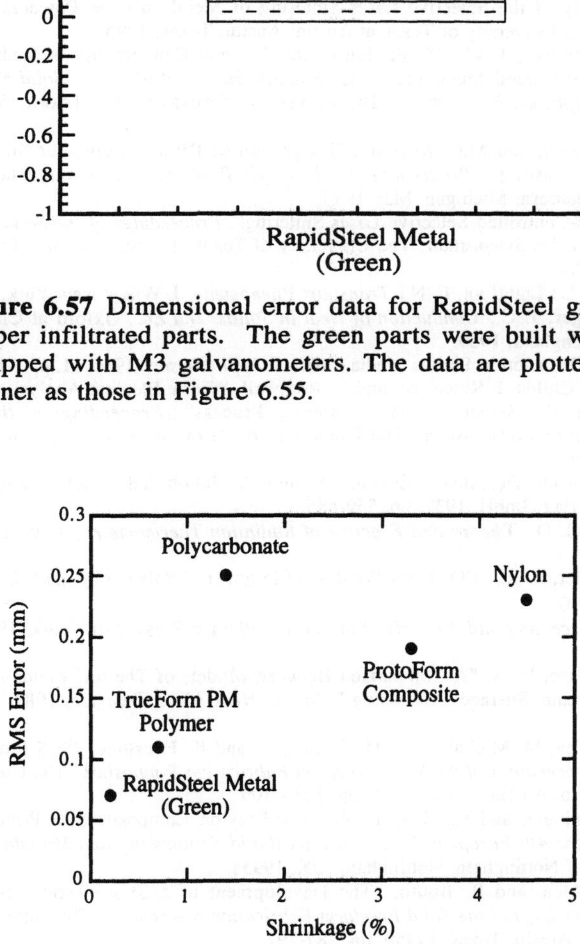

Figure 6.58 RMS error as a function of SLS material shrinkage for commercial materials. All the measured parts were built on a machine equipped with M3 galvanometers.

REFERENCES

Ames, W. F., *Numerical Methods for Partial Differential Equations*, 2nd Ed., Academic Press: New York, 1977.

ASTM, "Standard Method of Test for Thermal Conductivity of Refractories: ASTM C177-63," *1967 Book of ASTM Standards*, Part 14, 1967, 17-28; "Tentative Method of Test for Thermal Conductivity of Insulating Materials at Low Temperatures by Means of the Wilkes Calorimeter: ASTM C420-62T," *Ibid*, Part 14, 1967, 172-179; "Standard Method of Test for Thermal Conductivity of Materials by Means of the Heat Flow Meter: ASTM C518-67," *Ibid*, Part 14, 1967, pp. 230-238.

Badrinarayan, B., "Study of the Selective Laser Sintering of Metal-Polymer Powders," *Ph.D. Dissertation*, The University of Texas at Austin: Austin, Texas, 1995.

Badrinarayan, B. and Barlow, J. W., "Prediction of the Thermal Conductivity of Beds which Contain Polymer-Coated Metal Particles," Marcus, H. L., *et al*, Eds., *Solid Freeform Fabrication Symposium Proceedings*, The University of Texas:Austin, Texas, 1990, pp. 91-97.

Baldwin, M.D., C. Atwood, and M.C. Maguire, "Integration of RP and Computer Simulation into Investment Casting", *Proceedings of the Rapid Prototyping & Manufacturing '95 Conference,* Dearborn, Michigan, May 1995.

Benda, J., "Temperature-Controlled Selective Laser Sintering", *Proceedings of the Solid Freeform Fabrication Symposium*, The University of Texas at Austin, Austin, Texas, 1994, pp. 277 - 284.

Bird, R. B.; Stewart, W. L.; Lightfoot, E. N., *Transport Phenomena,* J. Wiley: New York, 1960

Carslaw, H. S. and Jaeger, J. C., *Conduction of Heat in Solids, 2nd Ed.*, Oxford at Clarendon Press: London, England, 1959.

Currie, J. A., "Gaseous Diffusion in Porous Media," *Brit. J. Appl. Physics*, 1960, 11, pp. 314-324.

Dalgarno, K.W., T.H.C. Childs, I. Rowntree, and L. Rothwell, "Finite Element Analysis of Curl Development in the Selective Laser Sintering Process", *Proceedings of the Solid Freeform Fabrication Symposium*, The University of Texas at Austin, Austin, Texas, 1996, pp. 546-553.

Damkohler, G., *Der Chemie-Ingenieur*, Euken, M. and A. Jakob, eds., 3(1), Leipzig, A Kademischer Verlag GmbH, 1937, pp. 359-485.

DeWitt, D. P.; Nutter, G. D., *Theory and Practice of Radiation Thermometry*, J. Wiley: New York, 1988.

Duley, W. W. and Mueller, R. E., "CO_2 Laser Welding of Polymers," *Polym. Eng. and Sci.*, 1992, 32(9), pp. 582-590.

Duley, W. W., *Laser Processing and Analysis of Materials*, Plenum Press: New York, 1983, pp. 19-23.

Festa, R.; Manca, O.; Naso, V. A. "A Comparison Between Models of Thermal Fields in Laser and Electron Beam Surface Processing," *Int. J. Heat Mass Transfer*, 1988, 31, pp. 99-106.

Forderhase, P., K. McAlea, M. Michalewicz, M. Ganninger, and K. Firestone, "SLS Prototypes from Nylon", *Proceedings of the Solid Freeform Fabrication Symposium*, The University of Texas at Austin, Austin, Texas, 1994, pp. 102 - 109.

Forderhase, P., M. Ganninger, and K. McAlea, Nylon and Nylon Composite SLS Prototypes", *Proceedings of the 4th European Conference on Rapid Prototyping and Manufacturing*, The University of Nottingham, Nottingham, UK, 1995a.

Forderhase, P., K. McAlea, and R. Booth, "The Development of a SLS Plastic Composite Material", *Proceedings of the Solid Freeform Fabrication Symposium*, The University of Texas at Austin, Austin, Texas, 1995b, pp. 287-297.

Frenkel, J., "Viscous Flow of Crystalline Bodies under the Action of Surface Tension," *J. Phys.* (USSR), 1945, 9, pp. 385-391.

German, R.M., *Particle Packing Characteristics*, Metal Powder Industries Federation: Princeton, New Jersey, 1989, chapt. 11.

German, R.M., *Powder Metallurgy Science*, 2nd Ed., Metal Powder Industries Federation: Princeton, N.J., 1994, Appendix B.

Ginnings, D. C., and G. T. Furukawa, "Heat Capacity Standards for the Range 14 to 1200°K", *J. Am. Chem. Soc.*, 1953, 75, pp. 522- 529.

GSI, DG Series: Scan Head Controllers, Users Manuel, General Scanning Incorporated, Watertown, MA, GSI PN 12P-DG, Rev #5, 1988.

Halpin, J.C., *A Primer on Composite Materials Analysis*, Technomic, Lancaster, PA, 1992.

Hobbs, S. Y., Dekkers, M. E. J., and Watkins, V. H., "Effect of Interfacial Forces on Polymer Blend Morphologies," *Polymer*, 1988, 29, pp. 1598-1602.

Ikeda, M., *et al*, "Basic Techniques of Laser Processing," *Precision Machinery*, 1985, 51(12), pp. 2253-2258. (in Japanese)

Incropera, F. P. and DeWitt, D. P., *Fundamentals of Heat and Mass Transfer*, 2nd Ed., John Wiley & Sons: New York, 1985, pp. 740-782.

Jacobs, P.F., *Rapid Prototyping and Manufacturing* , SME, Dearborn, Michigan, 1992, Chapter 11.

Jurand, R., ed., *Modern Plastics Encyclopedia*, McGraw Hill: New York, 1989, p. 595.

Juster N.P. and T.H.C. Childs, "A Comparison of Rapid Prototyping Processes", *Proceedings of the 3rd European Conference on Rapid Prototyping and Manufacturing*, University of Nottingham, Nottingham, United Kingdom, 1994.

Kuczynski, G. C., Neuville, B., and Toner, H. P., "Study of Sintering of PMMA," *J. Appl. Polym. Sci.*, 1970, 14, pp. 2069-2077.

Landolt-Börnstein: *Numerical Data and Functional Relationships in Science and Technology*, 6th Ed., Springer Verlag: Heidelberg, 1980.

Liu, Y. N. and Kannatey-Asibu, E., "Laser Beam Welding with Simultaneous Gaussian Laser Preheating," *Trans. ASTHMA*, 1993, 115, p. 34.

MacKenzie, J. K. and Shuttleworth, R. A., "A Phenomenological Theory of Sintering," *Proc. R. Soc.* (London), 1949, B62, 833-852.

McAdams, W. H., *Heat Transmission*, McGraw-Hill: New York, 1954.

Nelson, J. C. and Barlow, J. W., "Sintering Rates in the Selective Laser Sintering Process," *Solid Freeform Fabrication Symposium Proceedings,* The University of Texas:Austin, Texas, 1990, pp. 164-170.

Nelson, J. C., "Selective Laser Sintering: A Definition of the Process and an Empirical Sintering Model", *Ph.D. Dissertation*, The University of Texas at Austin, 1993.

Nelson, J. C., Xue, S., Barlow, Joel W., Beaman, J. J., Marcus, H. L., and Bourell, D. L., "Model of the Selective Laser Sintering of Bisphenol-A Polycarbonate," *Ind. Eng. Chem. Res.*, 1993, 32, pp. 2305-2317.

Nelson, J. C., Vail, N. K., Barlow, J. W., Beaman, J. J., Bourell, D. L., and Marcus, H. L., "Selective Laser Sintering of Polymer-Coated Silicon Carbide Powders," *Ind. Eng. Res.*, 1995, 34, pp. 1641-1651.

Nelson, J.C., K. McAlea, and D. Gray, "Improvements in SLS Part Accuracy", *Solid Freeform Fabrication Symposium Proceedings*, The University of Texas:Austin, Texas, 1995b, pp. 159-169.

Nielson, L.E., *Mechanical Properties of Polymers and Composites*, Volume 2, Marcel Decker, New York, 1974.

Pang, T.H., Stereolithography Epoxy Resins SL 5170 and SL 5180: Accuracy, Dimensional Stability, and Mechanical Properties", P*roceedings of the Solid Freeform Fabrication Symposium*, The University of Texas at Austin, Austin, Texas, 1994, pp. 204-224.

Perkin-Elmer Corporation, *Instructions: DSC-2 Differential Scanning Calorimeter*, 1976.

Press, W. H., Flannery, B. P., Teukolsky, S. A., and Vetterling, W. T., *Numerical Recipes in C, The Art of Scientific Computing*, Cambridge University Press: New York, 1988.

Rahim, W., "Rheological and Thermodynamic Properties," in *Principles of Polymer Processing*, *loc. cit.*(Tadmor and Gogos, 1979), Appendix A.

Ristic, M. M. and Dragojevic-Nesic, M. J., "A New Approach for the Description of the Initial-Stage Kinetics during Amorphous Materials Sintering," *Powder Technol.*, 1987, 49, pp. 189-190.

Rohsenow, W. M., Hartnett, J. P., and Ganic, E. N., Eds., *Handbook of Heat Transfer Fundamentals,* 2nd. Ed., McGraw-Hill: New York, 1985, pp. 3-113.

Ryder, G.J., M. Berzins, and T.H.C. Childs, "Modeling Simple Feature Creation in Selective Laser Sintering", *Proceedings of the Solid Freeform Fabrication Symposium*, The

University of Texas at Austin, Austin, Texas, 1996, pp. 554-561

Sawade, H., *J. Macromol. Sci*., 1969, C3, 313.

Scherer, G., "Sintering of Low Density Glasses: 1. Theory," *J. Am. Ceram. Soc*., 1977, 60, pp. 236-239.

Scherer, G., "Viscous Sintering under a Uniaxial Load," *Comm. Am. Ceram Soc*., 1986, 69(9), pp. 206-207.

Scherer, G., "Viscous Sintering of a Bimodal Pore Size Distribution," *J. Am. Ceram. Soc*., 1984, 67, pp. 709-715.

Schmidt, L.D., "A Benchmarking Comparison of Commercial Techniques in Rapid Prototyping", *Proceedings of the Fifth International Conference on Rapid Prototyping*, Dayton, Ohio, 1994, p. 333.

Schmidt, L.D., *Report Number rptcomp/bc*, Chrysler Motor Corporation, 1992.

Schroder, J., "Apparatus for Determining the Thermal Conductivity of Solids in the Temperature Range from 20-100 °C," *Rev. Sci. Instr.*, 1963, 34(6), pp. 615-621.

Sih, S. S. and Barlow, J. W., "The Prediction of the Thermal Conductivity of Powders," *Solid Freeform Fabrication Symposium Proceedings*, The University of Texas:Austin, Texas, 1995, pp. 397-401.

Sih, S. S. and Barlow, J. W., "Measurement and Prediction of the Thermal Conductivity of Powders at High Temperatures", *Solid Freeform Fabrication Symposium Proceedings*, The University of Texas:Austin, Texas, 1994, pp. 321-329.

Sih, S. S. and Barlow, J. W., "Emissivity of Powder Beds," *Solid Freeform Fabrication Symposium Proceedings*, The University of Texas:Austin, Texas, 1995, pp. 402-408.

Sih, S. S., The Thermal Properties of Powders, *Ph.D. Dissertation*, The University of Texas at Austin, 1996.

Subramanian, K, Vail, N. K., Barlow, J. W., and Marcus, H. L., "Selective Laser Sintering of Alumina with Polymer Binders," *Rapid Prototyping J*., 1995, 1(2), pp. 24-35.

Sun, M. M., Beaman, J. J., and Barlow, J. W., "Parametric Analysis of the Selective Laser Sintering Process," *Solid Freeform Fabrication Symposium Proceedings*, The University of Texas:Austin, Texas, 1990, pp. 146-154.

Sun, M. M., Physical Modeling of the Selective Laser Process, *Ph.D. Dissertation*, The University of Texas at Austin: Austin, Texas, 1991.

Swift, D. L., " The Thermal Conductivity of Spherical Metal Powders Including the Effect of Oxide Coating", *Int. J. Heat Mass Transfer*, 1966, 9, pp. 1061- 74.

Tadmor, Z. and Gogos, C., *Principles of Polymer Processing*, John Wiley & Sons: New York, 1979.

Taniguchi, N., Ikeda, M., Miyamoto, I., and Miyazaki, T., *Energy Beam Processing of Materials*, Clarendon Press, Oxford, 1989.

Vail, N. K., Badrinarayan, B., Barlow, J. W., and Marcus, H. L., "A Thermal Model of Polymer Degradation During the Selective Laser Sintering of Polymer-Coated Ceramic Powders", *Rapid Prototyping J.*, 1996, 2(3), in press.

Vail, N. K., and Barlow, J. W., "Modeling of Polymer Degradation in SLS," *Solid Freeform Fabrication Symposium Proceedings*, The University of Texas: Austin, Texas, 1994, pp. 330-338.

Vail, N. K., Barlow, J. W., Beaman, J. J., Marcus, H. L., Bourell, D. L., "Development of a Poly(Methyl Methacrylate-co-n-Butyl Methacrylate) Copolymer Binder System," *J Appl. Polym. Sci*., 1994, 52, pp. 789-812.

Vail, N. K., "Preparation and Characterization of Microencapsulated, Finely Divided Ceramic Materials for Selective Laser Sintering," *Ph.D. Dissertation*, The University of Texas at Austin: Austin, Texas, 1994.

Vail, N. K. and Barlow, J. W., "Ceramic Structures by Selective Laser Sintering of Microencapsulated, Finely Divided Ceramic Materials," *Solid Freeform Fabrication Symposium Proceedings*, The University of Texas:Austin, Texas, 1992, pp.124-130.

Vail, N. K. and Barlow, J. W., "Effect of Polymer Coatings as Intermediate Binders on Sintering of Ceramic Parts," *Solid Freeform Fabrication Symposium Proceedings*, The University of Texas:Austin, Texas, 1991, pp.195-204.

Van Putte, D.A., "A Brief Benchmarking Study of Rapid Prototyping Processes", *Proceedings of the Third International Conference on Rapid Prototyping*, Dayton, Ohio, 1992, p. 251.

Wakao, N. and Kato, K., "Effective Thermal Conductivity of Packed Beds," *J. Chem. Eng. Japan*, 1969, 2(1), pp. 24-32.

Williams, J., D. Miller, and C. Deckard, "Selective Laser Sintering Part Strength as a Function of Andrew Number, Scan Rate, and Spot Size", *Solid Freeform Fabrication Symposium Proceedings*, The University of Texas at Austin, Austin, Texas, 1996, pp. 537-545.

Wu, S., "Calculation of Interfacial Tension in Polymer Systems," *J. Polym. Sci*. Part C, 1971, 34, pp. 19-30.

Xue, S. S. and Barlow, J. W., "Thermal Properties of Powders," *Solid Freeform Fabrication Symposium Proceedings*, The University of Texas:Austin, Texas, 1990, pp.179-185.

Xue, S. S. and Barlow, J. W., "Models for the Prediction of the Thermal Conductivities of Powders," *Solid Freeform Fabrication Symposium Proceedings*, The University of Texas:Austin, Texas, 1991, pp. 62-69.

Xue, S. S., "The Thermal Properties of Powders," *M. S. Thesis*, The University of Texas at Austin, 1991.

Yagi, S. and Kunii, D., "Studies on the Effective Thermal Conductivities in Packed Beds," *J AIChE*, 1957, 3[3], pp. 373-381.

Zehner, P. and Schlunder, "Warmeleitfahigkeit von Schuttungen bei maßigen Temperaturen," *Chemie-Ing.-Tech.*, 1970, 42[14], pp. 933-941.

7

DIRECT SLS FABRICATION OF METALS AND CERAMICS

In the following, direct fabrication of high-density metal and ceramic structures is discussed. This discussion is primarily centered on the SLS (Selective Laser Sintering) process, but a brief discussion of other direct techniques is given at the end of this chapter. All of these techniques (including SLS) are still in the research stage.

7.1 SLS BINDING MECHANISMS

In Selective Laser Sintering, binding of adjacent powder particles may take place by viscous flow, particle wetting/infiltration and/or curvature effects. The operation of these mechanisms depends on the material system. In SLS, localized heating of a small volume of powder is laser induced, shown schematically in Figure 7.1. The duration of the laser beam at any powder particle is short, typically between 0.5 and 25 ms. Therefore, the thermally-induced binding reactions must be kinetically rapid. Three such mechanisms are viscous flow when the powder has the appropriate temperature-dependent viscosity (e.g., polymers above the softening point), melting with particle wetting and infiltration, and, in cases of ultrafine or nanocrystalline particles, a curvature effect.

As discussed in Chapter 6, viscous sintering has been described by Frenkel (1945) using a two-sphere model. For early stages of particle binding, the neck of radius x joining two spheres of radius R grows according to

$$\left(\frac{x}{R}\right)^2 = \frac{3}{2}\left(\frac{\sigma t}{R\eta}\right) \tag{7.1.1}$$

where σ is particle surface tension, η is thermally activated particle viscosity, and R is particle radius. Similar relations have been developed for spheres considering various atomic rearrangement mechanisms to accommodate densification (Swinkel and Ashby, 1981). While the exact form varies considerably depending on the unit cell geometry being considered, the neck growth rate is always directly proportional to σ and is inversely related to η. Experimentally, the densification rate in early stage sintering of cylinders and spheres often reveals an inverse viscosity dependence (Scherer, 1977, 1986).

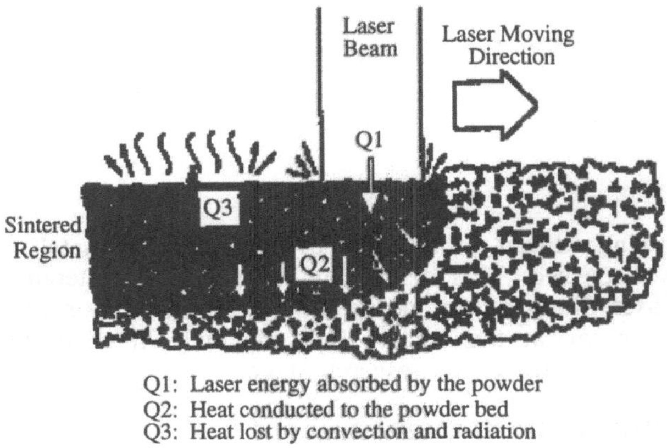

Figure 7.1 Schematic of the laser beam/powder bed interaction zone (Sun, *et al*, 1991).

Sun, *et al.* (1991), have derived a sintering rate equation for SLS based on Frenkel's assumptions and assuming a "shrinking cubic" geometry shown in Figure 7.2. Their sintering rate is given as

$$\dot{x} = -\frac{3(1-\rho)\pi\sigma R^2}{24\eta\rho^3 x^3}\left[R-(1-\xi)x+\left(x-\left(\xi+\frac{1}{3}\right)R\right)\frac{9(x^2-R^2)}{18Rx-12R^2}\right] \quad (7.1.2)$$

where $2x$ is the cube side dimension defined from Figure 7.2, R is sphere radius, and ρ is apparent density. The variable ξ, ranging between 0 and 1, represents the probability of forming a neck between two arbitrary particles. From a statistical thermodynamics perspective, it is related to the energy status of the contact region. The reciprocal viscosity dependence of sintering rate is still maintained.

Materials with strongly temperature-dependent viscosities (or high activation energy for viscous flow) have been readily processed using SLS. This group includes a number of thermoplastic polymers such as polycarbonate, poly(vinyl chloride) and rubber modified poly(styrene-co-acrylonitrile).

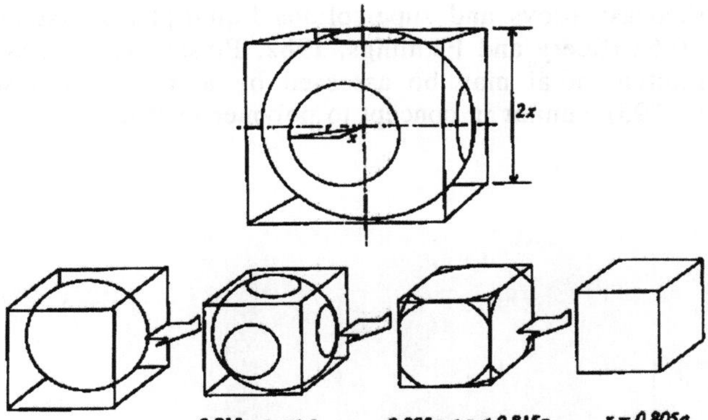

Figure 7.2 Unit cell and process of densification.

Early attempts to SLS process single-phase metals such as lead, zinc or tin were unsuccessful (Haase, 1989, Manriquez-Frayre and Bourell, 1990). For this reason, a unique melting/solidification approach has been developed (Bourell, *et al*, 1990). When melting was induced by the laser beam, the molten powder quickly consolidated into a sphere of diameter approximately equal to the laser beam diameter rather than consolidating into the previous layer. Figure 7.3 shows this "balling" phenomenon for single layer SLS processing of tin at various laser beam scan speeds (top) and powers (bottom). The laser scan direction is vertical. At lowest energy densities the powder did not sinter and remained loose. Once melting occurred, spheroidization resulted independent of scanning rate and power density. Sphere diameter increased with decreasing scan speed and increasing power.

In an attempt to mollify the tendency to form spheres, a two-phase powder approach was developed. Similar to liquid-phase sintering (German, 1985), the powder bed consists of at least two different powders of significantly disparate melting points. The laser beam heats the powder bed locally, inducing melting only of the low-melting-point species. The liquid wets and binds the high-melting-point solid. From the supersolidus liquid-phase sintering and rheocasting literature, the viscosity of a solid-liquid mixture η_{sl} is related phenomenologically to the viscosity of the liquid phase in the absence of any solid η_o and the relative solid content Φ (German, 1990):

$$\eta_{sl} = \eta_o \left[1 - \left(\frac{\Phi}{\Phi_m}\right)\right]^{-2} \qquad (7.1.3)$$

where Φ_m is a critical volume fraction above which the mixture has infinite viscosity (Metzner, 1985, Wildemuth and Williams, 1984). Φ_m has been measured in rheocast alloys and supersolidus liquid-phase systems to be approximately 0.62 (Suery and Flemings, 1982, Pinsky, *et al*, 1984). The viscosity of molten metal may be assessed by a melt-flow viscometer (Bunnell, *et al*, 1995), similar in concept to polymer melt indexers.

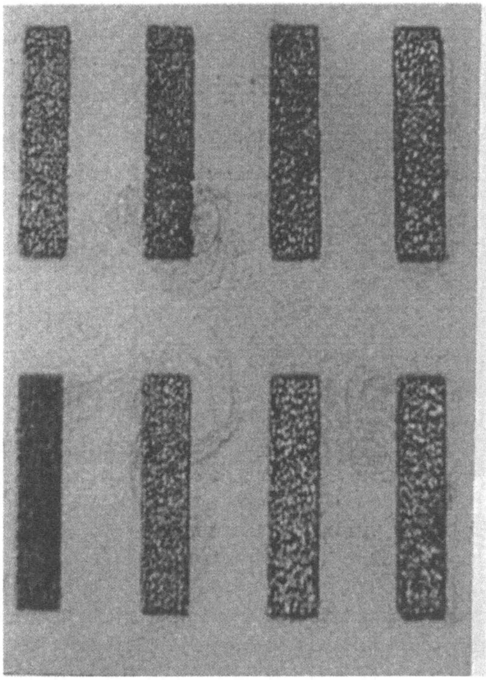

Figure 7.3 Single layers of tin powder processed using Selective Laser Sintering. Top row shows samples processed with different scanning speeds (6.8, 4.5, 3.4, 2.7 cm/s, left to right) at 24W laser power. Bottom row shows samples with different laser powers (8, 18, 22, 26W) at 3.4 cm/s scanning speed. Rectangle dimensions are 6 mm x 25 mm.

In this two-phase powder approach, particle bonding is driven by the liquid viscosity η_o, assuming the materials are suitably chosen for interparticle wetting to occur. However, the tendency for balling is controlled by η_{sl}, which is increased substantially by the retention of solid particles in the melt assemblage. Therefore, by adjustment of the volume fraction of the high-melting-point phase, it becomes possible to induce binding without balling.

This two-phase powder approach has been used successfully in SLS processing of metals and ceramics. Examples of metallic and ceramic systems include Cu-Sn, Cu-(70 Pb-30 Sn) solder, bronze-Ni, Ni-Sn, alumina-ammonium dihydrogen phosphate, SiC-Si, WC-Co and alumina-boron oxide. A variation of this approach has been demonstrated in prealloyed 89Cu-11Sn bronze (Prabhu and Bourell, 1993) by SLS processing in the supersolidus regime. The metal is only partially molten in a two-phase zone between the solidus and liquidus temperatures. Solid and liquid metal are in thermodynamic equilibrium, and this serves the same function as the binary powder approach.

A critical aspect of the two-phase, solid-liquid approach to binding in SLS is wetting of the solid phase by the liquid. Wetting of metals by liquid metals reveals a time dependence to wetting often on the order of minutes and hours. Transient wetting consideration is necessary in SLS due to the short time duration of the thermal exposure. A wetting balance may be used to assess the transient wetting behavior of superheated liquids on a solid substrate (Bunnell, *et al*, 1994). A pendant drop test has been used to assess surface tension of molten metal at the melting point (Bunnell, *et al*, 1995), and a sessile drop test provides information on substrate wetting at or near the melting point. In SLS, wetting is enhanced by atmosphere control to prevent oxidation. Depending on the system, nitrogen, argon, a nitrogen-hydrogen mixture and hydrogen have been used as protective/reducing atmospheres. Additionally, various types of fluxes have been used to enhance the wettability of SLS material systems. Included are metal chlorides and phosphates used in brazing or soldering. Based on the work of Bunnell, binary metal systems for optimum density SLS processing should have the following characteristics: for the solid phase, fine particle size, high surface energy, good laser coupling; for the liquid phase, high surface energy, low viscosity, low solubility of the solid phase, no volatile components. The melting points of the solid and liquid phase materials should be large so substantial superheat can be achieved in the molten metal. The volume fraction of liquid must be greater than about 30 percent for complete densification by infiltration wetting of interparticle regions of the solid phase. Liquid volume fractions exceeding 60 to 70 percent preclude either balling up due to the inadequacy of the solid framework to hold the liquid in the plane or cracking due to the extensive shrinkage. Atmosphere control is important in minimizing oxygen pick up in the liquid and oxide formation on the liquid and solid surfaces. Oxygen pick up lowers the liquid surface energy. The metal oxides have much lower surface energy than the associated metal.

Curvature effects also provide a driving force for binding and densification. Ashby (1990) has derived a relation for the driving stress P_s arising from curvature effects at the onset of sintering as

$$P_s = \frac{3\sigma}{R}\frac{\Delta^3}{1-\Delta} \tag{7.1.4}$$

where R is particle radius, σ is particle surface energy and Δ is the relative or fractional density (ρ/ρ_{th}). Taking an approximate value of 1 J/m^2 for surface energy, a relative density at the onset of sintering equal to 0.4 and a particle size 2R equal to 20 nm, the curvature pressure is 32 MPa. This significant driving stress arises solely from curvature effects and is available to drive sintering in nanocrystalline materials. Densification operates rapidly in nanocrystalline materials relative to their coarse-grained counterparts, and it appears theoretically possible to sinter a large variety of nanocrystalline materials to near full density at reasonable temperatures and times of the duration associated with SLS.

7.2 SLS PROCESSING OF BINARY METALLIC POWDER BLENDS

Selective Laser Sintering is the one of the few SFF technologies that allows direct production of metallic parts without the necessity of additional post processing. Binary metal powders investigated to date have been the low-melting-point systems Cu-(70 Pb-30 Sn) solder and Cu-Sn. A Q-switched Nd:YAG laser was used: λ = 1.06 µm, nominal maximum power output = 50 W, beam size = 0.5 mm. Starting powders included 99.0 % Cu (-100 mesh), 99.5 % Sn (-100 mesh) and a 70 Pb-30 Sn solder mixture (-100 mesh). Maximum particle size was 150 µm. Atmosphere control or flux additions were necessary to prevent surface oxidation and facilitate wetting. When a flux was applied, crystals of zinc chloride ($ZnCl_2$) were dissolved in alcohol and metallic powder was added. The solution was evaporated under vacuum at temperatures between 60°C and 90°C, and the remaining powder was sieved to produce a fluxed coated powder blend. Powder was SLS processed in air or using a reducing atmosphere of 96 N_2-4 H_2. Laser output power ranged between 8 W and 26 W with laser beam surface scan speeds between 3.4 cm/s and 6.8 cm/s.

The effect of fluxing is demonstrated in Figure 7.4 for the 50 Cu-50 solder system SLS processed at 25 W, 3.4 cm/s and using the reducing atmosphere. The final scanned surface is shown in unfluxed (Figure 7.4a) and fluxed (Figure 7.4b) conditions. The solder melted and flowed in both conditions, but the wettability was clearly improved for the fluxed material. This fluxed 50 Cu-50 solder system was used to prepare the first directly generated SLS metal part, a 72 layer gear. The conditions were identical to those described previously, excepting that the powder bed was heated to a temperature of

90°C before SLS processing.

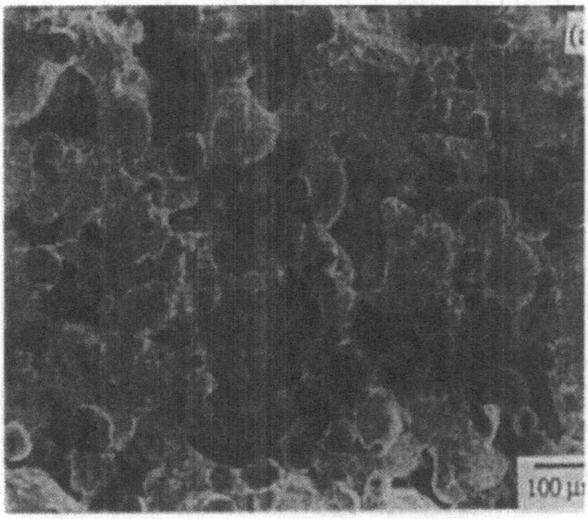

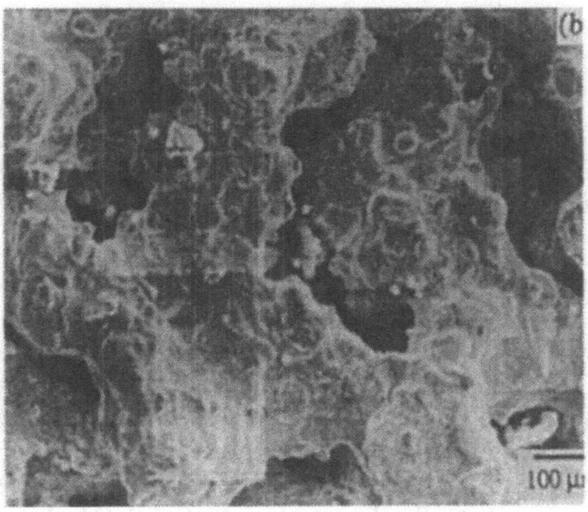

Figure 7.4 SEM of surface sintering characteristics (a) unfluxed and (b) $ZnCl_2$ fluxed 50 Cu-50 (70Pb-30Sn) solder powder after processing. Shown are the surfaces produced after four and six layers, respectively.

Density is an important factor in SLS. Figure 7.5 shows the variation of density with Φ, the volume fraction of high-melting-point phase, for fluxed Cu-(70 Pb-30 Sn) solder and Cu-Sn. Also shown is the relative viscosity η_{sl}/η_o from equation 7.1.3, with Φ_m assumed to be 0.62. Density of the SLS processed part is improved by decreasing the viscosity of the molten mass heated by the laser beam. For most powder systems, Φ is limited to between approximately 0.25 and 0.6. At low solid contents below 0.25, the "balling" phenomenon occurs (cf., Figure 7.3). At large solid contents, insufficient bonding takes place due to the inadequate supply of low-melting-point phase, and the part does not hold together.

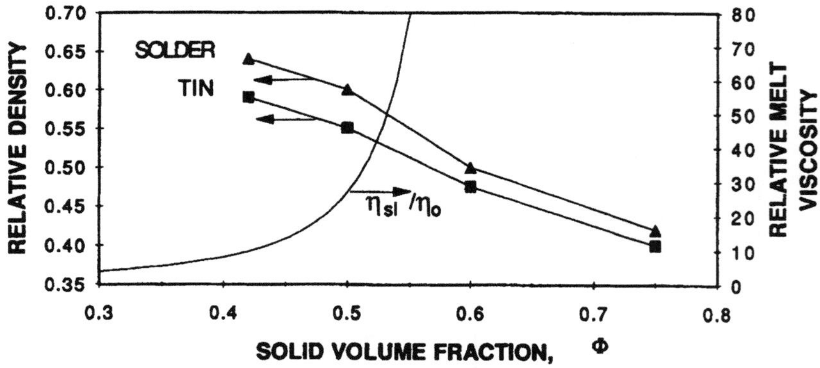

Figure 7.5 The influence of volume fraction of high-melting-point phase Φ on the relative density of selective laser sintered parts. Curves are shown for Cu-(70Pb-30Sn) solder and Cu-Sn. Also shown is the relative viscosity η_{sl}/η_o of the molten mass as calculated using equation 7.1.3 with T_m = 0.62.

Prealloyed 90Cu-10Sn bronze mixed with Ni powder has been SLS processed successfully (Zong, *et al*, 1991, 1992, Agarwala, *et al*, 1993, 1994). Figure 7.6a shows a typical microstructure of the materials after SLS processing. This system consists of bronze, nickel and copper phosphate. The nickel is the high temperature component, bronze forms the liquid phase during SLS processing, and copper phosphate serves as a flux. This mixture is a match to the Electrolux material (Nyrhila, 1991) which is sometimes used in SLS systems (Sindel, *et al*, 1994). It is possible by post-SLS conventional sintering to further densify and homogenize the material, Figure 7.6b, since copper, tin and nickel form a solid solution. Some residual bronze is visible, but, based on microstructural, Xray and electron diffraction, the majority phase is the ternary solid solution.

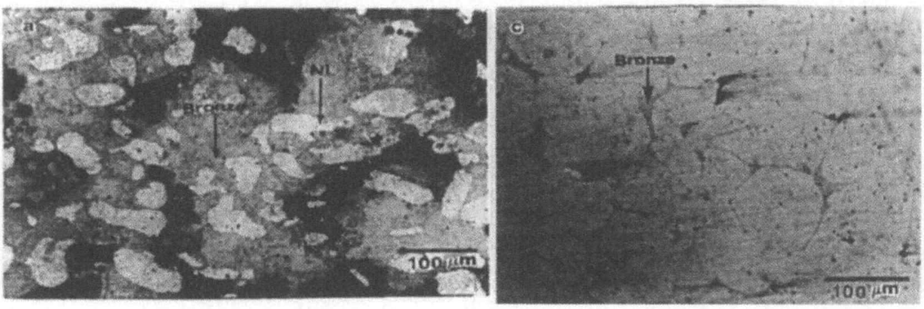

(a) (b)

Figure 7.6. Optical micrographs of (a) SLS bronze-nickel part (56W, 0.3 in/sec) showing bronze and nickel as separate phases and high porosity, and (b) post-SLS liquid phase sintered bronze-nickel part (1030°C, 1 hr) showing nearly complete homogenization and low porosity.

Supersolidus liquid phase SLS of prealloyed, -150 mesh, 89Cu-11Sn bronze has been performed. This material begins to melt at approximately 850°C, but melting is not complete until a temperature in excess of 1000°C is reached. Within this temperature range the material coexists as solid and tin-rich liquid. For SLS, a CO_2 laser was used, of beam size 4.5 mm and 100 W nominal power. The powder bed was heated to 300-500°C in a nitrogen protective atmosphere. Figure 7.7 shows the surface/top view and side/build view.

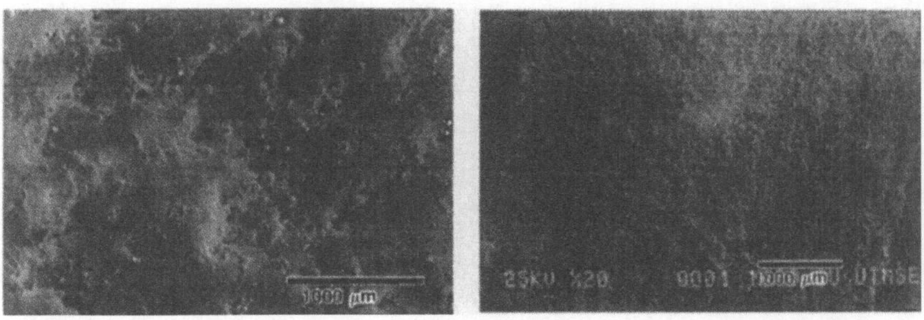

(a) (b)

Figure 7.7 Selective Laser Sintering of prealloyed bronze (89Cu-11Sn). Bed temperature was 350°C, 100 W laser power, 2.5 cm/sec scan speed. (a) Top surface, (b) End view. Layers are barely distinguishable.

7.3 SLS PROCESSING OF CERMETS

Selective Laser Sintering can directly process cermets when a two-phase metal-ceramic powder system is used. This approach has been used for production of cermet composite turbine blade sealing components (Fuesting, *et al*, 1996a, 1996b). Primary considerations for SLS processing were reuse of high-cost raw materials, reduced labor compared to the standard tape casting techniques, repeatability in production, binderless processing, reduction or elimination of a low-melting point metallic phase, and the ability to produce a complex shape directly for production runs of several hundred parts.

The materials system consisted of a titanium-coated proprietary ceramic grit, a nickel-based superalloy and a cobalt-based brazing alloy. Early evaluation using indirect processing of powder with an added binder were successful. Blade tip geometries were also created using direct SLS processing, Figure 7.8. Porosity was minimized at energy densities of approximately 2000-3000 J/cm^2, Figure 7.9. Test samples produced by direct SLS processing were successfully evaluated for stress rupture to 1150°C and wear resistance according to guidelines set for turbine blade sealing components.

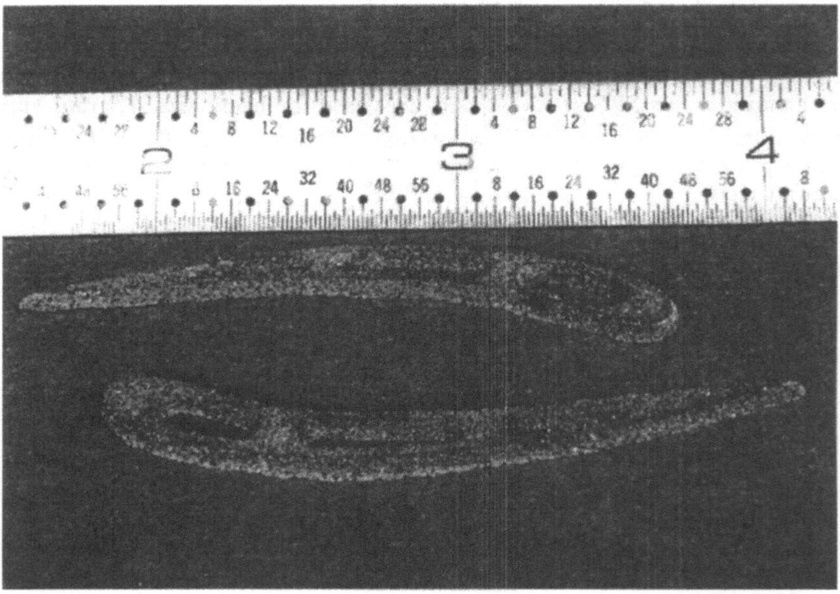

Figure 7.8 Cermet turbine engine abrasive blade tip fabricated by SLS process (Fuesting, *et al*, 1996).

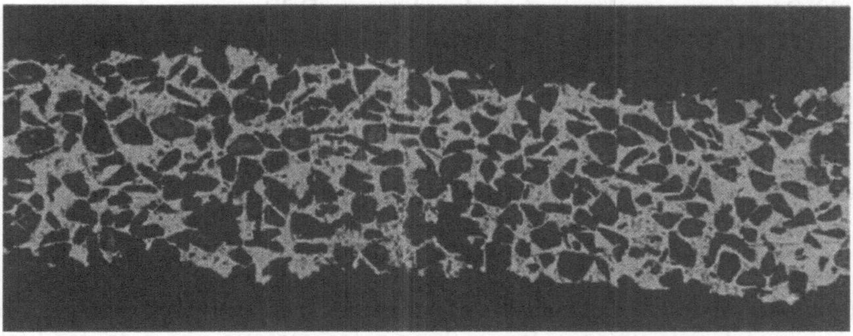

Figure 7.9 Micrograph of cermet blade tip show above (Fuesting, *et al*, 1996).

Prealloyed tungsten carbide-cobalt powder has been SLS processed with Co and Ni in the weight fraction 70(88WC-12Co)-20Co-10Ni (Zong, *et al*, 1992a, 1992b). This was one of the first attempts to use SFF techniques to directly generate a part from a conventional tool material. The material was SLS processed in a high-temperature workstation (1.1 kW CO2 laser, nitrogen atmosphere). The surface texture of the SLS processed part was rough, but extensive regions were almost fully dense, Figure 7.10.

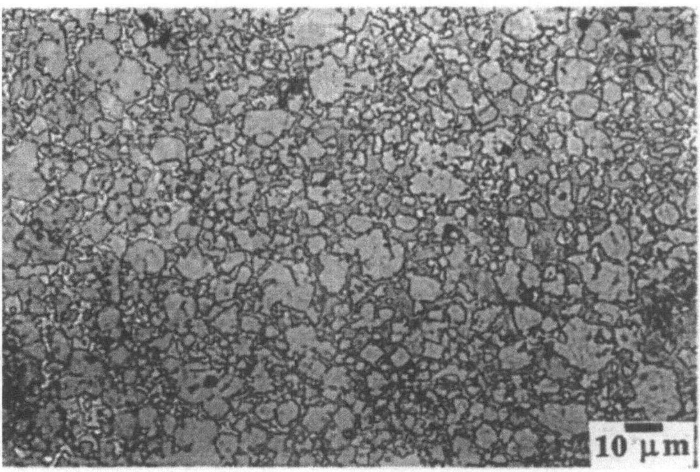

Figure 7.10 Optical micrograph of selective laser sintered 70 (88WC-12Co)-20 Co-10 Ni.

SLS of a nanocrystalline 50 alumina-50 cobalt system was also studied (Manthiram, *et al*, 1993a, 1993b). The powder mixture was synthesized using sol-gel processing to produce aluminum and cobalt oxides 50 nm in diameter. The mixture was partially reduced by annealing in hydrogen at 1000°C to form an alumina/cobalt mixture. Following agglomeration in nitrogen at 1350°C, single layer SLS scans of the powder produced a promising, smooth layer, Figure 7.11. Another nanocrystalline system studied was 20 nm to 50 nm tungsten carbide particulate SLS processed after mixing with 15 percent cobalt (Cottle, *et al*, 1995). The wear rate of the SLS processed material was approximately 5 times lower than that for traditional WC-15Co under identical test conditions.

7.4 DIRECT SLS PROCESSING OF CERAMIC POWDER BLENDS

The binary material approach described earlier has been applied to SLS of ceramic materials using a low-melting-point ceramic phase to facilitate bonding under thermal activation of the scanning laser beam (Marcus, *et al*, 1990, Lakshminarayan and Marcus, 1991, Lakshminarayan, *et al*, 1990). The components were low-purity alumina ($95Al_2O_3$, $3TiO_2$, $0.75SiO_2$, $0.46Fe_2O_3$, $0.25MgO$; $T_m > 2000°C$) and ammonium dihydrogen phosphate ($NH_4H_2PO_4$, $T_m = 190°C$) . A number of reactions occur on heating (Baumgart, *et al*, 1984), but the binding reaction involves formation of a glassy phase by reaction with the alumina at or below 200°C according to:

$$2\ NH_4H_2PO_4 \Rightarrow P_2O_5 + 3\ H_2O(g) + 2\ NH_3(g)\ .$$

Higher temperature reactions result in the formation of $Al(PO_3)_3$ by the following

$$Al_2O_3 + 3\ P_2O_5 + 6\ H_2O \Rightarrow 2\ Al(H_2PO_4)_3$$

$$Al(H_2PO_4)_3 \Rightarrow Al(PO_3)_3 + 3\ H_2O(g)\ .$$

Figure 7.11 SEM of 50 mol% Al_2O_3 - 50 mol% Co nanocomposite after firing at 1350°C in hydrogen for 8 hr followed by selective laser sintering.

Powder blends of alumina-ammonium dihydrogen phosphate were SLS processed under the conditions: pulsed Nd:YAG laser, 40 kHz frequency, 20 W to 25 W incident power, 2 cm/s to 6 cm/s scanning speed, 0.5 mm beam diameter, 21 to 60°C powder bed temperature. The ammonium dihydrogen phosphate phase melted and wet the alumina, but, based on X-ray diffraction results, formation of

$Al(H_2PO_4)_3$ was not extensive.

The SLS processed alumina-ammonium dihydrogen phosphate samples were fired in air after processing, based on the high-temperature transformation:

$$Al_2O_3 + 2\ NH_4H_2PO_4 \Rightarrow 2\ AlPO_4 + 2\ NH_3(g) + 3\ H_2O(g)\ .$$

Specimens were heated at 100°C/hr to 450°C and held 2 hr. They were then heated at the same rate to 850°C and held 6 hr before cooling at 30°C/hr. Shrinkage was minimal due to evolution of the gaseous phases, and the structure was porous.

Spherical alumina (15 μm) and aluminum (20 μm) have been mixed in weight fractions between 15 and 35 percent Al, and SLS processed (Subramanian, *et al*, 1992). The aluminum melted as the binding phase during SLS. The microstructure consisted primarily of a dense network of 200 μm sized regions of alumina in an aluminum matrix. Part apparent density was on the order of 50 percent of theoretical. Similar results were obtained by substitution of silicon carbide powder of comparable shape and size for the alumina.

Supersolidus liquid phase sintering of the ceramic, high-temperature superconductor Y2BaCu3O7-x has been performed (Agarwala, *et al*, 1993). This ceramic melts incongruently and therefore may be directly SLS processed without the need of a second constituent phase. Porous preforms were produced by SLS. The preforms were largely noncrystalline, but an 8-10 hr restoration anneal in oxygen at 930°C recovered the crystalline superconducting phase. After infiltrating with molten silver in a post-processing step, SQUID magnetometer measurements confirmed an onset critical temperature between 86 and 93°K depending on the volume fraction of silver.

7.5 REACTIVE SINTERING OF BINARY POWDER

The two-powder approach was initially developed as a binding mechanism for SLS, Section 7.1. Under certain conditions, chemical reactions may take place between the two phases during or after SLS processing to produce a new compound. This is termed *reactive sintering*. Reactive sintering has been demonstrated for production of intermetallic compounds and spinels. Section 8.4 treats a variation of reactive sintering, Selective Laser Reactive Sintering (SLRS). In SLRS, the sintering atmosphere reacts with the powder bed during laser scanning to produce a new phase

The reactive sintering approach affords the possibility to produce net-shape components of intermetallics if the elemental metals comprising the intermetallic have appropriately disparate melting points and good wetting characteristics (Bourell, *et al*, 1992, Weiss, 1991, Weiss and Bourell, 1993). This concept was tested by SLS processing a 5 % ZnCl2 fluxed powder mixture of 59 Ni-41 Sn. Both metal powders were roughly spherical and commercially obtained. The nickel was 99.9 % pure and -100 mesh; the tin was 99.5 % pure and -325 mesh. The selected weight fraction corresponds to the stoichiometric intermetallic compound Ni3Sn which is stable under equilibrium conditions to over 950°C (Massalski, 1990). Cylindrical specimens were prepared by SLS processing using the parameters: Nd:YAG laser, 17 W to 25 W power, 3 cm/s to 6 cm/s scan speed, 150°C powder bed temperature, 0.5 mm beam diameter, nitrogen atmosphere. The tin melted and wet the nickel particles, and residual porosity is evident. Relative density was measured to range between 60 percent and 70 percent depending on the process parameters used. Based on Xray diffraction and energy dispersive spectroscopy, the SLS processed materials consisted of only elemental nickel and tin (no intermetallic phases) with residual zinc and chlorine from the flux.

SLS processed specimens were post-process annealed in hydrogen. The structure was less dense than before annealing, supported by macroscopic swelling. Conversion to the intermetallic was incomplete, the phases detected by Xray diffraction being Ni3Sn and Ni3Sn2 with less than 5 percent Sn present. No zinc or chlorine was detected. Density was measured to be 42 percent of theoretical, and the fired parts tended to be friable.

This initial investigation of SLS processing of binary metals followed by subsequent conversion to an intermetallic compound has shown that the concept is correct and possible. Melting of the low-melting-point phase to bind the high-melting-point phase was accomplished without reactive transformation under the laser beam, and the transformation anneal converted the metallic phases into intermetallics without loss of part shape.

Nanocrystalline composites produced by sol-gel synthesis routes have been sintered and annealed to produce a homogeneous phase reactively]. Table 7.1 lists numerous potential nanocomposite systems which appear amenable to both sol-gel processing (with or without subsequent reduction of one or more constituents) and Selective Laser Reactive Sintering (or SLRS). Presently, two systems have been pursued experimentally, $75Al_2O_3$-$25Co_3O_4$ (mole percent) and $50Al_2O_3$-50NiO (mole percent) (Manthiram, *et al*, 1993a). The crystallized powder was less than 100 nm in diameter after extended anneals as high as 700°C although agglomeration into approximately 100 µm superparticles occurred. Upon complete reaction, stoichiometric spinels $CoAl_2O_4$ and $NiAl_2O_4$ are the respective anticipated final phases. Complete conversion to the spinel occurred at about 1000°C, Figure 7.12. SLRS of these nanocomposites was limited to single layer testing. An advantage of the Al_2O_3-Co_3O_4 system is a color change from black to bright blue as the material reacts to form the spinel. This conversion occurred during 200°C SLRS of ball milled nanocomposite. Unlike the binary intermetallic mentioned above, reactive sintering actually takes place as the material was SLS processed.

Table 7.1 Examples of Nanocomposite Systems Designed for SLS or SLRS

System	High *T* Compo-nent	T_{melt} (°C)	Low *T* Compo-nent	T_{melt} (°C)	Possible Product	Process
Ceramic-Ceramic						
Al_2O_3-FeO_x	Al_2O_3	2,070	FeO	1,370	$Fe_xAl_{3-x}O_4$	SLRS
Al_2O_3-Co_3O_4	Al_2O_3	2,070	CoO	1,795	$Co_xAl_{3-x}O_4$	SLRS
Al_2O_3-NiO	Al_2O_3	2,070	NiO	1,985	$Ni_xAl_{3-x}O_4$	SLRS
Al_2O_3-CuO	Al_2O_3	2,070	CuO	1,325	$Cu_xAl_{3-x}O_4$	SLRS
Al_2O_3-MgO	MgO	2,850	Al_2O_3	2,050	Mg Al_2O_3	SLRS
Al_2O_3-ZrO_2	ZrO_2	2,700	Al_2O_3	2,070	ZrO_2-Al_2O_3	SLS
Y_2O_3-ZrO_2	ZrO_2	2,700	Y_2O_3	2,410	$Zr_{1-x}Y_xO_{2-0.5x}$	SLRS
Al_2O_3-SiO_2	Al_2O_3	2,070	SiO_2	1,710	$3Al_2O_3 \cdot 2SiO_2$	SLRS
Ceramic-Metal						
Al_2O_3-Fe	Al_2O_3	2,070	Fe	1,535	Al_2O_3-Fe	SLS
Al_2O_3-Co	Al_2O_3	2,070	Co	1,495	Al_2O_3-Co	SLS
Al_2O_3-Ni	Al_2O_3	2,070	Ni	1,455	Al_2O_3-Ni	SLS
Al_2O_3-Cu	Al_2O_3	2,070	Cu	1,085	Al_2O_3-Cu	SLS
WC-Co	W C	2,870	Co	1,495	WC-Co	SLS
YBCO*-Ag	YBCO	1,000	Ag	960	YBCO-Ag	SLS
Metal-Metal						
Cu-Ni	Ni	1,455	Cu	1,085	$Cu_{1-x}Ni_x$	SLRS
Fe-Co	Fe	1,535	Co	1,495	Fe-Co	SLS
W-Mo	W	3,410	Mo	2,610	$W_{1-x}Mo_x$	SLRS

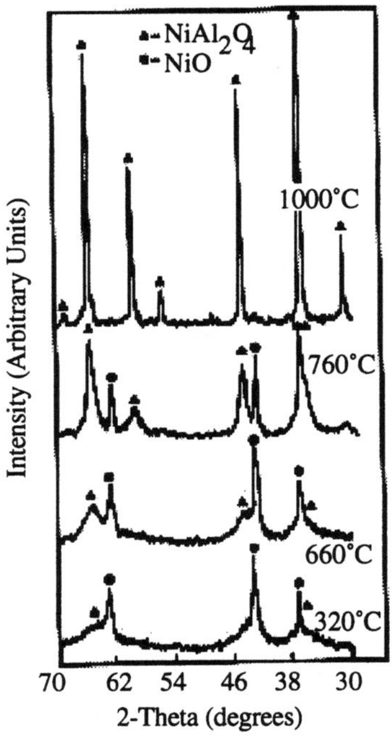

Figure 7.12 Xray diffraction pattern of 50 mol% Al_2O_3 - 50 mol% NiO (Al:Ni ratio = 2:1) nanocomposite powder after firing at various temperatures for approximately 20 hr.

7.6 DENSITY OF SLS PROCESSED PARTS

An important feature of SLS processed materials is the part apparent density. In powder processing, density strongly influences mechanical and physical properties (German, 1984). For this reason, density and its control are critical to successful SLS, particularly if direct production of structural parts is to be realized.

In broad terms, the apparent density of SLS processed ceramic materials falls between 40 and 60 percent; for metals, density can approach 100%. Density of SLS processed nanocrystalline composites falls in the 30 to 50 percent range. Figure 7.13 shows the effect of laser scan speed and power on part relative density for a bronze-nickel mixture (Agarwala, *et al*, 1993). Density increases with increase in thermal exposure, accomplished either by increasing laser power or the duration of the beam at any point on the surface (inverse of scan speed). Numerous methods have been employed to increase part density without sacrifice of the freeform advantage. Increasing the powder bed density before SLS is one approach. Based on SLS of polymer coated soda-lime glass (Vail and Barlow, 1991), the density of the powder bed is not much greater than the untapped powder density. One method of improving powder bed density has been to increase the intrinsic powder tap density by conventional means. These include use of spherical particulate, bimodal distributions of powder size, agglomeration of fine powder and optimizing the volume fraction of combustible binder. Figure 7.14 illustrates the role of binder volume fraction and particle size on the density of SLS processed alumina with ammonium dihydrogen phosphate binder. Similar to conventional powder processing, increasing particle size increases density. As processed density also increases with binder volume since more material is available to flow under thermal activation of the laser. This is a liability to post-processed density since the binder is combusted which increases porosity.

The second approach to increased powder bed density is to increase the effectiveness of the powder distribution process in SLS in increasing powder bed density. Examples are the use of magnetic fields (Melvin and Beaman, 1992) and electrostatics (Melvin and Beaman, 1991) as well as mechanical agitation of the powder and/or powder bed (Bunnel, 1995, Rao, *et al*, 1995)].

In numerous applications, porosity in a structural part is important to its proper function. An important application for porous ceramic parts is ceramic molds for metal casting. SLS processing of ceramics easily allows direct production of single-piece or multiple-piece ceramic molds and has been called "Lost-Lost-Wax" processing (Lakshminarayan, *et al*, 1990). The limited shrinkage of the ceramic part during firing assists in the initial computer-assisted mold design.

Once the part is SLS processed, post-processing methods are available to improve density so long as part shape is not adversely affected. Conventional sintering and hot isostatic pressing are two common methods for increasing part density. With respect to metal and ceramic powder coated with polymer binders, an initial burnout anneal similar to metal and ceramic injection molded parts is needed prior to sintering. In the case of material systems amenable to SLRS, post-process annealing may also result in further conversion of precursor phases into a new resultant phase or phases. Examples are production of homogeneous alumina from SLS-processed alumina-aluminum (Subramanian, *et al*, 1992), formation of intermetallic compounds from SLS-processed nickel/tin, and formation of spinels from binary oxide precursors.

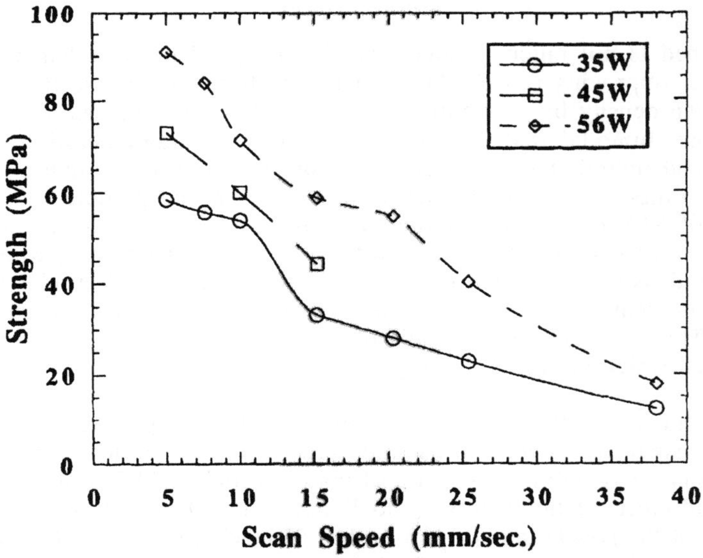

Figure 7.13 Fractional density of selective laser sintered bronze-nickel parts as a function of scan speed and laser power.

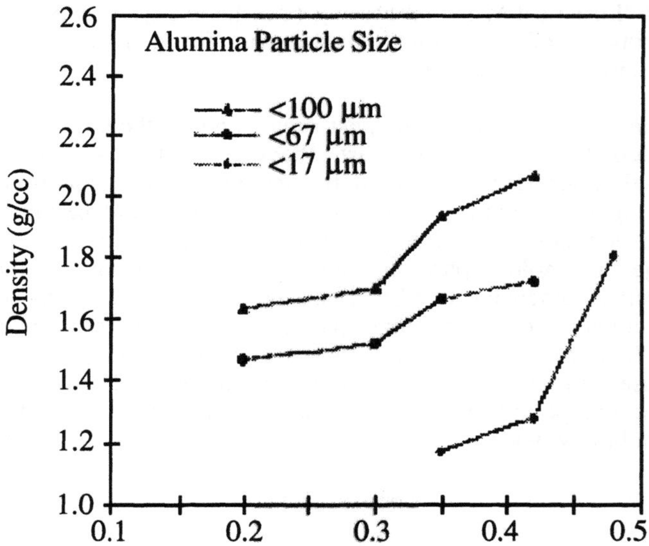

Figure 7.14 Variation of density with alumina powder size and ammonium dihydrogen phosphate binder volume fraction before firing

Electrolux Material

A bronze-nickel-copper phosphate system has been developed by Electrolux (Nyrhila, 1991) for shrinkless mold applications. Direct SLS processing of this material was previously described in Section 7.2. Post processing above the solidus temperature (840°C) resulted in increased or decreased density based on time at temperature (Agarwala, 1994, Agarwala *et al*, 1995). Liquid-phase sintering (LPS) in the bronze-nickel system occurs in three stages which follow.

Stage I. At temperatures above the solidus temperature, bronze melts partially, and above the liquidus (1,020°C) it is completely melted. In this stage of LPS, the liquid bronze flows into the pores and a rearrangement of the solid nickel particles takes place. This stage leads to rapid shrinkage and an increase in density and strength. Most of the densification in LPS is achieved during this stage. If sufficient amounts of liquid are formed with low viscosity to free flow, the final density at the end of Stage I is independent of the starting density, which is determined by laser power and scan speed. The amount of liquid phase formed and the viscosity of the liquid formed depend on the sintering temperature. Increasing the sintering temperature causes a larger amount of liquid phase to form with lower viscosity. A very high sintering temperature can lead to a large amount of liquid phase with very low viscosity, which can result in loss of shape. Solid phase particle size and the rate of heating also affect densification at this stage.

Stage II. In this stage of LPS, normal densification is accompanied by chemical homogenization between bronze and nickel. Bronze, predominantly copper (90 per cent weight), and nickel homogenize at high temperatures, either by interdiffusion if the homogenization is between solid bronze and solid nickel, or by solution reprecipitation if the homogenization is between liquid bronze and solid nickel, to form a homogeneous solid solution. In either case, the bronze phase disappears completely, leaving behind pores in its place and an expanded solid solution of Cu in Ni. This occurs as a result of the Kirkendall effect, a mass shift due to the more rapid diffusion of Cu into Ni than of Ni into Cu. The pores created at this stage are referred to as Kirkendall porosity. In Cu-Ni sintered compacts, the greatest shrinkage is observed when sintering only one of the pure components: copper or nickel. With increasing nickel content in copper, conversely, a decrease in shrinkage is observed and even growth of sintered parts may occur under certain sintering conditions. Similar growth or swelling as opposed to shrinkage has been observed in other metallic systems such as Cu-Zn, and age and in several ceramic systems also, such as MgO-Al_2O_3, MgO-Fe_2O_3, MgO-SiO_2 and MgO-TiO_2. In the ceramic systems, swelling occurs because of unequal diffusion rates of the cations at the phase boundary. In Stage II, normal densification is dominated by the growth or swelling of the part as a result of Kirkendall porosity.

It is this phenomenon that accounts for the "shrinkless" behavior of the material system. The reader should note that this is at the expense of decreased density (and therefore strength). More importantly, this effect occurs over a time controlled by Cu diffusion (i.e., hours, see Figure 7.15) and therefore does not occur during laser processing (i.e., time scale of seconds).

Detailed investigations of LPS have shown that similar material transport mechanisms prevail during Stage I as well as Stage II. Therefore a sharp distinction between these stages is not justified. Whether Stage I or Stage II predominates the process depends on the particular circumstances. Powder particle size, starting density, sintering temperature, time and rate of heating are most influential in determining this. Figure 7.6 shows the microstructural changes associated with two different stages of LPS as found during SLS processing. Figure 7.6a depicts the microstructure of the material after SLS and before LPS. Nickel is present as a light phase surrounded by bronze. Porosity is evident by dark contrast. LPS at a high temperature (1,030°C) resulted in chemical homogenization with some untransformed bronze present (Figure 7.6b). However, at 950°C (Figure 7.15), Stage II homogenization occurred, with formation of fine Kirkendall porosity.

Figure 7.15 Optical micrograph of post-processed bronze-nickel sample showing Kirkendall porosity (K).

Stage III. Once homogenization has occurred, only normal shrinkage due to sintering occurs, but now the sintering occurs in the solid state as the temperature is below the solidus of the newly formed solid solution of Cu-Ni (T_{melt} ~1100°C). Therefore, even for prolonged periods of time, shrinkage or densification during this stage is small, leaving an overall expansion or shrinkage in the samples, depending on the extent of Stage I densification or Stage II expansion.

Figures 7.16 shows that at post-processing sintering temperatures below 1,000°C, the density of SLS parts decreases irrespective of starting density. The density reaches a minimum and then increases slowly to nearly the starting density. At such low temperatures, the amount of liquid phase formed is very small with high viscosity, which prevents any significant densification normally associated with Stage I LPS. Instead, Stage II dominates the LPS initially and then Stage III continues until the end of the process.

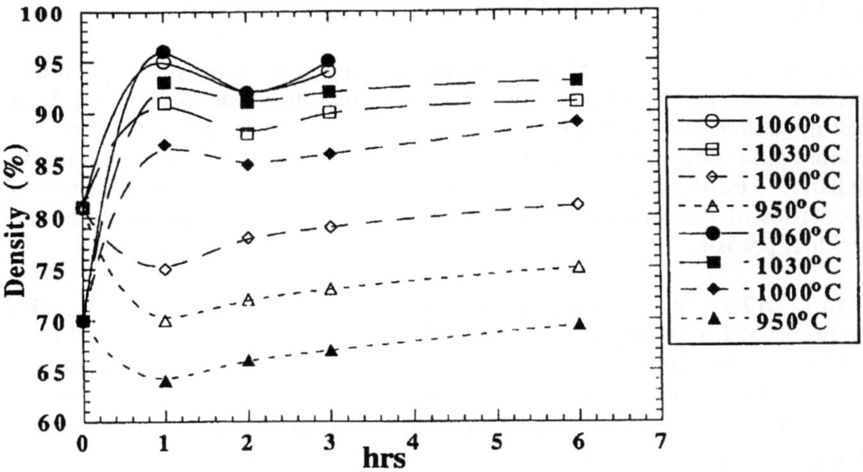

Figure 7.16 Fractional density vs time of post-SLS liquid phase sintered bronze-nickel parts showing the effect of time, temperature, and starting density.

At 1,060°C, a temperature well above the liquidus of bronze, bronze is completely in the liquid state and has sufficiently low viscosity to allow Stage I LPS to dominate. The density of the SLS parts increases rapidly at short times of one hour. Following this rapid gain in density and strength, there is a slight loss of density and strength owing to Stage II LPS formation of Kirkendall porosity. Some of this loss in density and strength is regained in Stage III of LPS as a result of solid state sintering of the product Cu-Ni solid solution. At this high temperature, the flow of liquid in Stage I is relatively free. Therefore, the final density at the end of Stage I is independent of starting density and strength. At temperatures above 1,060°C, SLS parts exhibited loss of shape even at short times.

At the liquidus temperature of 1,030°C, liquid bronze is more viscous than at 1,060°C. Flow of liquid into pores and rearrangement of solid nickel particles is less dominant. The density in Stage I at 1,030°C was lower than at 1,060°C. In a high-starting density part, the rigid network of solid nickel particles restricts the flow further and prevents their rearrangement. In a lower-starting density part, the liquid is more free to flow and rearrange the solid particles. Also, the interdiffusion length in a high-starting-density part is smaller, thereby promoting Stage II LPS and resulting in lower final strength and density than that for a lower-starting-density part. This effect is more pronounced at 1,000°C when Stage II homogenization completely dominates over Stage I densification for a higher-starting-density part. Therefore, at 1,000°C, the high-starting-density part exhibits expansion, while the low-starting-density part exhibits improvements in density.

7.7 MECHANICAL BEHAVIOR OF SLS PROCESSED PARTS

Mechanical properties of SLS processed and post-processed specimens includes tension testing, compressive testing, three and four point bend testing and hardness testing. It appears that the primary microstructural feature affecting mechanical behavior of SLS parts is porosity or relative density. Further, this effect is similar and predictable to mechanical properties of conventional processed powder metallurgy parts which contain residual porosity.

Figure 7.17 shows the effect on density on part strength for a bronze-nickel system as-SLS processed (Figure 7.17a) and post-SLS processed (Figure 7.17b). The linear relationship on these log-log plots suggests a strength-density relationship of the form, $\sigma = C\sigma_o\rho^m$, where C and m are constants, σ_o is the strength of fully dense material and ρ is the relative density. This form is commonly used to describe the strength of conventionally sintered metals (German, 1984).

Figure 7.18 shows the dependence of the modulus of rupture in a four-point bend test on binder volume fraction for two alumina particle size distributions (Lakshminarayan and Marcus, 1991, Bourell, *et al*, 1992). The effects of particle size and binder volume fraction on the modulus of rupture are similar to their effect on part density.

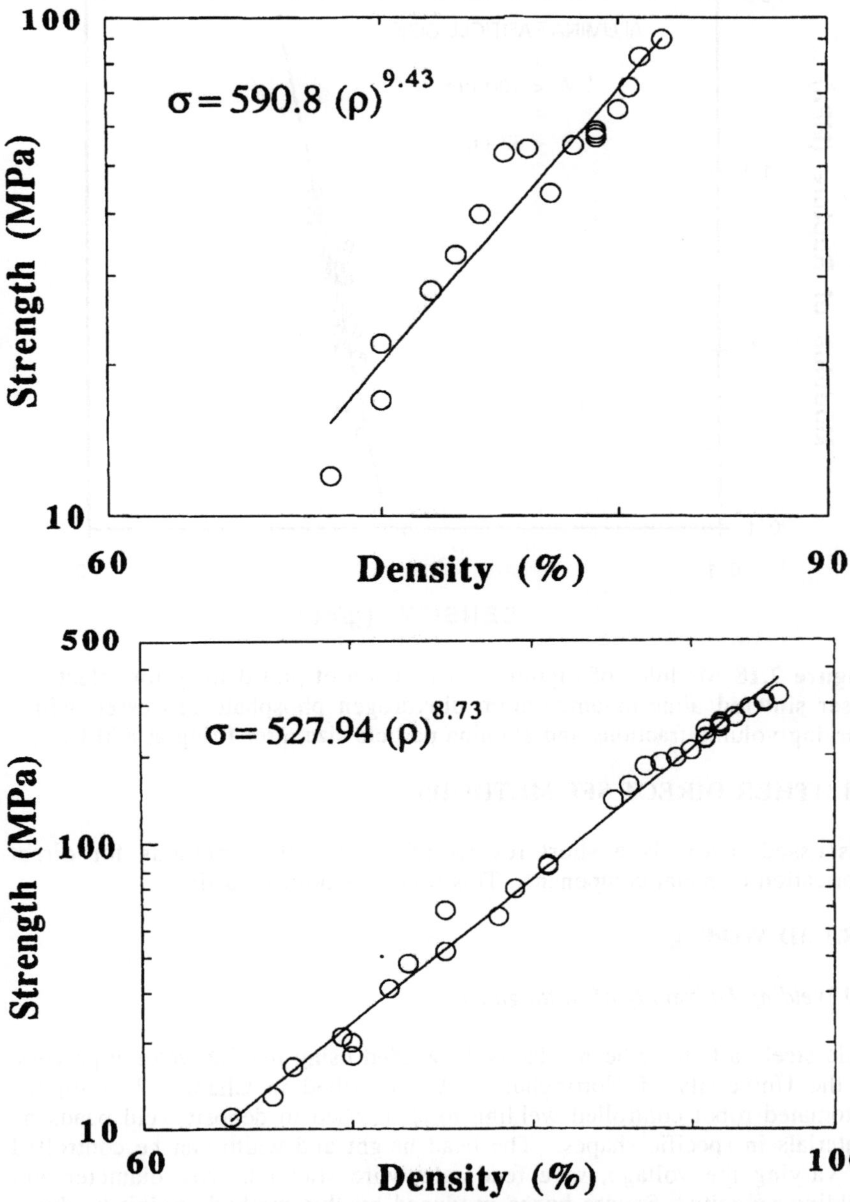

Figure 7.17 Strength and density relationships for selective laser sintered bronze-nickel parts (top) without post processing, (bottom) with post processing by liquid phase sintering.

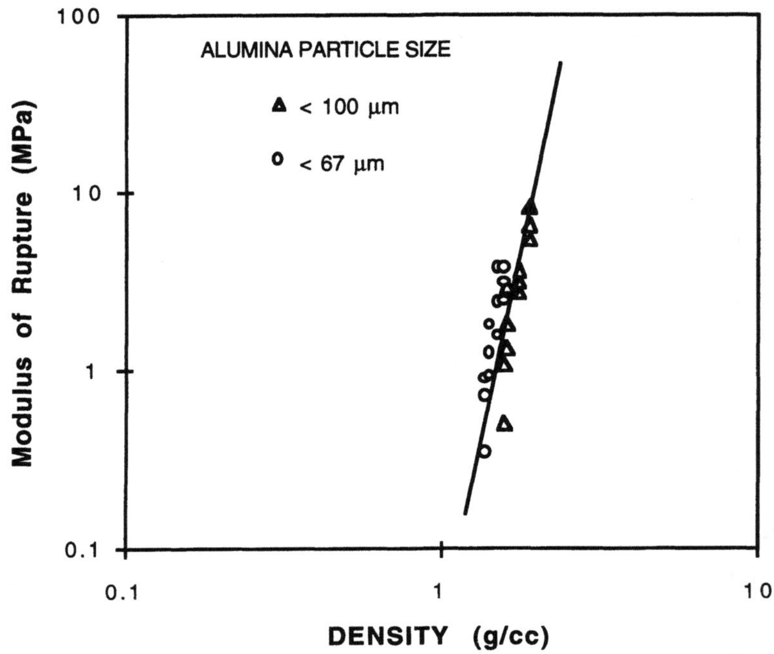

Figure 7.18 Modulus of rupture as a function of part density for selective laser sintered alumina-ammonium dihydrogen phosphate processed with varying volume fractions and alumina particle size after firing at 850°C.

7.8 OTHER DIRECT SFF METHODS

Discussed below is a short review of a few other methods for direct fabrication of metal components. This review is not exhaustive.

7.8.1 3D Welding

3D Welding-University of Nottingham

Mild steel parts have been selectively welded using the 3D welding process at the University of Nottingham. As described in Chapter 2, a highly automated robot controlled welding torch is used to deposit weld beads of materials in specific shapes. The bead height and width can be controlled by varying the voltage, wire feed rate, wire stickout, wire diameter and welding velocity. Square boxes produced by this method exhibit hardness values of 146.3 VHN (base) and 172.6 VHN (top). Tensile strengths of 490 MPa and a density that is 99.5% of the equivalent wrought material were obtained. Microstructure observations indicate mainly large equiaxed grains with an average size of 60 µm. No oxidation was observed. Part accuracy is at best +- 0.2 mm, and surface roughness is poor. Heat build up can occur during deposition, causing previous layers to remelt and distort the part. Solid layers of metal have not been built without producing gaps in the material. This process has been used to produce a thermostat housing for a Ford automobile, shown in Figure 7.19. (Dickens, *et al*, 1992).

Figure 7.19 Welded thermostat housing made in 3D welding process (Dickens, *et al*, 1992).

Electron Beam SFF (EBSFF)

This technique is a variant of 3D welding based on E-Beam welding being developed at MIT. The addition of metal in each layer is achieved by feeding wire into a melt pool that is maintained on the surface of the previous layer by an electron beam. A positioning system controls the motion of the melt pool. Research done by UTC/Pratt & Whitney (East Hartford, CT) and the Materials Science and Engineering Department at MIT has been focused on metals used in nautical service, specifically stainless steels, rotor steels, bronzes, brasses, aluminum and titanium alloys, but the process can potentially be used for any material that can be melted at reasonable power levels. Temperature gradients set up in the part as a consequence of the heat input cause residual stresses that must be eliminated by pulsing the electron beam during deposition (theoretically) or through post processing. Spatial resolution is limited to approximately the diameter of the feedstock wire. For large part production, the wire dimension must be relatively large, thus dimensional accuracy is low.

Mechanical properties are expected to be superior to cast parts made of the same material but inferior to those of forgings. Segregation issues are expected to be alleviated since the extent of segregation is limited to less than the bead cross-section. Parts of 304 and 308 stainless steels and titanium have been made. In preliminary research, a layer of 308L stainless steel was deposited onto a stainless steel mandrel. A micrograph of the radial section is shown in Figure 7.20. Individual weld beads are visible due to different solidification environments, and porosity is very low. In the titanium parts, porosity was found between the substrate and the first layer of weld beads. (Davé, *et al*, 1995)

Figure 7.20 Micrograph of radial section of Type 308L stainless steel deposit, 50x made in EBSFF (Davé, *et al*, 1995).

7.8.2 Shape Deposition Manufacturing

Shape deposition manufacturing is a rapid prototyping technique that combines microcasting, 5-axes milling, and stress relieving (shot peening) to produce functional metal shapes with complex geometry, superior material properties, good surface quality and low residual stresses. A description of the process was provided in Chapter 2. During microcasting, small droplets of metal are deposited in the desired shape. The arc that forms between the plasma welding torch and the feedstock wire is not transferred to the part. This prevents the heat build up and layer remelt that can occur in 3D welding. Part accuracy is limited by the droplet diameter, which is on the order of 1 to 10 mm. Stainless steel parts have been produced by the Robotics Institute and the Engineering Design Research Center at Carnegie Mellon University and by Stanford University. (Merz, *et al*, 1994). A complex part made of 308 stainless steel is shown in Figure 7.21. This piece was embedded in a sacrificial copper support material.

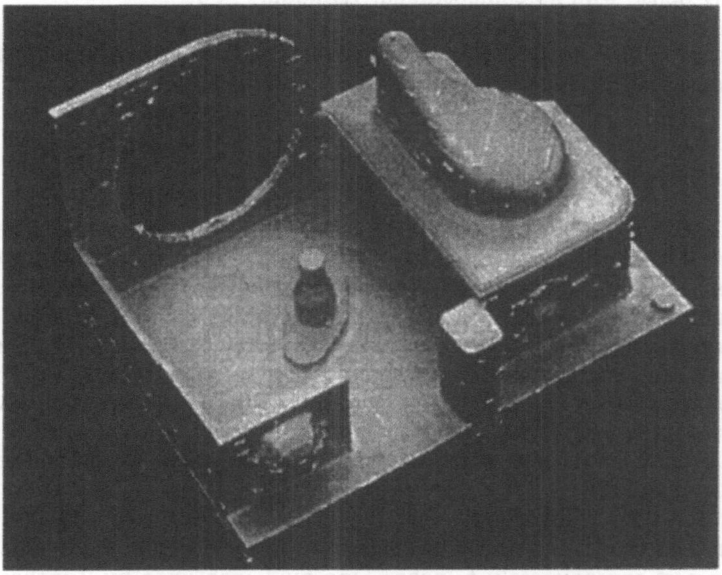

Figure 7.21 Complex 308 stainless steel part made by Shape Deposition (Aubin, 1991).

The material properties of 308 stainless steel parts have been found to be superior to their weldment counterparts. Average tensile strength is reported as 677.2 MPa, average 0.2% offset yield strength as 481.1 MPa and percent elongation as 44.8%, versus 597.2 MPa, 399.9 MPa and 35% elongation, respectively, for 308 stainless steel weldments. The 5-axes milling step eliminates the stair-step effect found in many other rapid prototyping methods.

A process similar to Shape Deposition is under development by the Fraunhofer IPT, Aachen, Germany.

7.8.3 Laser Cladding

Sandia National Laboratories has developed a laser cladding technology known as Laser Engineered Net Shaping (LENS), to fabricate parts directly in metals (Griffith, 1996). A process schematic appears as Figure 7.22.

Metallic powder is injected into a focused laser beam which is incident on a metal substrate. The laser beam creates a weld pool into which laser-heated powder is injected and melted. Metal build-up occurs by continued powder deposition on a three-axis positioning system. Parts such as those shown in Figure 7.23 have been produced using 316 stainless steel powder, nickel-based powder and H13 tool steel powder. Feasibility studies are underway to assess tungsten and tungsten carbide. Parts made using LENS are fully dense with no compositional variation.

A process similar to LENS is under development at Los Alamos National Laboratory called Directed Light Fabrication (DLF).

7.8.4 LaserCast of Titanium

LaserCast is a variant of Selective Laser Sintering developed at The John Hopkins University (House, *et al*, 1996). A schematic of the LaserCast process is shown as Figure 7.24. The process steps are to load the target into target chamber and purge with argon; load the powder feeder, heat and purge with argon; align the laser to the target; monitor the oxygen level; initiate the laser run; charge the target chamber with powder feeder; fluidize the powder using target chamber gas to deposit powder onto the target; level the powder on the target; run the laser over the target in a predefined path; continue until the desired shape is completed; cool under argon and remove from the chamber.

To date, two shapes have been produced as shown in Figure 7.25. These are a linear part approximately 6-inches long by 1-inch wide by 2-inches high, and a cylindrical part approximately 4-inches outside diameter by 2-inches inside diameter by 2-inches high. A box beam starter target of titanium ASTM B265-93 Grade 2 was used for the linear part. LaserCast Ti parts exceed ASTM B367 specifications for composition and mechanical properties.

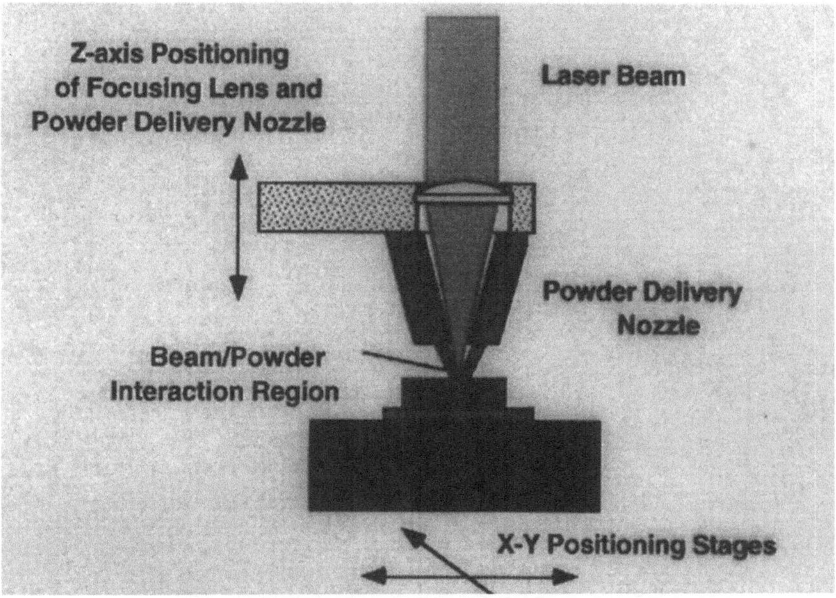

Figure 7.22 Schematic of LENS system (Griffith, *et al*, 1996)

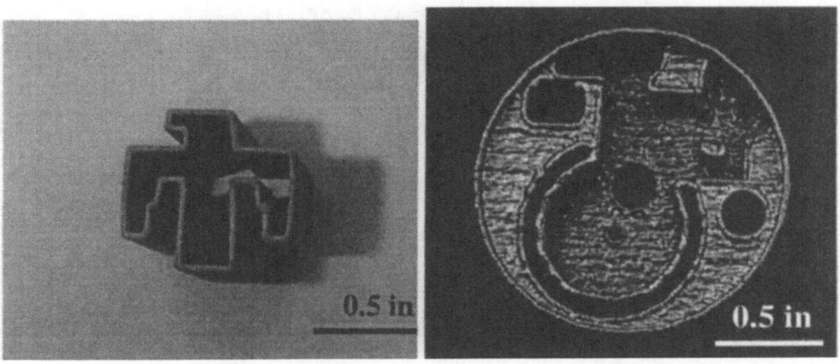

(a) (b)

Figure 7.23. (a) Thin-walled geometry fabricated from H13 tool steel. (b) Solid geometry fabricated from 316 stainless steel (Griffith, *et al*, 1996).

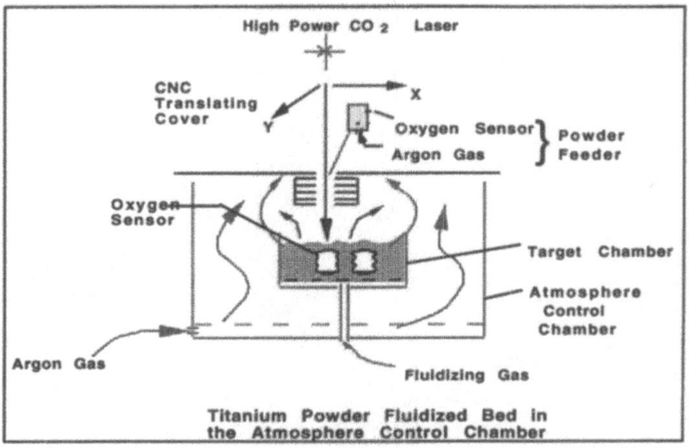

Figure 7.24 Schematic of LaserCast process (House, *et al*, 1996)

Figure 7.25 Titanium parts produced in LaserCast process (House, *et al*, 1996).

7.8.5 Nano-Liter Droplet Deposition

A droplet-based solid freeform fabrication technique for the fast manufacture of metallic structural components directly from CAD information is under development in the Department of Mechanical and Aerospace Engineering at the University of California at Irvine (Orme and Huang, 1995). In droplet-based solid freeform manufacturing, metallic structures are fabricated by depositing precisely controlled molten nano-liter droplets onto a temperature controlled substrate.

The unique feature of the technique under development is the fact that the molten droplet streams can be precisely controlled and manipulated. Precise control refers to droplet streams which have an angular stability of the order of 1 micro-radian and speed dispersions as low as 3x10-7 times the average stream speed. Droplet generation can be manipulated to produce ultra-uniform droplets at frequencies which would otherwise yield highly nonuniform droplet streams with conventional droplet formation methods (i.e., Rayleigh breakup). Moreover, precise control refers to the ability to "customize" the droplet stream configuration according to the desired material property (i.e., gradient materials) and three-dimensional geometry of the structural component. Customized droplet streams may include streams with two or more droplet sizes and/or separations. The sizes and separations can be predicted (in order of exact occurrence) from a numerical model based on conservation of momentum which requires knowledge of the applied disturbance given to the molten fluid prior to droplet formation.

A conceptual schematic of the controlled droplet deposition manufacturing technique is shown in Figure 7.26. Molten droplets are ejected from a droplet generator into a controlled environment which may be either a vacuum or a nonreactive gas. As the droplets are formed, they acquire a charge by passing through a charge electrode (not shown). As the charged droplets travel through the deflection electrodes their trajectories are precisely controlled. The charged droplets can be deflected into a gutter for rapid turn-on/turn-off procedures, or they may be deflected onto a substrate for the deposition of fine detailed structures. As the droplets travel they convect heat. The degree of convection is governed by the initial temperature of the droplets, droplet speed and the environment background pressure and temperature. Upon arrival at the substrate, they undergo a "splatting" action which entails simultaneous spreading and solidification. Under certain circumstances when the droplets have partially pre-solidified due to convection prior to impact, the solidified droplet profile will be nearly spherical. In other realizations the solidification time may be long compared to the spreading time, and the solidified droplet profile will approximate the shape of a short cylinder. Successive droplet deliveries build the 3d structural component. Methods of controlling the thermal conditions in order to insure that the thermal energy content of incoming droplets is sufficient to remelt a thin layer of the previously solidified deposit are necessary. The remelting scheme is advantageous because it enables a layer on the surface of the previously deposited and solidified material to unite in the molten phase with incoming material so that individual splat boundaries will become obliterated, leading to a homogeneous structural component.

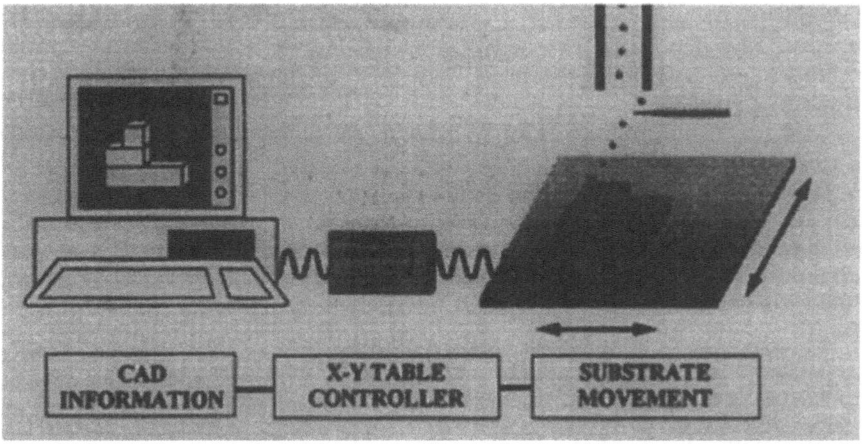

Figure 7.26 Droplet system under development by Orme (1995).

REFERENCES

Agarwala, M., D. Bourell, J. Beaman, H. Marcus, J. Barlow, "Post-Processing of Selective Laser Sintered Metal Parts", *Rapid Prototyping Journal*, vol. 1 #2, 1995, pp. 36-44.

Agarwala, M. K. , David L. Bourell, Joseph J. Beaman, Benny Wu and H.L. Marcus, "Structurally Sound Metal Parts by Selective Laser Sintering", *Proceedings of the 123rd TMS-AIME Annual Meeting*, Gary Warren, ed., San Francisco CA, 1994, pp. 833-851.

Agarwala, M.K. , "Synthesis, Selective Laser Sintering and Post Processing of Metal and Ceramic Composites", *Ph.D. Dissertation*, University of Texas at Austin, 1994.

Agarwala, M., D.L. Bourell, B. Wu and J.J. Beaman, "An Evaluation of the Mechanical Behavior of Bronze-Nickel Composites Produced by Selective Laser Sintering", *Solid Freeform Fabrication Proceedings*,The University of Texas, Austin, Texas, 1993, pp. 193-203.

Agarwala, M., D.L. Bourell, A. Manthiram, B.R. Birmingham and H.L. Marcus, "Synthesis, Selective Laser Sintering and Infiltration of High T_c Dual Phase $Y_2BaCu_3O_{7-x}$ Superconductor Composites", *Solid Freeform Fabrication Proceedings*,The University of Texas, Austin, Texas, 1993, pp. 339-349.

Ashby, M.F. , *Sintering and Isostatic Pressing Diagrams*, 6; , Cambridge, England CB2 1PZ, Department of Engineering, Trumpington St, 1990.

Baumgart, W. , A.C. Dunham and G.C. Amstutz, *Process Mineralogy of Ceramic Materials*,, Stuttgart, Germany, Ferdinand Enke Verlag, 1984.

Bourell, D.L. , H.L. Marcus and Wendy L. Weiss: "Selective Laser Sintering of Parts by Compound Formation of Precursor Powders", US Patent #5,156,697, issued October 20, 1992.

Bourell, D. L. , H. L. Marcus, J. W. Barlow and J. J. Beaman, "Selective Laser Sintering of Metals and Ceramics", *Intl. J. Powd. Metall.*, 1992, 28#4, pp. 369-381.

Bourell, D.L. , H.L. Marcus, J.W. Barlow, J.J. Beaman, C.R. Deckard, "Multiple Material Systems for Selective Beam Sintering", Patent # 4,944,817, issued July 31, 1990.

Bunnell, D.E. , S. Das, D.L. Bourell, J.B. Beaman and H.L. Marcus, "Fundamentals of Liquid

Phase Sintering During Selective Laser Sintering", *Solid Freeform Fabrication Proceedings*,The University of Texas, Austin, Texas,1995, pp.440-447.

Bunnell, D.E. , D.L. Bourell and H.L. Marcus, "Fundamentals of Liquid Phase Sintering Related to Selective Laser Sintering", *Solid Freeform Fabrication Proceedings*,The University of Texas, Austin, Texas,1994, pp. 379-386.

Bunnell, D.E., "Fundamentals of Selective Laser Sintering of Metals", *PhD Dissertation*, The University of Texas at Austin, Austin, Texas, 1995.

Cottle, R.D. , Britton Birmingham, Zwy Eliezer and H.L. Marcus: "Selective Laser Sintering, Preparation and Tribological Testing of Nanostructured Tungsten Carbide-Cobalt Composites", *Solid Freeform Fabrication Proceedings*,The University of Texas, Austin, Texas, 1995, pp.369-373.

Davé, V.R., J.F. Matz, and T.F. Eagar, "Electron Beam Solid Freeform Fabrication of Metal Parts", *Solid Freeform Fabrication Proceedings*,The University of Texas, Austin, Texas, 1995, pp. 64-71.

Dickens, P.M., M.S. Pridham, R.C. Cobb, I. Gibson, and G. Dixon, "Rapid Prototyping Using 3D Welding", *Solid Freeform Fabrication Proceedings*,The University of Texas, Austin, Texas, Austin, Texas, 1992, pp. 280-90.

Frenkel, J., "Viscous Flow of Crystalline Bodies under the Action of Surface Tension," J. Phys. (USSR), 1945, 9, pp. 385-391.

German, R.M., *Liquid Phase Sintering*, 1985, New York, Plenum Press.

German, R.M., *Powder Metallurgy Science*, 1984, Princeton NJ, Metal Powder Industries Federation, pp. 234-252.

German, R.M., "Supersolidus liquid phase sintering. II Densification theory", *Int'l. J. Powder Metall.*, 1990, 26#1, 35-43.

Griffith, M.L. , D.M. Keicher, C.L. Atwood, J.A. Romero, J.E. Smugeresky, L.D. Harwell, D.L. Greene, "Free Form Fabrication of Metallic Components Using Laser Engineered Net Shaping (LENS)", *Solid Freeform Fabrication Proceedings*,The University of Texas, Austin, Texas, 1996, pp. 121-128.

Haase, P. : "Selective Laser Sintering of Metal Powders", *MS Thesis*, University of Texas at Austin, 1989.

House, M. A. , E. J. Whitney, D. G. Krantz, and F. G. Arcella, "Rapid Laser Forming of Titanium Near Shape Articles: LaserCast", *Solid Freeform Fabrication Proceedings*,The University of Texas, Austin, Texas, 1996, pp. 232-241.

Lakshminarayan, U. , S. Ogrydziak and H.L. Marcus: "Selective Laser Sintering of Ceramic Materials", *Solid Freeform Fabrication Proceedings*,The University of Texas, Austin, Texas, 1990, pp. 16-26.

Lakshminarayan, U. and H.L. Marcus: "Microstructural and Mechanical Properties of Al_2O_3/P_2O_5 and Al_2O_3/B_2O_3 Composites Fabricated by Selective Laser Sintering", *Solid Freeform Fabrication Proceedings*,The University of Texas, Austin, Texas, 1991, pp. 205-212.

Manriquez-Frayre, J.A. and D.L. Bourell: "Selective Laser Sintering of Binary Metallic Powder", *Solid Freeform Fabrication Proceedings*,The University of Texas, Austin, Texas, 1990, pp. 99-106.

Manriquez-Frayre, J.A. and D.L. Bourell: "Selective Laser Sintering of Cu-Pb/Sn Solder Powders", *Solid Freeform Fabrication Proceedings*,The University of Texas, Austin, Texas, 1991, pp. 252-260.

Manthiram, A. , F. Chi, L.F. Johnson, B.R. Birmingham and H.L. Marcus: "Development of Nanocomposites for Solid Freeform Fabrication", *Solid Freeform Fabrication Proceedings*,The University of Texas, Austin, Texas, 1993a, pp. 4-73.

Manthiram, A. , D.L. Bourell and H.L. Marcus, "Nanophase materials in solid freeform fabrication", *JOM*, 1993b, 45#11, 66-70.

Marcus, H.L. , J.J. Beaman, J.W. Barlow and D.L. Bourell: *Am. Ceram. Soc. Bull.*, 1990, 69#6, 1030-1031.

Massalski, T.B. , ed., *Alloy Phase Diagrams*, 2 edn,, Materials Park OH, ASM International, 1990, p. 2864.

Melvin III, L.S. and J.J. Beaman: "The Application of an Artificial Body Force to the Selective Laser Sintering Process", *Solid Freeform Fabrication Proceedings*,The University of Texas, Austin, Texas, 1992, pp. 118-123.

Melvin III, L.S. and J.J. Beaman: "The Electrostatic Application of Powder for Selective Laser

Sintering", *Solid Freeform Fabrication Proceedings*,The University of Texas, Austin, Texas, 1991, pp. 171-177.

Metzner, A.B., "Rheology of suspensions in polymeric liquids", *J. Rheology*, 1985, 29, 739-75.

Nyrhila, O.,"Manufacture of dimensionally precise pieces by sintering", US Patent 5,061,439, 1991.

Orme, M.E., C. Huang, "Thermal design parameters critical to the development of Solid Freeform Fabrication of structural materials with cintrolled nano-liter droplets", *Solid Freeform Fabrication Proceedings*,The University of Texas, Austin, Texas, 1995, pp. 88-95.

Pinsky, D.A. , P.O. Charreyron and M.C. Flemings, "Compression of semi-solifd dendritic Sn-Pb alloys at low strain rates", *Met. Trans. B*, 1984, 15, p. 173.

Prabhu, G. and D.L. Bourell: "Supersolidus Liquid Phase Selective Laser Sintering of Prealloyed Bronze Powder", *Solid Freeform Fabrication Proceedings*,The University of Texas, Austin, Texas, 1993, pp. 317-324.

Rao, T.S. , D.L. Bourell and H.L. Marcus: "Densification Behavior of SLS Processed Al_2O_3/Al Composite", *Solid Freeform Fabrication Proceedings*,The University of Texas, Austin, Texas, 1995, pp. 374-380.

Scherer, G.W. , "Sintering of low-density glasses. I. Theory", *J. Am. Ceram. Soc.*, 1977, 60#5-6, 236-239.

Scherer, G.W. , "Viscous sintering under a uniaxial load", *J. Am. Ceram. Soc.*, 1986, 69#9, C206-207.

Sindel, M., T. Pintat, M. Greul, O. Nyrhila, and C. Wilkening, "Direct laser sintering of metals and metal infiltration for near net shape fabrication of components", *Solid Freeform Fabrication Proceedings*,The University of Texas, Austin, Texas, 1994, pp. 94-101.

Subramanian, P.K. , G. Zong, and H.L. Marcus: "Selective Laser Sintering and Reaction Sintering of Ceramic Composites", *Solid Freeform Fabrication Proceedings*,The University of Texas, Austin, Texas, 1992, pp. 63-71.

Suery, M. and M.C. Flemings, "Effect of strain rate on deformation behavior of semi-solid dendritic alloys", *Met. Trans. A*, 1982, 13, pp. 1809-19.

Sun, M.M. , J.C. Nelson, J.J. Beaman, and J.W. Barlow: "A Model for Partial Viscous Sintering", *Solid Freeform Fabrication Proceedings*,The University of Texas, Austin, Texas, 1991, pp. 46-55.

Swinkels, F.B. and M.F. Ashby, "Mechanisms of hot-isostatic pressing", *Acta Metall.*, 1981, 29, pp. 259-281.

Vail, N.K. and J.W. Barlow: "Effect of Polymer Coatings as Intermediate Binders on Sintering of Ceramic Parts", *Solid Freeform Fabrication Proceedings* ,The University of Texas, Austin, Texas, 1991, pp. 195-204.

Weiss, W.L. : "The Formation of Intermetallics by Selective Laser Sintering", *MS Thesis*, University of Texas at Austin, 1991.

Weiss, W. L. and D.L. Bourell, "Selective Laser Sintering of Intermetallics", *Metallurgical Transactions A*, 24A, 1993, pp. 757-759.

Wildemuth, C.R. and M.C. Williams, "Viscosity of suspensions modeled with a shear-dependent maximum packing fraction", *Rheologica Acta*, 1984, 23, p. 627.

Zong, G. , Y. Wu, N. Tran, I. Lee, D.L. Bourell, J.J. Beaman and H.L. Marcus: "Direct Selective Laser Sintering of High Temperature Materials", *Solid Freeform Fabrication Proceedings*,The University of Texas, Austin, Texas, 1992a, pp. 72-85.

Zong, G. , Y. Wu, H.L. Marcus, D.L. Bourell, and J.J. Beaman: "Direct Solid Freeform Fabrication of High Temperature Materials Using Selective Laser Sintering", in P*roc. of the Symposium on Processing and Fabrication of Advanced Materials for High Temperature Applications - II*, (V.A. Ravi and T.S. Srivatsan, eds), Chicago IL, Nov. 1-5, 1992b, The Metals, Minerals and Mining Society, pp. 503-518.

8

SOLID FREEFORM FABRICATION USING GAS PHASE PRECURSORS

In addition to Selective Laser Sintering two other Solid Freeform Fabrication approaches that involve precursor gas phases are being explored at The University of Texas. The first is a variation of the Selective Laser Sintering process, where a layer of powder is spread and the laser scans under the identical conditions described earlier, but the action of the scanning laser is to heat the powder surface to a temperature where it will react with the gas phase to create a compound. This is another version of Selective Laser Reactive Sintering (SLRS) defined earlier (Birmingham, 1992; Birmingham, 1993; Birmingham, 1995a; Bourell, 1992; Subramanium, 1992) for powder/powder reactive sintering. Typical precursor gases include O_2, CH_3, C_2H_2, tetramethylsilane, N_2, and NH_4. An example is Si powder in the presence of CH_4 forming SiC. In this case, Si was locally melted by the scanning laser beam and reacted with the CH_4 gas. In a similar case, Si powder under the laser beam reacts with a precursor of ammonia to produce shapes of Si_3N_4. Finally, a mixture of Si_3N_4 and Si powders, where Si reacts in the presence of the CH_4 gas, results in a composite of SiC and Si_3N_4.

The second approach using gas phase precursors is Selective Area Laser Deposition (SALD) (Lee, 1995, Marcus, 1991, Marcus, 1992a, Marcus, 1992b, Maxwell, 1993, Maxwell, 1995a, Thissell, 1991, Thissell, 1994a, Tompkins, 1993, Tompkins, 1994, Zong, 1990a, Zong, 1991a, Zong, 1991b). This approach results in direct formation of shapes from the gas phases such as, CH_4 and C_2H_2, using a computer controlled laser to define the deposition shape made by thermal and/or photo decomposition of the gas phase. Some of the key problems associated with SALD are maximizing deposition rate with desired microstructures, determination of the actual growth during the SALD processing, and appropriate computer sectioning models to allow the direct formation of complex shapes. The gas phase approach has the potential to be applied to many elements and to multi-element composite deposition by using multiple laser activation where both photo-assisted and thermal decomposition may occur over mixed or sequentially delivered gases. Although SALD has been used primarily for depositing stand alone structures, many additional applications will certainly be developed. For example, SALD can be used for ceramic joining in which a layer of ceramic material is deposited from the gas phase to bond together two adjacent ceramic materials (Tompkins, 1995).

8.1 SELECTIVE AREA LASER DEPOSITION

Laser chemical vapor deposition (LCVD) is a proven technique (Arnone, 1986; Kodas, 1987) for single-step local deposition or direct writing of thin films of metals, semiconductors or insulators. There are basically two interactions between the laser and the gas/substrate system. The first is a laser-induced pyrolytic reaction where the substrate is locally heated. The second is photolytic absorption of the laser energy by the gas phase at the point at which the modified gas phase precursor will be deposited. In pyrolytic LCVD, the deposition rates are high so that three-dimensional structures can be produced.

An experimental set-up is schematically shown in Figure 8.1 (Tompkins, 1992). The TEM mode of a pulsed CO_2 (λ =10.6 μm) laser (modulation 5 kHz, maximum output of 25W) was used to locally heat a substrate. To vary its size, the beam was focused by a ZnSe lens with a focal length, f, of 50mm. For accurate adjustment of the laser focus and monitoring of the deposition process, a video camera was used. The beam waist w_0 equals $2f\lambda/\pi a$ or 112.5 μm, where a is the diameter of the beam, 0.33 mm, on the lens of focal length *f*. The focal depth is given by $L = 2\pi w_0/\lambda$ = 7.5mm. A helium-neon laser beam, running collinearly, assisted in visualizing the path of the CO_2 laser beam. The reaction chamber rested on a *x-y-z* mechanical translator and was constructed from a stainless steel cylinder with an outside diameter of 6 inches and a height of 5 inches. A ZnSe window was installed on the top of the chamber to transmit the CO_2 laser beam. The chamber could be evacuated to 0.1 Torr with a rotary pump prior to introduction of the source gas. A static fill method over a range of pressures was employed. For safety considerations, the gas supply, laser sources, and reaction chamber were installed and operated under an exhaust hood. Additional laser sources such as a Nd:YAG are used to obtain a wider range of laser power and wavelength operating parameters.

SALD is a very complex process with respect to both chemical and physical aspects. A deposition material, carbon, will be chosen as an example for discussion of these aspects. Chemical vapor deposition (CVD) of the diamond form of carbon has received a great deal of research attention in the last ten years and is a worthy research deposition goal for SALD. The rate and deposit quality of traditional broad area CVD of diamond depends strongly on the substrate temperature, gas phase chemical mole fractions above the substrate, the process pressure, the nature of the atomic hydrogen generation source, and transport phenomena within the reactor. The rate and quality of deposition from SALD also depend on these parameters, but they are also significantly influenced by the extremely complex transport phenomena within the transient laser heated spot deposition area. Amorphous carbon on the other hand has less critical process control demands than exist for diamond deposition.

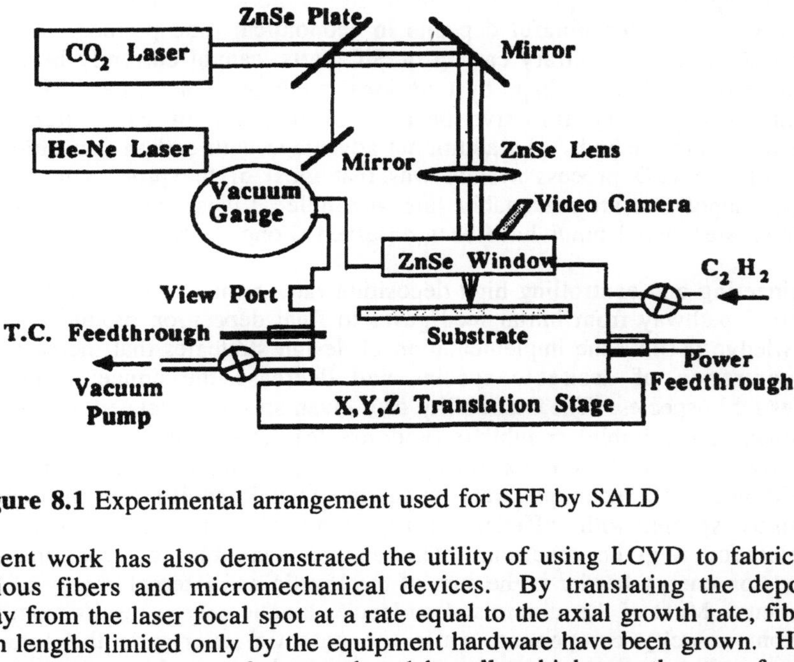

Figure 8.1 Experimental arrangement used for SFF by SALD

Recent work has also demonstrated the utility of using LCVD to fabricate various fibers and micromechanical devices. By translating the deposit away from the laser focal spot at a rate equal to the axial growth rate, fibers with lengths limited only by the equipment hardware have been grown. High precursor gas pressures of greater than 1 bar allow high growth rates of more than 0.3 mm/s for carbon fibers from ethylene gas (Wallenberger, 1993). Multiple gas precursors can be used simultaneously in the reaction chamber to yield fibers with tailored compositions. For example, mixtures of silane (SiH_4) and ammonia were used to grow fibers of silicon nitride (Si_3N_4) (Wallenberger, 1994). By combining translation of the stage/deposit with rotational motion relative to the laser spot, structures of greater complexity can be fabricated. In one example, a silicon fiber was deposited by LCVD from silane gas. A helix of tungsten was then deposited around the silicon core from a mixture of H_2/WF_6, forming a microsolenoid (Westberg, 1993). The examples discussed to this point have involved only the growth of rods and coils by LCVD in which the growth direction is parallel to the laser beam axis. Recently it was demonstrated that deposition from the gas phase can be achieved at the intersection of two overlapping laser beams. The growth of the deposit follows the path of this intersection, not the path of a single beam, so direct writing in free space in nearly any direction is possible (Lehmann, 1995). With these additional degrees of freedom, structures of virtually any conceivable complexity can be fabricated.

8.2 CHEMICAL KINETICS AND SALD

The complicated chemical kinetics involved in the CVD process in general, and for SALD in particular, are discussed in some detail in several recent references (Thissell, 1994b; Thissell, 1996a; Thissell, 1996b). Gas phase SFF using SALD requires fast deposition kinetics in order to achieve macroscopically meaningful deposits in economical time periods. Certain materials, such as arbitrary carbon or SiC parts, cannot currently be easily made into an arbitrary shape by traditional methods. Therefore, SFF using SALD is especially attractive because of the lack of easy alternative manufacturing methods. An additional advantage with carbon deposition as the initial SALD process system was that most of the process gases for carbon deposition are reasonably safe to handle. Linear carbon rod growth rates greater then 1 mm/s have been observed (Zong, 1990a).

Engineering and controlling high deposition rates requires knowledge of the reaction pathway from initial feed gases to final deposition products. This knowledge permits the implementation of design strategies that increase the concentration of desirable species and decrease the concentration of undesirable species. The desirability of a given species is related to whether it enhances (desirable) or inhibits (undesirable) the rate of deposition either directly or indirectly simultaneously producing the required microstructure. Traditional CVD techniques rely on thermodynamic equilibrium to produce chemical species with sufficient energy to overcome the activation energy barrier for a given reaction. The energy available to overcome the activation energy barrier is the sum of the kinetic and internal energy of the reactants. Most of the atomic and molecular kinetic energy results from the random velocity component, while the potential energy is the sum of vibrational energy, rotational energy, excited metastable energy, and unpaired electrons (free radicals). Kinetic energy may be transferred to internal energy via inelastic collisions, while potential energy may be transferred to kinetic energy via superelastic collisions. The distribution of random velocity for processes at or very near equilibrium and not under applied external forces is given by a local Maxwell-Boltzmann distribution function integrated over a differential volume sphere in velocity space (Bittencourt 1988), which defines the concept of temperature. Chemical reactions can be assumed to have rate constant dependencies that can be expressed as:

$$k_{f_i} = k_{oi}\left[\frac{T}{T_o}\right]^{\beta_i} e^{\frac{-E_{ai}}{RT}}$$

As an example, let us look at the kinetics of pyrolytic carbon deposition from acetylene. A first order rate constant for this reaction has the following rate constants in the temperature range of 820-1375°K (Tesner, 1984):

$$k_o = 17.1 \times 10^{-6} \left[\frac{\text{kg}}{\text{m}^2\ \text{s Pa}}\right]$$

$$\beta = 0$$

$$E_a = 138 \left[\frac{\text{kJ}}{\text{mole}}\right]$$

Pyrolytic carbon deposition from acetylene is not a one-step reaction, but includes many individual reactions. Overall reaction rate equations are process and temperature range specific. The above constants are likely fixed by the rate limiting step in the particular reaction pathway. Very high temperatures are required for an appreciable deposition rate at an acetylene partial pressure of 10^5 Pa due to the high activation energy for the reaction. To summarize, for thermal CVD processes, such as pyrolytic carbon deposition from acetylene, the deposition rate in the kinetic rate limited deposition regime increases exponentially with temperature and linearly with the partial pressure of acetylene.

Plasma deposition is more complicated than thermal CVD, but it can provide great increases in deposition rate when compared to thermal CVD at a given substrate temperature, feed gas partial pressures and flow rates. Electron, atomic and molecular species can exist with two very different kinetic energy distribution profiles below about 26-40 kPa of pressure, due to the great difference between the electrons and the others in mass, collision cross sectional area, and ability to accelerate through an electric field. Electrons cannot accelerate to high enough velocities between inelastic collisions at pressures higher than about 300 Torr to maintain a kinetic energy distribution profile significantly higher than the neutral and ionic species. However, the concentrations of ions is very low, about 10^{-5} times the neutral species concentrations in plasmas of interest for deposition, particularly for SALD. This category of plasma is called weakly ionized. A further characteristic of these plasmas is that the average kinetic energy of the free electrons is typically below about 9.64×10^5 KJ/mole (10 eV = 1.16×10^5 °K for a Maxwell-Boltzmann distribution), while the ions and neutrals are typically in thermal equilibrium in the temperature range of 400-600°K. The average electron kinetic energy is below the ionization threshold of most diatomics and nobles. The electrons with higher energy than particular bond strengths in neutrals and ions may cause bond breakage upon an inelastic collision with a neutral or ion, or excitation to long lived metastable states. Therefore, a weakly ionized plasma can be used to transfer internal energy from energetic free electrons to an ionic and neutral specie population having a relatively low random velocity distribution ("cold"). This can create much higher concentrations of species having sufficient internal energies to overcome the activation energy barriers of many reactions than exist in the case of thermal equilibrium alone. In fact, it is possible to create weakly ionized plasmas where most of the chemical species are radicals and atomic species if recombination reactions are properly controlled. This allows for the creation of internal energy transport regimes in reactors where excited states can transfer internal energy absorbed through inelastic impact with free electrons to a zone where the deposition process gases are introduced. Isolation of the plasma process from the SALD process can be thus achieved (Thissell, 1991). An additional problem with using a plasma in SALD is that wall recombinations associated with desired atomic species, such as hydrogen and oxygen in carbon deposition, greatly reduce the concentration within the plasma resulting in reduced kinetics (Fite, 1969, Greaves, 1959).

8.3 ORGANOMETALLIC PRECURSORS FOR SALD

Due to their strong covalent bonding, ceramic materials generally have a large negative free energy of formation from precursor gases. This indicates that ceramic materials might grow relatively quickly from precursor gases in SALD applications. An example material is silicon carbide, which has numerous structural, chemical, and high temperature applications. Tetramethylsilane, $Si(CH_3)_4$, is a safe and relatively inexpensive SiC precursor. Tetramethylsilane is a silane derivative which exists as a stable, high vapor pressure liquid (Bp 26°C), and is widely used as an NMR standard. It acts as a single source precursor according to the reaction: $Si(CH_3)_4 \rightarrow SiC + 3\ CH_4$. For SALD this reaction occurs in a thermal plasma within a focused infrared laser beam. The temperature is estimated to be approximately 1000-1300°C within the beam area, and at a pressure of 120 Torr of tetramethylsilane, growth rates greater than 1.5 mm/min have been observed. This is approximately 150 times the reported rate for purely pyrolytic 1500°C CVD of silicon carbide using tetramethylsilane (Figueras, 1991, Figueras, 1992). It is hypothesized that the thermal plasma and the solid angle of gas feed when compared to normal CVD significantly enhances the growth rate. The point source allows the full 2π solid angle to contribute to both the gas feed and the reactant gas removal from the point of deposition. In contrast, normal CVD is a one dimensional process. In addition, it is possible that the excited species in the thermally created plasma directly absorb the infrared beam, giving a photolytic contribution to the growth rate. Evidence for this is the great enhancement of growth rate when the plasma is observed in the SALD process zone.

Other compounds which can provide stable, high vapor pressure single source precursors include diethylsilane (Bp 56°C), hexamethyldisilane (Bp 126°C), and dichloromethylsilane (Bp 42°C) for SiC. Other precursors offer the possibility of depositing silicon nitride and silicon carbide simultaneously from single or multiple gas precursors, such as ammonia combined with tetramethylsilane. This can result in either a mixing on an atomic scale or in a novel nanophase microstructure. These compounds include diethylaminotrimethyl silane (Bp 126°C), dimethylaminotrimethylsilane (Bp 86°C), trimethylsilylazide (Bp 95°C), and tetrakis(dimethylamino)silane (Bp 180°C). Several boron carbide precursors are dimethylboronbromide (Bp 31°C), pentaborane (Bp 63°C), and tributylboron (130°C, 20mm). These can be used in conjunction with silicon carbide precursors to give a novel material composed of silicon carbide and boron carbide combined on an atomic or microstructural scale.

SALD metal precursors have also been considered. Tetramethyltin (Bp 78°C) and tricarbonylnitrosylcobalt (Bp 48°C) can deposit tin and cobalt respectively. Potential aluminum precursors are particularly interesting. In addition to a pure metal, they might produce aluminum oxide or an aluminum-aluminum oxide composite. The presence of carbon in the organometallic precursor also suggests that aluminum carbide, either as monolithic material or as a second phase in an aluminum matrix might be formed. Candidates include trimethylaluminum diethylether (Bp 45°C, 4mm), hydrodimethylaluminum (Bp 25°C, 2mm), bromodimethylaluminum (Bp 75°C, 50mm), and trimethylaluminumethylmethylether (Bp 56°C, 10mm). SALD of TiO_2 and SiO_2 from $TiCl_4$ and $SiCl_4$ is presently being pursued (Jakubenas, 1995). The composition of the composite is controlled by adjusting the relative partial pressures of the gas precursors. SALD can also be used to deposit metal alloys. In one example, the precursors $Fe(CO)_5$ and $Ni(CO)_4$ were mixed, resulting in Ni-Fe alloys (Maxwell, 1995b).

8.4 SELECTIVE LASER REACTIVE SINTERING SYSTEM FOR GASES

To be successful, a SLRS system must accomplish three basic functions. These functions are: 1) supply controlled laser power to a selected area, 2) provide multiple layers of a powder bed having proper thickness and density, and 3) provide the appropriate gas environments including vacuum. An overall schematic of the SLRS system is seen in Figure 8.2 (Birmingham, 1992).

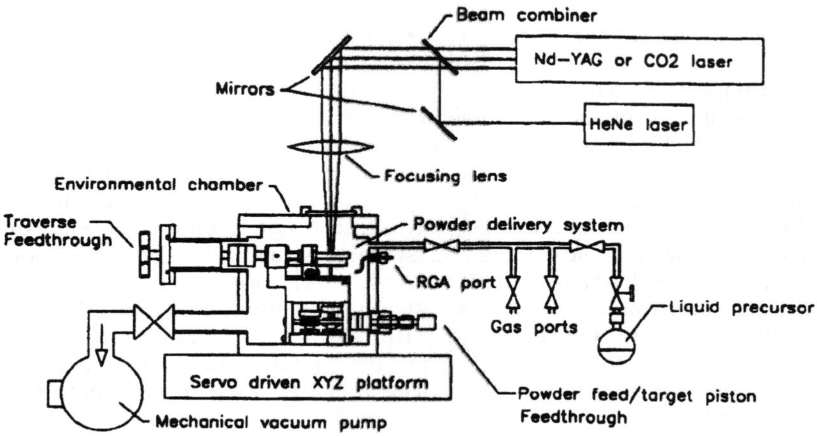

Figure 8.2 Selective Laser Reaction Sintering Workstation.

With the exception of the powder delivery mechanism, the basic arrangement of this system has been described earlier. The laser beam from a 25 watt CO_2 laser is directed onto a substrate that is located in an environmentally controlled chamber. A motor driven X-Y table moves the chamber/substrate under the stationary beam, effectively "scanning" the beam. Laser power fluctuations have been reduced to less than 5% of set point by using a pulse width modulation controller developed at the University of Texas (Tompkins, 1992). Power level can be adjusted manually or by computer. The positional accuracy of the table/beam is now better than 20 μm and repeatability better than 5 μm at scan speeds to 3 mm/s using a computer controlled dc servo system.

Environmental control is achieved by locating the powder delivery mechanism/substrate inside a vacuum chamber. The chamber has three independent gas inputs with the emphasis on flexibility due to the research nature of the system. The gas inputs have allowed for the introduction of a variety of gases including N_2, N_2/H_2, NH_3, Ar, H_2, O_2, CH_4, and C_2H_2. Mechanical pumping initially provides 10^{-3} Torr vacuum to minimize gas impurity content. Power and thermocouple feedthroughs provide the ability to heat substrates and monitor and control temperature using a PC based data acquisition system. A gas sampling port has also been included to enable connection to a residual gas analyzer. The powder delivery system is similar to the one described for SLS, but the feed system is 1.9 cm in diameter. This allows the use of small amounts of experimental powders for SLRS studies. When the materials systems have been proven on this scale they can easily be adapted to a larger system.

Laser reaction sintering has been used in a variety of ways to produce a variety of results. Standard pressureless sintering of oxides, nitrides, and carbides is difficult. However, formation of an oxide, nitride or carbide may be achieved by reaction sintering in the corresponding oxygen, nitrogen, or carbon rich atmosphere. One example is the sintering of an aluminum/alumina mixture in the presence of oxygen to form an alumina pre-form (Subramanium, 1992). These pre-forms can be infiltrated to create a metal-ceramic or ceramic-ceramic composite part. Composite structures can be produced directly by partial conversion of powder by reaction. A multi-layer Cu-TiN part was produced by SLRS using a Cu-10Ti alloy as base material (Birmingham, 1992). Reaction sintering in a nitrogen atmosphere caused the Ti to migrate to the surface of the alloy particles where it reacted with nitrogen and converted to TiN. The result was a copper matrix with a submicron TiN layer on each of the original powder particles. TiN presence was confirmed by X-ray diffraction, and the dispersion of Ti-rich areas(TiN) was mapped using EDS.

8.5 SELECTIVE AREA LASER DEPOSITION VAPOR INFILTRATION

A third technique involves SALD directly into layers of a powder bed resulting in simultaneous vapor deposition and infiltration. This approach has been named selective area laser deposition vapor infiltration (SALDVI). The hardware requirements are similar to that of the SLRS system discussed earlier, namely the capability to deliver sequential layers of powder in the presence of a reactive environment. One system currently under study (Birmingham, 1994) combines the pyrolytic formation of SiC from an organometallic gas precursor directly into a SiC base powder. The pyrolytically produced SiC infiltrates the SiC base powder and binds it together. Multi-layer shapes have been formed in this fashion. Figure 8.3 is an example of this.

8.6 MODELING AND SIMULATION

The nucleation, growth, and resulting microstructures of SALD and SLRS shapes are the result of very complex thermal, chemical and reaction processes. To enable the intelligent choice of operating parameters, process modeling and simulation are extremely valuable (Lee, 1995, Maxwell, 1993, Thissell, 1991, Zong, 1990b, Zong, 1991a, Zong, 1991b, Zong, 1991c). These studies are needed for SALD and gas phase SLRS of arbitrary three-dimensional shapes because these processes present unique problems and issues which are significantly different from those relevant to conventional CVD or SLS. A laser beam is used in these processes to heat the substrate locally to generate a spatially confined region in which the gas and substrate conditions are suitable for a chemical deposition/reaction to occur. Thus, SALD, SALDVI, and SLRS are complex maskless selective area nucleation and growth processes, which involve moving boundary heat transfer, "point" sink diffusion geometry (also called hemispherical boundary layers), as well as laser-gas and laser-solid interactions. The calculations can provide spatially resolved deposition rates, temperatures, and concentration fields.

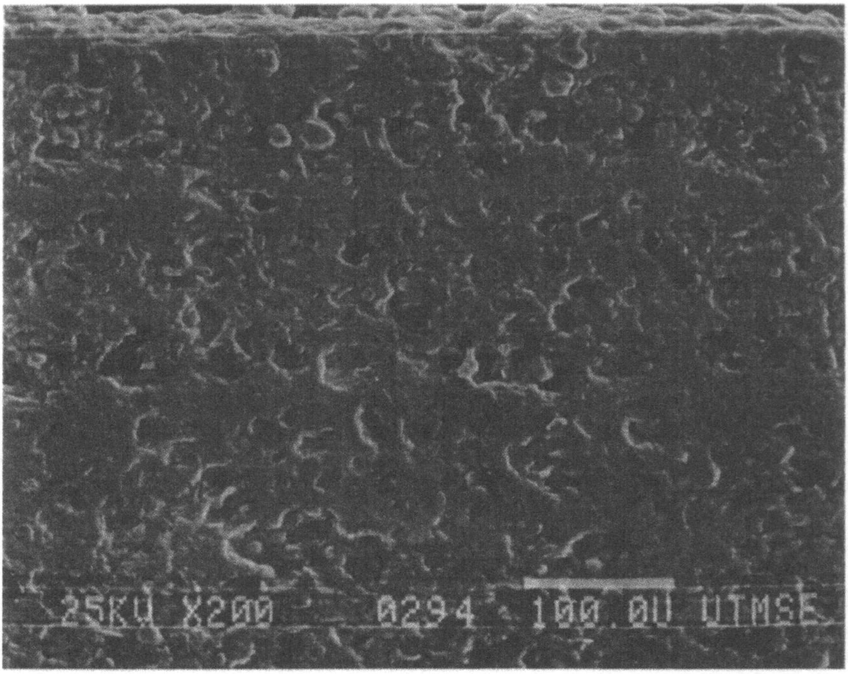

Figure 8.3 SALDVI of SiC from Tetramethylsilane into SiC Powder.

Transport phenomena within the reactors are studied because momentum transport through the reactor affects the mass transport through boundary layers that, in turn, affect deposition rates. The techniques for modeling transport phenomena in traditional CVD reactors are well developed (Jensen, 1983, Mills, 1988, Zhu, 1990). Heat, mass, and momentum transport within the reactor can be modeled in two and three dimensions. Modeling of the gas phase chemical kinetics is also required in order to calculate the concentrations and rates of production of species in both microwave and laser induced plasmas. The chemical kinetics analysis can be addressed using approaches like the CHEMKIN II software developed at Sandia National Laboratories (Kee, 1989), and data from the NIST chemical kinetics database and the literature on neutral-neutral reactions (Mallard, 1991).

Examples of the use of modeling to predict the behavior of the SALD systems are given in the references (Lee, 1995, Maxwell, 1995a, Zong, 1990b, Zong, 1991a, Zong, 1991b, Zong, 1991c). Another example is the use of thermodynamic data to predict the SALD of SiC/Si_3N_4 from a gas mixture of tetramethylsilane and ammonia. The higher stability of the Si_3N_4 makes it the choice of deposit over the SiC as the pressure of ammonia gas is increased. This was experimentally demonstrated. Many of the interactions between the laser, the powder, and the gas are described in a recent publication (Birmingham, 1995b).

8.7 SUMMARY

Methods of using gas precursors to produce solid freeform fabricated parts were described. These included direct deposition from the gas phase (SALD), reaction between the gas phase and a layered powder bed (SLRS), and vapor infiltration and deposition into a powder bed (SALDVI). All of the above have real promise as SFF approaches.

REFERENCES

Arnone, C., Rothschild, M., Black, J.G., and Ehrlich, D.J., *Appl. Phys. Lett.* 48, 1986, 14.

Birmingham, B.R., Tompkins, J.V., Zong, G., and Marcus, H.L., *Solid Freeform Fabrication Proceedings*,The University of Texas, Austin, Texas, 1992, pp. 147-153.

Birmingham, B.R., and Marcus, H.L., *Solid Freeform Fabrication Proceedings*,The University of Texas, Austin, Texas, 1993, pp. 308-316.

Birmingham, B.R., Tompkins, J.V., and Marcus, H.L., *Solid Freeform Fabrication Proceedings*,The University of Texas, Austin, Texas, 1994, pp. 348-355.

Birmingham, B.R., and Marcus, H.L., *Solid Freeform Fabrication Proceedings*,The University of Texas, Austin, Texas, 1995a, pp. 389-396.

Birmingham, B.R., Tompkins, J.V., Harrison, S.L., and Marcus, H.L., *Proceedings of the Second International Conference on Beam Processing of Advanced Materials*, Oct. 30-Nov. 2, 1995, Cleveland, OH, 105-117, 1995b.

Bittencourt, J.A., *Fundamentals of Plasma Physics*, New York: Pergamon Press, 1988.

Bourell, D.L., Marcus, H.L., and Weiss, W.L., "Selective Laser Sintering of Parts by Compound Formation of Precursor Powders" U.S. Patent Issued October 20, 1992, #5,156,697.

Figueras, A.S., Rodroguez-Clemente, R., Garelik, S., Santiso, J., Armas, B., Combescure, C., Mazel, A., Kihn, Y., and Sevely, J., *Journal de Physique IV*, Colloque C2 Supple. au Journal de Physique II, Volume I, 1991, pp C2-225-232.

Figueras, A.S., Garelik, J.S., Rodroguez-Clemente, R., Armas, B., Combescure, C., Berjoan, R., Saurel, J.M., and Caplain, R., *Materials Science and Engineering*, Volume B11, Nos 1-4, 1992, 83-88.

Fite, W.L., *Chemical Reactions in Electrical Discharges*, Edited by B. D. Blaustein, Advances in Chemistry Series, ACS 80, 1969.

Greaves, J.C., and Linnett, J.W., Transactions of The Faraday Soc 55, 1959, 1355-61.

Jakubenas, K., and Marcus, H.L., *Solid Freeform Fabrication Proceedings*,The University of Texas, Austin, Texas, 1995, pp. 381-388.

Jensen, K.F., and Graves , D.B., *J. Elec. Soc.*, 130, 9, 1983, pp. 1950-7.

Kee, R.J., Rupley, F.M., and Miller, J.A., Chemkin-II: A Fortran Chemical Kinetics Package for The Analysis of Gas-Phase Chemical Kinetics, *SANDIA Report: SAND89-8009 UC-401*, September 1989.

Kodas, T.T., Baum, T.H., and Comita, P.B., *J. Appl. Phys.* 62, 1987, p. 281.

Lee, Y.L., Tompkins, J.V., Sanchez, J.M., and Marcus, H.L., *Solid Freeform Fabrication Proceedings*,The University of Texas, Austin, Texas, 1995 , pp. 433-439.

Lehmann, O., and Stuke, M., *Science*, Vol. 270, Dec. 8, 1995, 1644-1646.

Mallard, W.G., *NIST Chemical Kinetics Database*, Version 3.0, NIST, Apr. 1991.

Marcus, H.L., "Gas Phase Selective Beam Deposition" U.S. Patent Issued May 21, 1991, #5,017,317, 1991.

Marcus, H.L., "Positioning, Focusing and Monitoring of Gas Phase Selective Beam Deposition" U.S. Patent Issued August 4, 1992, #5,135,695, 1992a.

Marcus, H.L., Zong, G., and Thissell, W.R., "Catalyst and Plasma Assisted Nucleation and Renucleation of Gas Phase Selective Laser Deposition", U.S. Patent Issued December 8, 1992, #5,169,579, 1992b.

Maxwell, J.L., Pegna, J., and Ostrogorsky, A., *Solid Freeform Fabrication Proceedings*,The University of Texas, Austin, Texas, 1993, pp. 253-270.

Maxwell, J., Pegna, J., and Hill, E., *Solid Freeform Fabrication Proceedings*,The University of Texas, Austin, Texas, 1995a pp. 143-150.

Maxwell, J.L., Pegna, J., DeAngelis, D., and Messia, D.,"Three-Dimensional Laser Chemical Vapor Deposition of Nickel-Iron Alloys", *Proceedings of the Materials Research Society*, Fall 1995 Meeting, Symposium B: Laser Chemical Processing, Nov. 27-Dec. 1, 1995, Boston, MA, 1995b.

Mills, P.L., and Ramachandran, P.A., Math. Comp. Mod. 11, 1988, 375-79.

Subramanium, P.K., Zong, G., and Marcus, H.L., *Solid Freeform Fabrication Proceedings*,The University of Texas, Austin, Texas, 1992 , pp. 63-71.

Tesner, P.A., *Chemistry and Physics of C.* Edited by Peter A. Thrower, 19, 1984, 65-161.

Thissell, W.R., Zong, G., Tompkins, J.V., Birmingham, B.R., and Marcus, H.L., *Solid Freeform Fabrication Proceedings*,The University of Texas, Austin, Texas, 1991, pp. 259-270.

Thissell, W.R., and Marcus, H.L., *Solid Freeform Fabrication Proceedings*,The University of Texas, Austin, Texas, 1994a, pp. 311-320.

Thissell, W.R., "Processing and Control of Selective Area Laser Deposition from Methane and Hydrogen", *Ph.D. Thesis*, University of Texas at Austin, December 1994b.

Thissell, W.R., and Marcus, H.L., "Design of a Closed Loop Computer Controlled System for Selective Area Laser Deposition-Part 1: Laser Systems, Gas Flow and Substrate Temperature Control", *Materials and Manufacturing Processes* 11, 1996a, p. 5.

Thissell, W.R., and Marcus, H.L., "Design of a Closed Loop Computer Controlled System for Selective Area Laser Deposition-Part 2: Remote Plasma Source and Pressure Control", *Materials and Manufacturing Processes* 11,1996b, p. 5 .

Tompkins, J.V., *Masters Thesis*, University of Texas at Austin, 1992.

Tompkins, J.V., and Marcus, H.L., *Solid Freeform Fabrication Proceedings*,The University of Texas, Austin, Texas, 1993, pp. 325-332.

Tompkins, J.V., Laabi, R., Birmingham, B.R., and Marcus, H.L., *Solid Freeform Fabrication Proceedings*,The University of Texas, Austin, Texas, 1994, pp. 412-421.

Tompkins, J.V., Birmingham, B.R., and Marcus, H.L., *Solid Freeform Fabrication Proceedings*,The University of Texas, Austin, Texas, 1995 , pp. 409-416,.

Wallenberger, F.T., and Nordine, P.C., *Science*, Vol. 260, Apr. 2, 1993, pp. 66-68.

Wallenberger, F.T., and Nordine, P.C., J. Mater. Res., Vol. 9, No. 3, Mar. 1994, pp. 527-530.

Westberg, H., Boman, M., Johansson, S., and Schweitz, J.-A., J. *Appl. Phys.* 73(11), Jun. 1, 1993, pp. 7864-7871.

Zhu, D., and Sahai, Y., *Journal of Materials Shaping Tech*. 8, 2, 1990, pp. 83-90.

Zong, G., Carnes, R., Wheat, H.G., and Marcus, H.L., *Solid Freeform Fabrication Proceedings*,The University of Texas, Austin, Texas, 1990a, pp. 83-90.

Zong, G., and Marcus, H.L., *Solid Freeform Fabrication Proceedings* ,The University of Texas, Austin, Texas, 1990b, pp. 74-82 .

Zong, G., Jacquot, Y., Thissell, W.R., and Marcus,H.L., *Plasma and Laser Processing of Materials*. Edited by K. Upadhaya, The Minerals, Metals, & Material Soc., 1991a, 23-48.

Zong, G., Tompkins, J.V., Thissell, W.R., Sajot, E., and Marcus, H.L., *Solid Freeform Fabrication Proceedings*,The University of Texas, Austin, Texas, 1991b, pp. 271-278.

Zong, G., and Marcus, H.L., *Solid Freeform Fabrication Proceedings* ,The University of Texas, Austin, Texas, 1991c, pp. 279-287.

9

SLS APPLICATIONS

9.1 BACKGROUND

Much of the growth in the freeform fabrication industry can be attributed to parallel advances in materials and process technologies. As it has become possible to produce accurate, durable parts from a variety of materials, the number of applications for rapid prototyping technologies has greatly increased. In general, early plastic-based parts had relatively poor mechanical properties and dimensional accuracy and were suitable mainly as visualization models. With material and process improvements, it is now possible to use these plastic parts as functional models as well as patterns for secondary operations. In particular, plastic patterns are widely used for making silicone rubber molds for urethane casting and for making ceramic shells for metals investment casting. More recently, as pattern-based secondary processes have become more sophisticated and metals-based rapid prototyping materials have become available, there has been an increasing industry focus on producing durable prototype tooling. In this chapter, rapid prototyping applications will be discussed from a SLS perspective. After a brief overview, two customer application stories are provided to illustrate current state-of-the-art. At the end of this chapter, a few advanced applications are discussed.

Advances in SLS materials and process technology were discussed in previous chapters. In particular, SLS plastics were discussed in Chapters 4 and 6 and SLS metals in Chapter 5. The commercial SLS materials available as of this writing and their general applications are summarized in Table 9.1.

Nylon-based polymers (Forderhase, *et al* 1994, Forderhase, *et al*, 1995a, Forderhase, *et al*, 1995b), polycarbonate and acrylic-based polymers, foundry sand, and carbon steel (McAlea, *et al*, 1996) have been commercialized by DTM Corporation. Nylon-based polymers, polystyrene polymer, foundry sand, and a nickel-bronze system have been commercialized by EOS (Sindel *et al* 1994, Serbin *et al* 1995). The nylon-based materials are generally used as functional models when durability or heat and chemical resistance are critical. The properties of nylon-based parts were discussed in Chapter 6. In some cases, these parts are also used as patterns, particularly if they must be used multiple times or when toughness and wear resistance are particularly important. In general, however, pattern-based applications are best addressed by the polycarbonate, polystyrene, and acrylic-based materials. In particular, the acrylic-based system offers superior dimensional accuracy, surface finish, and feature resolution - characteristics which are usually critical for replication processes. The porosity of these parts also allows them to be used as patterns for metals investment casting. The properties of these materials were also discussed in Chapter 6. The foundry sand allows direct production of sand casting molds and cores. The carbon steel (RapidSteel metal) and nickel-bronze system are used to produce metal-based tooling. The commercial processes for the carbon steel material were discussed in detail in Chapter 5.

Table 9.1 Current Commercial SLS Materials and General Applications

SLS Material(s)	General Application(s)
Nylon Polymer/ Glass Reinforced Nylon Polymer	Functional Prototypes Patterns
Polycarbonate Polymer/ Polystyrene Polymer/ Acrylic-Based Polymer	Patterns Investment Casting Patterns
Foundry Sand	Sand Casting Molds and Cores
Carbon Steel/ Nickel-Bronze System	Metal-Based Tooling

To date, a significant market focus of the industry has been plastics injection molding. This focus can be attributed to a number of factors: most rapid prototyping systems produce plastic-based parts which are best suited for use as plastic part models or for replication processes which also yield plastic part models or prototypes; the complexity of most injection molded geometries makes them well suited to layerwise manufacturing techniques; and the plastics injection molding market is very large. It is possible to produce plastic-based parts by a variety of routes with the SLS process. The most common routes are summarized in Table 9.2.

Table 9.2 Routes to Plastic Prototypes With SLS Materials

SLS Material(s)	Conversion Process	Final Part Material	Typical Number of Parts
Nylon Polymer Glass Reinforced Nylon Polymer	None	SLS Material	1-5
Acrylic-Based Polymer	Vacuum Casting-Silicone Rubber mold	Urethane	20-50
Carbon Steel/Epoxy Nickel-Bronze/Epoxy Carbon Steel/Copper	Injection Molding-Metal-Based Mold	Choice	50 - 200 ? 50-50k+

The first two options produce final parts which are really plastic models in the sense that they are not made with the production process or material. The last option allows "true" prototypes to be injection molded in the material of choice. Durable plastic models can be produced from nylon-based materials very quickly since the parts are used directly from the SLS process with no conversion step. Typically, this approach is best suited when a relatively small number (< 5) of plastic models is needed. If more models are required, it is generally more economical to produce urethane castings from a silicone rubber mold made from a SLS pattern. Typically, 20 - 50 castings can be made from a single silicone rubber mold. Urethane systems with a range of mechanical properties and colors are available. The steel-based materials offer the user the ability to quickly make a durable metal-based mold. For an epoxy infiltrated steel mold, approximately 200 parts can be molded, while for a copper infiltrated mold, as many as 50,000+ parts can be made. The lead times are approximately 3 days for the epoxy infiltrated tooling and 10 days for the copper infiltrated tooling. It should be pointed out that other pattern-based approaches for making prototype tooling have been the focus of much attention in the last several years. A number of secondary processes, including investment casting (Denton and Jacobs 1994), spray metal (Fussell 1994), and Keltool have been used to produce metal tooling from plastic patterns, particularly those made with the SLA process. While it is also possible to produce metal tooling from SLS patterns using such processes, more attention has been given to the more direct metal powder-based methods. The ability to produce nylon parts, urethane castings, and injection molded parts quickly with a single process allows users to address the prototype requirements of this market in a comprehensive fashion. It is common to produce nylon or urethane models early in the design cycle and then injection mold parts with a SLS metal-based mold once initial design changes have been made.

The market for prototype metal parts has not been addressed to the same degree as that for prototype molded plastic parts. Much of the focus has been on limited run investment casting and, more recently, sand casting. The most common routes for producing metal parts with the SLS process are summarized in Table 9.3.

Table 9.3 Routes to Metal Parts With SLS Materials

Metals Production Process	SLS Material(s)	Application for SLS Parts
Investment Casting	Acrylic-Based/Polystyrene/ Polycarbonate Polymers	Expendable Patterns
	Carbon Steel/Nickel	Tooling for Wax Pattern Molding
Sand Casting	Foundry Sand	Molds and Cores
	Nylon-Based Polymers	Limited Run Pattern
Die Casting	Carbon Steel/Copper	Limited Run Tooling
Forming	Carbon Steel/Copper	Limited Run Tooling

For investment casting, direct pattern production for a limited number of parts is possible as was discussed in Chapter 5. It is also possible to produce production quantities of injection molded wax patterns utilizing SLS metal-based tooling. For sand casting, nylon-based patterns have been used as short run patterns for sand molds, and foundry sand material can be used to directly manufacture sand cores as well as molds. These processes were also described in Chapter 5. Finally, initial trials have demonstrated that copper infiltrated carbon steel molds can be used for short run die casting and forming. It is anticipated that as more metals-based materials become available, the options for producing metal parts via the SLS process will increase dramatically. There are three broad areas of interest: metal-based "models", that is, metal parts which are not made in the production material or process, but which can be used for some functional testing and evaluation early in the design cycle; durable tooling for processes such as die casting and forming; and parts made from materials which can not be shaped using traditional processes such as machining.

It is also anticipated that there will be some specialized SLS ceramics development for medical (Lee and Barlow 1994) and aerospace (Deckard and Claar 1993, Griffin and McMillin 1995) applications.

In the next two sections, current applications of SLS technology are illustrated through two customer inputs. In keeping with the application trends discussed above, the focus of both discussions is prototype tooling. The authors would like to thank Richard Gee of Sunstrand Aerospace and Tim Gornet of The University of Louisville for their contributions.

9.2 PLASTIC PART PROTOTYPING AT THE SPEED SCIENTIFIC SCHOOL, UNIVERSITY OF LOUISVILLE

The process by which we develop and manufacture products has changed dramatically over the last ten years. The advent of new technologies has added many new options to the engineer's toolkit. The spread of breakthrough technologies such as computers, parametric and variational solid modeling, finite element analysis, injection molding simulation, and rapid prototyping has given companies the chance to change the development process from serially based sequential design to parallel or concurrent engineering design.

The serial design cycle can be simplified into the following steps: concept, design, prototype, test, and manufacture. The problem with the serial design cycle is twofold. First, the cycle time is long because each function does not play a part in the development cycle until the previous function's work is complete. Second, 85% of the total project dollars are committed when only 5% have been spent. This is illustrated schematically in Figure 9.1.

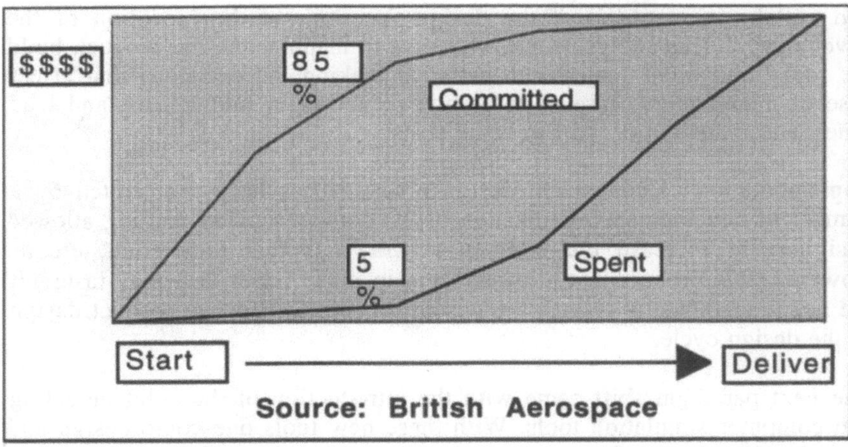

Figure 9.1 The impact of a serial design approach on project spending.

Therefore, decisions made in the first steps of a design are critical in controlling the product cost and time to market. The product cost can also increase dramatically if changes are made late in the design cycle. A change made during the manufacturing engineering phase is ten times more expensive to make than it is during the concept or design phase. This same design change is 100 times more expensive if it occurs after the part is in production. These relationships are shown in Figure 9.2. It is obvious from these numbers that if various "what-if" options can be examined early in the design cycle, a better product can be brought to market faster.

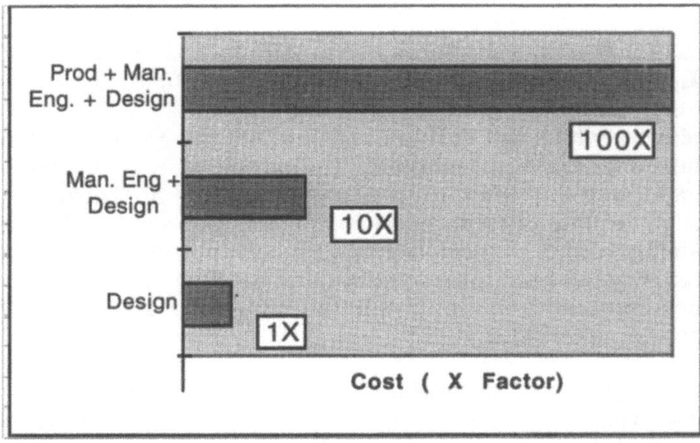

Figure 9.2 Cost multipliers associated with the time at which design changes are made.

This method of design - parallel or concurrent - involves the coordination and collaboration of all of the design functions at the inception of the development. This allows people who have to manufacture the product, build the tools, and order material to assist in making the decisions that affect ease of manufacture, ease of assembly, reduction of tooling cost and lead times, and reduction of part material costs.

Implementation of concurrent design practices has been accelerated by a number of developments. Utilization of 2D computer aided drafting allowed designers to increase the number of drawings that they could create. However, this process just allowed them to make paper drawings faster. It did not fundamentally change the design process or improve product design or the design cycle.

The next paradigm shift came with the introduction of the solid modeling and computer simulation tools. With these new tools one could design and visualize 3D objects. Parametric and variational solid modeling allow parts to be designed using features rather than edges as the building blocks. Solid models provide an unambiguous design with much more information - volume, surface area, and inertial properties - than a wireframe model. Assemblies can be related and constraints and dimensions added to give a part the necessary design intent. These tools allow checking assemblies for interferences, calculating part weights, and constructing a bill of materials. The designs can easily be modified by varying dimensions, parameters, or relations. In addition, finite element analysis can be used to determine application suitability, for making material choices, and for optimizing designs to minimize part weight or wall thicknesses.

The manufacturing process can also be simulated. If the part is to be injection molded, process simulation can be performed to evaluate part manufacturability. This allows one to optimize the process parameters to decrease the molding cycle time. The simulation can also provide qualitative shrinkage and warpage information. Thus, gating issues, material selections, and design concerns can be addressed before the part is in production. Generative machining software provides the capability to create NC tool paths automatically even for complex parts. Automatic fixture avoidance, feed and speed optimization for various materials, and tool path verification are also built in.

Engineering documentation is easily derived from the solid model. It is an outgrowth of the design, not the driving factor. In addition, all of these modules are bidirectionally associative. Changes made at any level are carried through the entire design cycle.

The introduction of rapid prototyping (RP) was a timely addition to the new solid modeling tools, even though the early parts were quite fragile. Models of objects could be produced quickly. Instead of pouring over a large stack of prints in a design review, the model could be passed around and discussed. Such parts provide an excellent communications tool for engineers, manufacturing engineers, and toolmakers. They can be used to evaluate form, fit, and design for manufacture/assembly issues. Despite these advantages, engineers and designers wanted more from RP parts. They wanted to be able to test actual function, drill screws, mount solenoids and switches, and assemble RP parts into a product for functional testing as well as produce material of choice parts from prototype tooling.

It is now possible to produce such functional parts and prototype tooling using DTM's Sinterstation 2000 and SLS materials. The application of SLS nylon-based functional prototypes and metal-based tooling to plastic part manufacturing are discussed below.

9.2.1 Nylon-Based Functional Prototypes.

As was discussed in Chapter 6, nylon and glass reinforced nylon (ProtoForm Composite) offer end users functional RP parts with good stiffness, elongation, and temperature capabilities. Parts made with these materials have been a tremendous boon to designers of plastic parts due to the long lead times normally required to get parts from a tool. This lag time obviously increases the time to market. It also limits the number of design iterations that can be undertaken due to new product introduction constraints.

Product development can be accomplished on an almost real time basis with nylon-based materials. The Colgate-Palmolive Co. plant in Jeffersonville, Indiana produces Colgate toothpaste in large 8 ounce tubes. The product is filled from the cap area using a filling head. The filling head is positioned inside the open tube and two different color pastes are extruded as the filling head is backed out of the tube. The bottom of the tube is then sealed. It is imperative that the two different color formulations do not mix. They wanted to address a new market with a travel size (1.3 ounce) container of two color toothpaste. However, the time frame for product delivery was critical and they wanted to be able to use existing filling equipment to accomplish the task.

After contacting Speed School Dean Thomas Hanley, Colgate representatives, including an engineer flown in from the company's worldwide technology center in Piscataway, New Jersey, quickly collaborated with the engineering school's Rapid Prototype Facility to modify the factory dispensing nozzle. Sitting down with a graduate engineer, they quickly developed the first iteration of the nozzle. Prototypes were built from SLS nylon. These inserts were placed into the filling equipment and tested. Several iterations were made and also tested. In less than two weeks, a final design that had a high flow rate of material with no mixing of the two components of the toothpaste was completed. Jim Humphrey, personal care products manager at the plant, estimated that it saved six weeks of development time. The company's conservative estimate on cost savings was "in the hundreds of thousands of dollars" due to capital equipment expenditures which were avoided.

Another example in which nylon-based parts proved to be critical was in the development of a contractor grade screw gun. Paul Scherer of Intec, Inc. of Lexington, Kentucky was the primary design engineer for this project for Quik-Drive, Inc of Nashville, Tennessee. The assembly, consisting of 30 parts, is a mechanical engineer's dream. The design takes the high speed rotary motion of a commercial screw gun and transfers it through a shaft to a square bit drywall or decking type screw. The product utilizes banded screws that come in rolls that fit into a magazine. This allows a contractor to set over 100 screws without reloading. Depth set is controllable, and there are multiple guide tubes for various screw lengths. Over 5,000 screws were run with glass filled nylon parts for evaluation purposes.

Over the development cycle time, numerous prototypes were built for all of the components to evaluate and fine tune the performance of the screw gun.

Feedback from customers and Quik-Drive guided Intec in this process. The prototypes also provided a basis for cost reduction discussions and manufacturing cost calculations.

Many companies produce such RP parts just to check tooling and provide guidance for the tool vendor. This has provided some people tens of thousands of dollars in tooling cost avoidance due to rework or even the total scrapping of a tool. One example concerns a product where two halves of an electronics housing were designed by separate people. The chips were already being cut on the production tooling. An RP part was built to show management. One small problem - one of the engineers had moved a snap fit on the cover without informing the designer of the base to move the mating through holes! Upon discussions with the tool vendor, it was determined that discovering this error before the area in question had been machined saved the company over $50,000 in tooling change costs.

9.2.2. Rapid Tooling Using the RapidTool Process.

As was discussed in Section 9.1, RP parts have been used as masters for secondary tooling since the infancy of rapid prototyping. All of these methods require one or two secondary processes to obtain the final tool. With the introduction of DTM's RapidTool Process, one can create injection mold core and cavity inserts in a more direct fashion.

In July of 1995, the University of Louisville's Rapid Prototype Facility and Xerox Corporation's Manufacturing Technology Center became Beta sites for this process. The RapidSteel metal is run in the same machine as the nylon-based plastics. The time required to switch between them is short (0.5 -1 hour).

A description of the RapidTool Process for producing core and cavity mold inserts can be found in Chapter 5. Preparing the solid model for the RapidTool Process requires some up front evaluation. This initial work is greatly simplified if mold design tools such as those found in some of the variational/parametric modeling software are available. First, the parting line must be determined to split the mold into the core and cavity insert plates. These plates usually conform to a standard pocket size for a Master Unit Die (MUD) assembly. The core and cavity dimensions must be adjusted for injection molding process shrinkage as well as for any post machining and finishing stock which is required. For parting lines, an additional 0.020 to 0.030 inch of stock is typically added to allow for machining the parting surfaces. For areas with critical dimensions and tight tolerances, 0.030 - 0.040 inch additional material is allocated for machining to the exact dimensions. Cooling lines can be added to the inserts as well. This includes curved cooling lines that follow contours of parts. The diameter of these must be at least 0.25 inch to allow the metal powder to be blown out of the lines before the polymer infiltration. For ejector pins and other through holes, the option exists to put them in the model at true size, slightly undersized for locating purposes, or leave them for secondary machining. In most cases they are designed undersized and brought to dimension later.

Once the resulting RapidTool inserts are built, they are ready for finishing and placement into a MUD assembly for injection molding. The finishing required to create the final parts obviously varies depending upon part geometry and complexity. All conventional tooling practices have been evaluated on the infiltrated metal material. The material has the strength of steel, but the machining properties of aluminum. Surface grinding, drilling, tapping, wire and plunge EDM, milling, turning, and MIG and TIG welding have all been performed. These processes have been accomplished with no special setups or constraints. The parts polish quite well to a very metallic shine. Polishing is done using conventional techniques. A 300 to 400 paper finish can be achieved with a minimal amount of bench time. In addition, the parts take plating quite well due to their excellent conductivity. By polishing the parts, plating, and then polishing the parts again, a very durable mold can be realized.

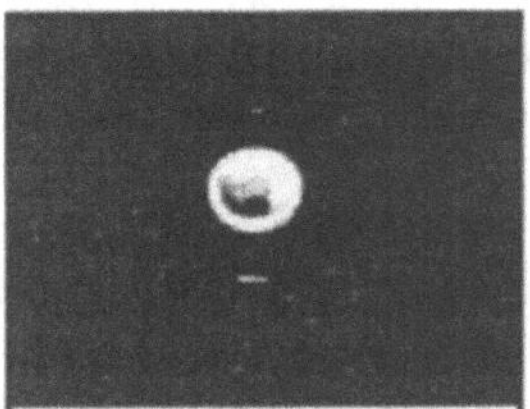

Figure 9.3 A plastic "star wheel" pump impeller assembly

As an example, a RapidTool insert set was used to evaluate the feasibility of a new product design for Grindmaster Corporation, Louisville, Kentucky. The part, a pump impeller assembly, consisted of a flexible vane pump wheel insert molded about a rigid "star wheel", as shown in Figure 9.3.

Due to the design requirements, nylon-based rapid prototype parts could not be used to simulate the assembly for product testing. The new product hinged upon the function of the vane pump wheel. RapidTool mold insert sets were completed and placed into a MUD base. One of the inserts is shown in Figure 9.4.

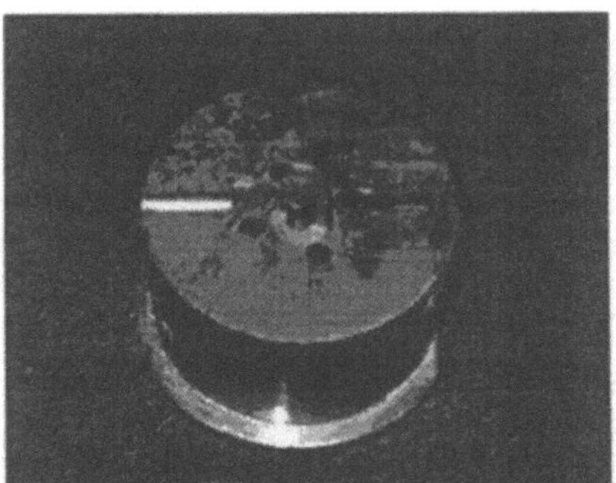

Figure 9.4 A RapidTool insert for the part shown in Figure 9.3.

The mold designs were straightforward. For the vane pump insert the ejector pin holes were simply marked with a dimple. Due to the flexibility of the material, a 1/16" ejector pin was placed at each vane. The ejector holes were machined conventionally as a secondary operation.

Approximately 50 "star wheel" parts were molded out of Delrin. These parts were then inserted into the vane pump mold and Texin was molded about it. Table 9.4 shows the costs and savings associated with this project.

Table 9.4 Cost and Time Savings For the "Star Wheel" Pump Impeller Assembly

	Star Wheel	Vane Pump
Insert Size	3"x4"	3"x4"
Design Time	3 hrs	4 hrs
RapidTool Build Time	6 days	6 days
Polishing/Machining Lead-time	2 days	2 days
Polishing/Machining Cost	$1,600	$2,500
Conventional Mach. Lead-time	6-8 weeks	6-8 weeks
Conventional Machining Cost	$3,800	$4,400
Material	Delrin	Texin
Lead-time Savings	4 -6 weeks	4 - 6 weeks
Cost Savings	$2,200	$1900

The RapidTool Process has been marketed primarily as a prototype tooling solution. Experience, however, is beginning to show that there are additional applications for this new tooling technology. These applications include interim tools, "disposable" production tools, and die casting tools. In the case of interim tooling, a company can utilize the inserts to accelerate the time to market for a product. The production tool is still critical for long term production, but it can be completed without the high costs associated with premium tool delivery. In addition, the part can go into production quickly without the long lead-time delay for tooling.

In the following case, an interim tool was required to fulfill production demands until the production tool could be built. The parts, shown in Figure 9.5, were made from unfilled polypropylene.

Figure 9.5 Polypropylene parts molded with a RapidTool mold.

It was expected that approximately 10,000 parts would be required. The larger part was designed and built in two inserts for each half so that it could be interchanged in the future. The core for the larger part used to create the internal geometry was a simple turned section. It was machined conventionally on a turning center. The inserts are shown in Figure 9.6 and the assembled molds in Figure 9.7. This tool now has a history of over 60,000 shots with no appreciable wear. It is believed that it could last for another 50,000 to 60,000 shots.

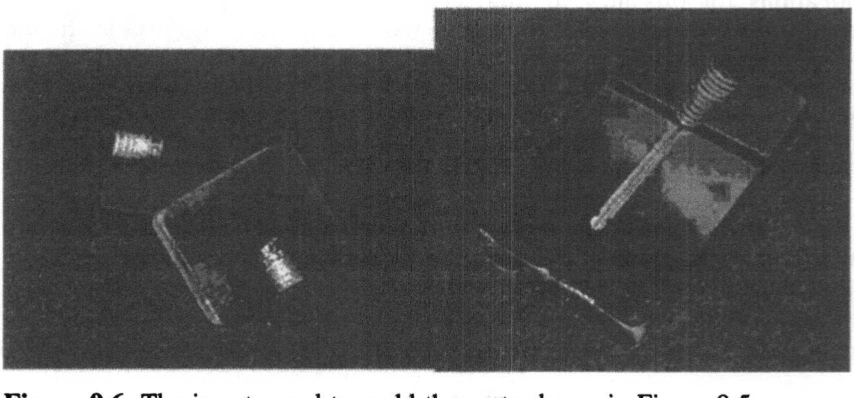

Figure 9.6 The inserts used to mold the parts shown in Figure 9.5.

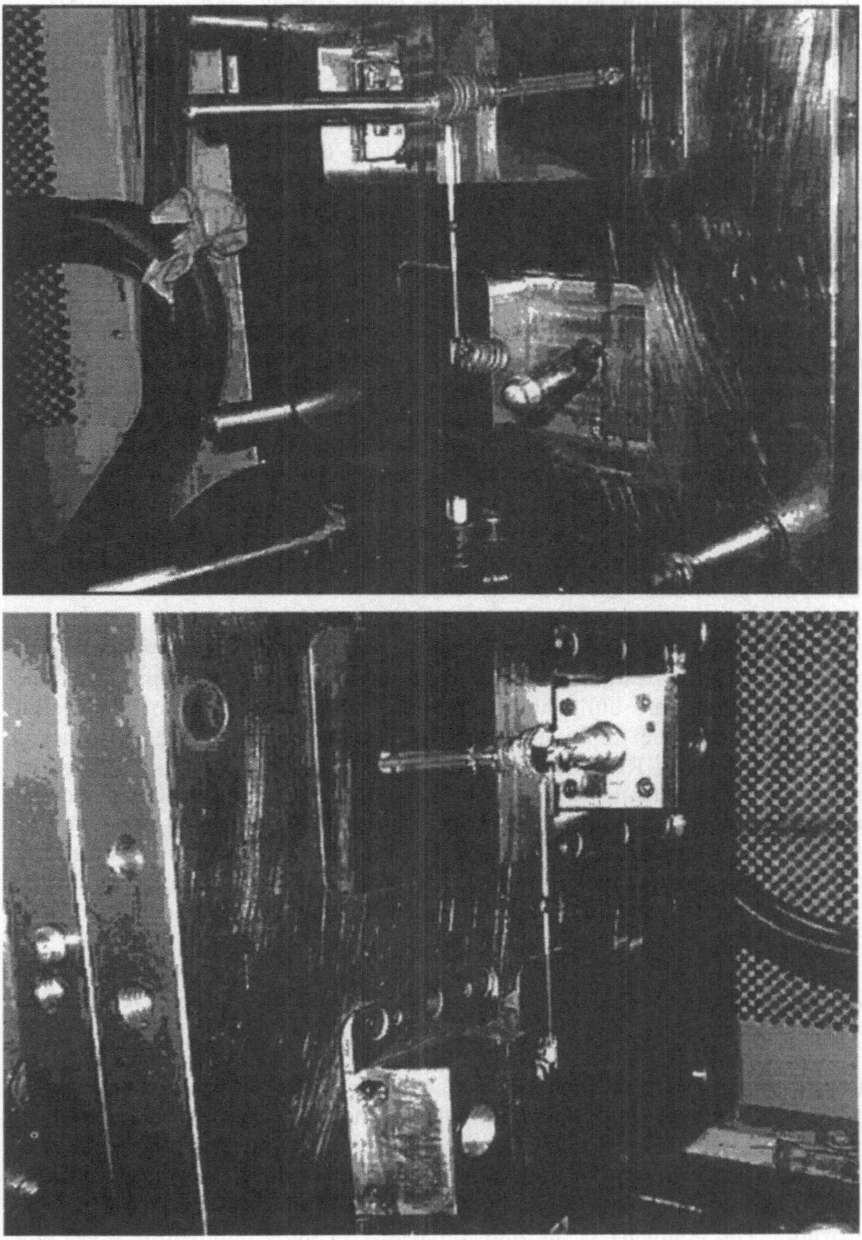

Figure 9.7 The assembled molds used to produce the parts shown in Figure 9.5.

The time to produce the RapidTool mold was 5 weeks and the total cost was $16,375. An aluminum mold would have taken 8 weeks and cost $23,750.

It is expected that the RapidTool Process will be utilized in conjunction with conventional tooling methods. RapidTool inserts for a complex geometry can be created with a slight steel safe condition and a final finish NC cut or finish EDM burn used to produce the final tool. Only critical features or dimensions would have to be made steel safe. Essentially, the Sinterstation would function as an unattended five axis mill. Therefore, a machine shop could get 80% - 95% of the way to a tool with little manpower or NC equipment time tied up in a project.

This synergy can include software as well. Solid modeling tools are rapidly maturing so that core and cavity mold halves can be created semi-automatically. In addition, injection molding simulation software packages are becoming more adept at allowing optimized gate and runner placement as well as cycle conditions and times to be determined. Some injection molding simulation software can even automatically create a mesh with defined wall thicknesses from a .STL file. Imagine combining such software with the RapidTool Process - finish a part design, create the mold halves, optimize injection placement and parameters, and start the RapidTool part build - all in the same day!

It is important to understand where prototyping makes the largest contribution to the design cycle. As mentioned earlier, most of the costs and design constraints are built into the product in the first stages of the design cycle. Therefore, the iterative process of evaluating various designs should be completed using RP parts made from nylon-based materials. This gives the designer much faster feedback on the part quality. The designer and the toolmaker or manufacturing engineer should be working together at this stage to minimize the complexity of the mold. Reducing or eliminating undercuts and features that would require cams, slides, or loose piece inserts should be tackled up front. This allows for simpler mold design, easier construction, and lower cost.

9.3 RAPID TOOLING BENCHMARKING AT SUNSTRAND AEROSPACE

Sundstrand Aerospace (Rockford, IL) has been in the aerospace business for fifty years, serving both the commercial and military sectors. Primary products include aircraft electric power generators and control electronics, control surface hydraulic actuators, electric motors, ram air turbines (RATS), jet engine fuel pumps, hydraulic pumps and electric power system software. The company is the industry leader in these application areas and is recognized as a preferred supplier.

Sundstrand has been very active in using rapid prototyping technologies over the past five years. This effort has been complemented by the addition of solid modeling tools and a desktop mill, as well as the implementation of supplier partnering and Integrated Product and Process Development (IPPD). The focus of the rapid prototyping effort has been to shorten the engineering development cycle and to achieve integration with chip cutting plant sites. To date, rapid prototyping parts have been used successfully as design verification models, marketing aids, non-functional mockups, and functional prototypes. There has been increasing interest in applying new rapid prototyping mold making capabilities to production part molding. In this project, the potential of several of these "rapid tooling" processes was examined. The application of new processes such as these has been a major factor in Sundstrand's ability to maintain its leadership position and increased profitability.

9.3.1 Benchmarking Selections.

Sundstrand uses a relatively small number of molded plastic parts each year. The production quantities of each of these parts is typically 1,000 or less. It was determined that a simple, existing production part, manufactured in relatively large quantities, should be utilized for the benchmarking exercise. A terminal board part used to make electrical connections was found to meet these criteria. A production part is shown in Figure 9.8.

Figure 9.8 A production terminal board molded from a fiberglass reinforced thermoset polymer

This part is molded from a mineral and fiberglass reinforced thermoset polyester, a material typically used in aircraft environments. The parts produced with the rapid tooling molds were made by the transfer molding process. This material and molding process require a durable mold material. To best meet this requirement, two rapid tooling approaches were chosen: the DTM RapidTool Process and the Cast Tool Steel (CTS) Process.

A current Sunstrand supplier, Kurz-Kasch (Wilmington, OH), was contracted to obtain the rapid tooling mold inserts and mold the parts. Kurz-Kasch utilized D&C Industries (Livonia, MI) to manufacture the RapidTool mold inserts and the SLS casting patterns. Sarcol, Inc. (Chicago, IL) was selected to produce the cast metal inserts from the SLS patterns.

9.3.2 Mold Making.

To facilitate use of the rapid tooling processes, a solid model of the terminal board was created using Unigraphics™V10.5.1. An IGES surfaced model file was then generated and sent to D&C Industries. Molds were created in parallel using both the DTM RapidTool and CTS processes.

A flowchart which illustrates the steps required to produce molded parts via the RapidTool process is shown in Figure 9.9.

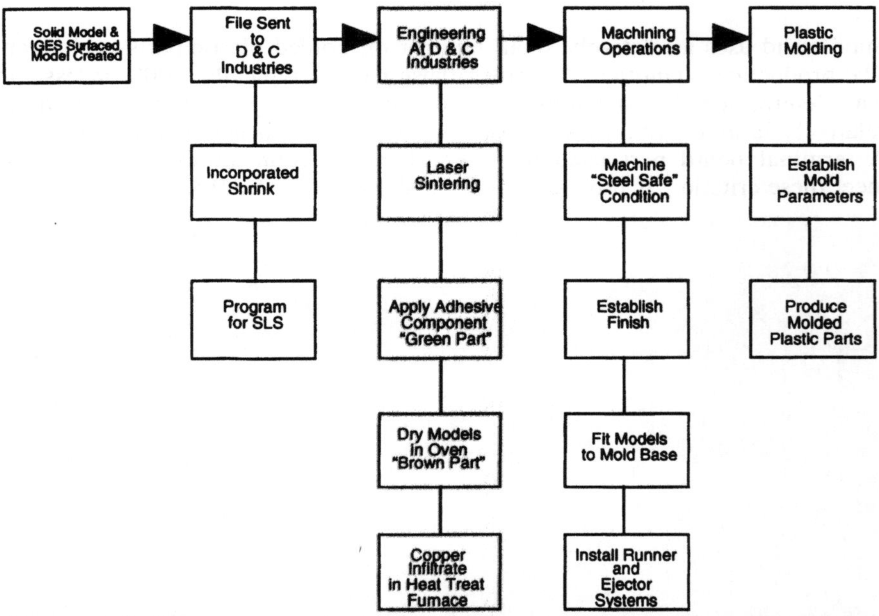

Figure 9.9 The steps required to produce molded parts via the RapidTool process.

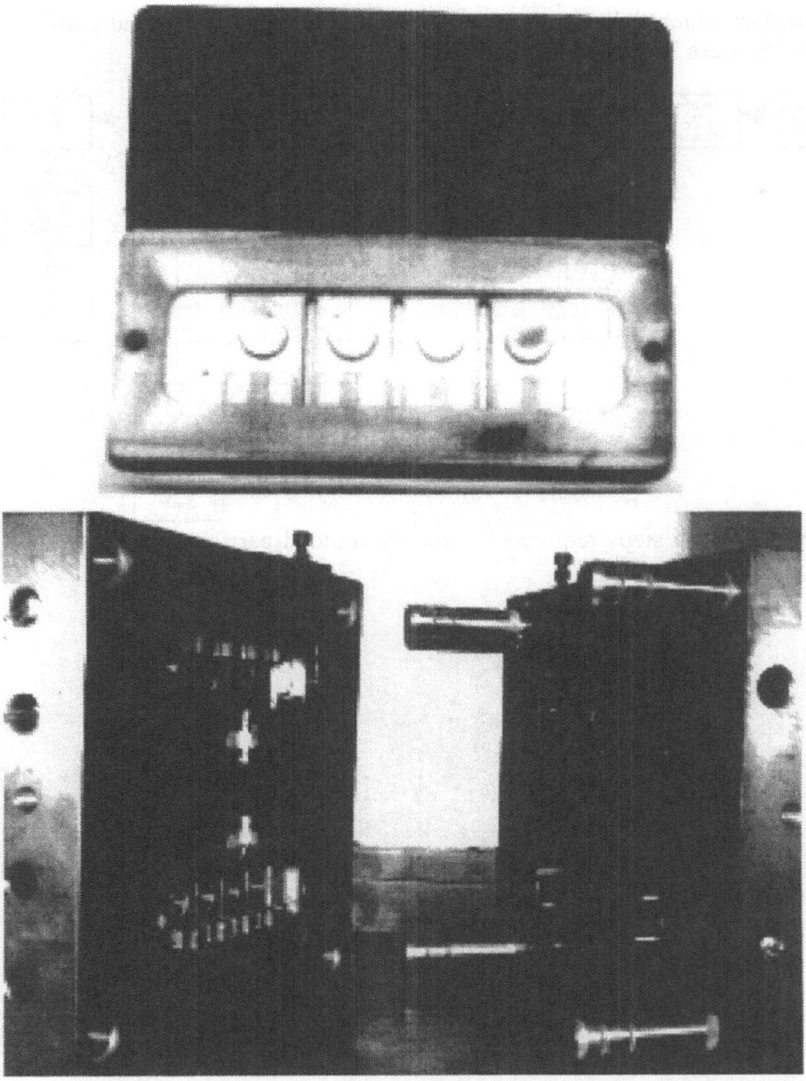

Figure 9.10 RapidTool inserts (top) and the completed mold (bottom).

D&C Industries utilized CAMAXTM CAM software to create the core and cavity inserts and parting line location from the IGES surfaced model. A mold base was then created to accommodate the inserts. Next, the RapidTool mold inserts were produced from scaled .STL files as described in Chapter 5. These inserts were machined to final configuration with the runner and ejector systems and installed in the mold base. The inserts and completed mold are shown in Figure 9.10.

A flowchart which illustrates the steps required to produce molded parts via the CTS process is shown in Figure 9.11.

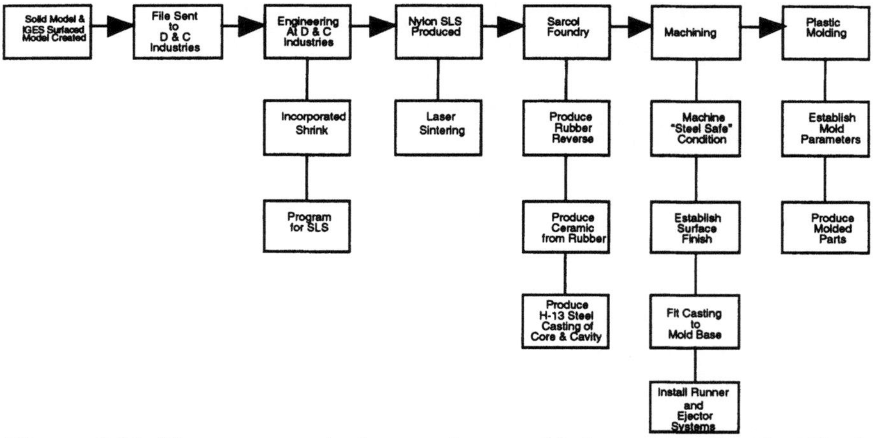

Figure 9.11 The steps required to produce molded parts via the Cast Tool Steel (CTS) process.

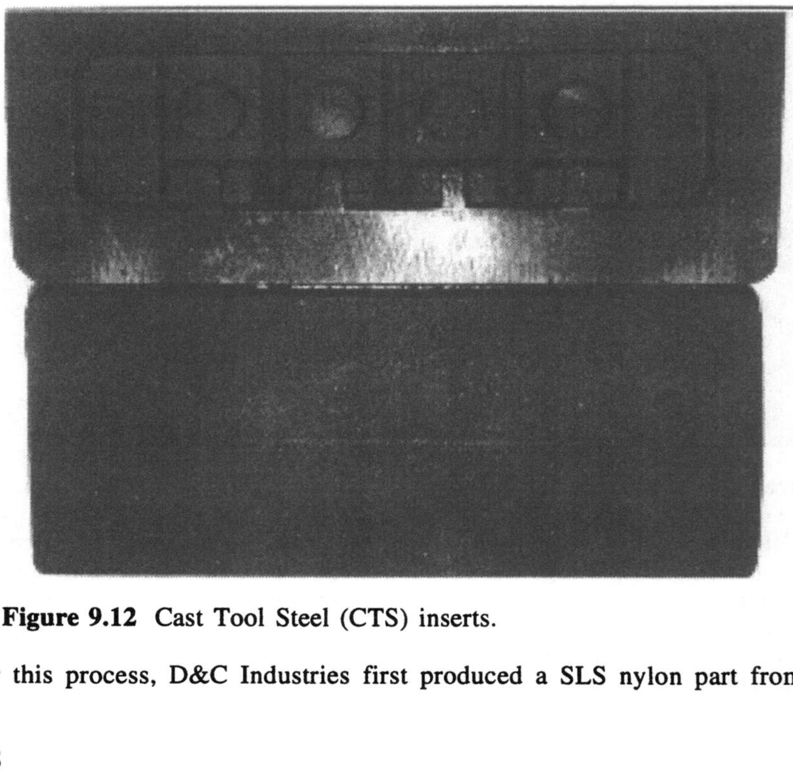

Figure 9.12 Cast Tool Steel (CTS) inserts.

For this process, D&C Industries first produced a SLS nylon part from a scaled .STL part file. Sacrol used this pattern to produce a ceramic mold via a silicon rubber negative. The core and cavity inserts were then cast from H-13 tool steel. These castings were annealed to R_c 26 to facilitate machining, but no further heat treating was performed in order to minimize the potential for distortion. The inserts were machined to final configuration with the runner and ejector systems and installed in the mold base. The inserts are shown in Figure 9.12.

9.3.3 Part Molding.

The process conditions used to mold parts with the rapid tooling molds and the production mold are given in Table 9.5. The conditions were nearly identical for all three runs.

Table 9.5 Terminal Board Molding Parameters.

	Mold Heat Top Half	**Mold Heat Bottom Half**	**Transfer Pressure**	**Clamp Pressure**	**Hold Time**	**Post Cure**	**Mtl.**	**Shot Wgt.**
DTM Rapid Tool[1]	290° F	310° F	900 psi	2300 psi	2 Minutes	2 Hours @ 250° F	IDI 44-10 HGM	220 gm
H-13 Cast Tool[1]	290° F	310° F	900 psi	2300 psi	2 Minutes	2 Hours @ 250° F	IDI 44-10 HGM	220 gm
Production Hard Tool[2]	290° F	290° F	900 psi	1800 psi	2 Minutes	2 Hours @ 250° F	IDI 44-10 HGM	220 gm

Secondary operation required: Minor flash removal and gate removal performed at the press during the molding cycle.
[1] Transfer mold process [2] Compression mold process

Parts made with the RapidTool and CTS molds are shown in Figures 9.13 and 9.14.

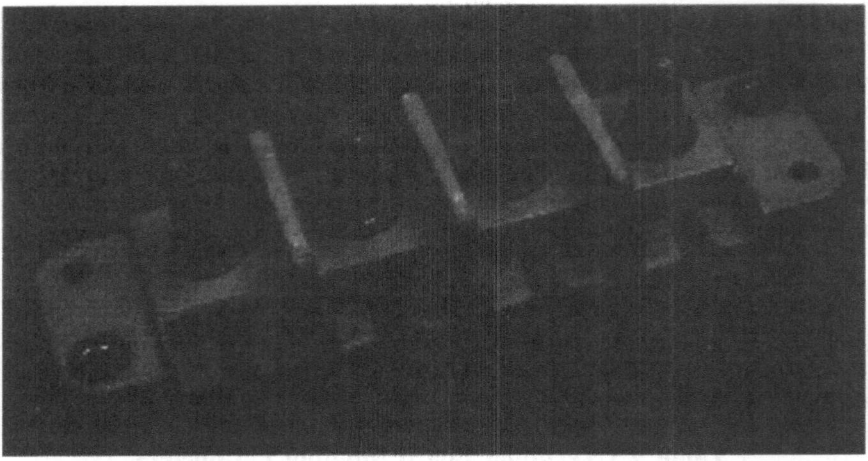

Figure 9.13 Terminal board molded with a RapidTool mold.

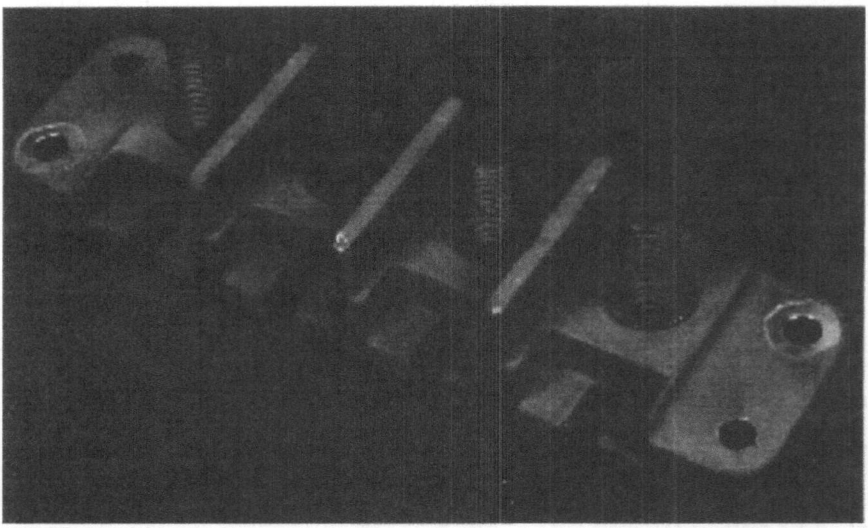

Figure 9.14 Terminal board molded with a Cast Tool Steel (CTS) mold.

Table 9.6 Kurz-Kasch First Article Inspection Report

No.	Print Dimensions	DTM Metal Insert	H-13 Cast Insert	Comments
1	5.812 ± .010	5.812	5.785	Overall Length
2	1.530 ± .010	1.533	1.542	Overall Width
3	5.188 ± .005 (2X)	5.187 5.190	5.187 5.193	
4	4.281 ± .010	4.286	4.290	
5	3.156 ± .010	3.160	3.164	
6	2.031 ± .010	2.027	2.036	
7	.906 ± .010	.898	.910	
8	.906 ± .005 (2X)	.909 .907	.904 .905	
9	.625 ± .010	.624	.622	
10	.312 ±.010 (2X)	.308 .316	.310 .310	
11	.005 +.025/ - .000 (2X)	.007 .010	.005 .006	
12	.228 ±.005 (2X)	.224 .225	.231 .233	
13	.258 ± .032 (2X)	.250 .255	.261 .263	
14	.620 ± .010	.614	.595	Body Thickness
15	4.312 ± .010	4.318	4.283	Width Across Studs
16	.500 ± .010	.502	.507	
17	.156 ±.010 (2X)	.158 .161	.168 .174	Lug Relief Thickness
18	.156 ± .010 (3X)	.158 .167 .170	.162 .161 .159	Rib Thickness

Flatness was checked across the bottom of the part. The DTM metal insert and the H13 cast insert produced parts flat from .006 inch to .009 inch.

Overall, the accuracy of the parts produced with the RapidTool mold was judged to be excellent, while the accuracy of the parts produced with the CTS mold was judged to be good (Table 9.6). Several dimensions were found to be "out-of-print" for parts made with both rapid tooling molds, particularly in the rib area. The parts molded with the cast tool inserts had a greater number of "out-of-print" dimensions. Part functionality was evaluated by conducting mechanical and electrical properties tests to determine stiffness, stud torque-out and dielectric strength. These part properties were compared to those of the production parts. The results of these property evaluations are given in Table 9.7. From this standpoint, parts made with both rapid tooling molds were judged to be excellent.

Table 9.7 Molded Part Functionality

FUNCTIONALITY TEST[1]			
Process	**Stiffness (Lbs/in. Deflection)**	**Stud Torque-out[2] (Ft.-lbs.)**	**Dielectric (Volts/Mil)**
Production Hard Tool[3]	7,080 AVE	62.6 AVE	290 - 300
DTM RapidTool[4]	6,381 AVE	57.3 AVE	290 - 300
Cast Tool Steel[4] (CTS)	6,480 AVE	56.9 AVE	290 - 300

1 Functionality tests conducted on three sample parts
2 In-service spec requires 144-165 In.-Lbs. (12-13.75 Ft.-Lbs.) torque
3 Parts produced via compression mold process
4 Parts produced via transfer mold process

Despite these positive results, it was evident that some mold modifications were needed in order to produce production quality parts. Parts made with both molds had distorted, poorly filled ribs caused by the lack of venting along the top of the ribs. In addition, the parts lacked the "molded-in" reference designations required for production parts. The tool modifications necessary to correct these problems could be made quite easily.

It might be possible to produce a more accurate CTS mold in less time. To accomplish this, castings would be made from SLS polycarbonate or TrueForm patterns, rather than a SLS nylon pattern. This would eliminate the need to produce a silicone rubber negative, since these materials can be used directly in the investment casting process.

The lead time and cost associated with the rapid tooling options are also critical benchmark criteria. A summary of the lead time and cost data is provided in Table 9.8. For the RapidTool process, the first parts were molded 6 weeks after the IGES file was delivered to D&C Industries. In the case of the CTS process, this time was 10 weeks, while for the production process, it was 16 weeks. The cost factors for both of the rapid tooling processes were approximately 1/3 that of the production tool.

Table 9.8 Lead Time and Cost: Rapid Tooling Options Versus Standard Hard Tooling

	Tooling Cost Factor	Lead Time	Cycles Before Repair	Are Slides Possible?	Cost For Slides	Added Lead Time
DTM Rapid Tool	35	6 Wks	* 1,200	** Yes	$1,500	3 Days
H-13 Cast Tool	27	10 Wks	1,200 +	** Yes	$1,500	3 Days
Standard Hard Tool	100	16 Wks	100 K+	** Yes	$3,000	1 Week

* DTM Corporation has compared the durability of the copper infiltrated steel to that of aluminum. Based on our experience, we can expect to see 1,200 cycles when processing reinforced thermoset molding compounds before wear occurs.

** Slide mechanism for adding reference designations.

NOTE: The figures in the "Cycles Before Repair" column have no historical data for representation.

9.3.3 Conclusions.

The overall project results were very favorable from the standpoint of tooling lead time and cost as well as part accuracy and functionality. Although the RapidTool and CTS processes need refinement, the molded part quality demonstrated that the concepts are sound. Based on the results to date the following conclusions were reached:

- Rapid tooling processes are viable for producing simple, thermoset plastic parts.
- Tooling lead time can be cut up to 50%.
- Tooling costs can be cut up to 50%.
- Part quality and repeatability were excellent for the DTM RapidTool process.
- Part quality and repeatability were good for the cast tool steel (CST) process.
- Part functionality was excellent (based on the tests conducted).

The next steps include modifying both tools to add venting in the rib area and a slide mechanism for the reference designations. The intent is to bring the molds up to "production" standard, and then mold and evaluate additional parts. New cast tool steel inserts will be made using SLS polycarbonate or TrueForm casting patterns in an effort to further refine the CST process. At this point in time, expectations are very high that these rapid tooling processes will be used to produce new thermoset plastic parts.

9.4 ADVANCED APPLICATIONS

9.4.1 Metal-Matrix Ceramic Composites

An interesting application of the indirect method of part fabrication is the production of Metal-Matrix Ceramic composites or MMC's. Lanxide Corporation, Newark, Delaware, and the University of Texas have combined technologies to demonstrate the production of MMC products. Polymer-bound Silicon Carbide green shapes were prepared by SLS (Vail, *et al*, 1993) and shipped to Lanxide, where the polymer was thermally removed and molten aluminum, in a nitrogen atmosphere at 750-850°C, was infiltrated by capillary action into the porous shape (Deckard and Claar, 1993). Silicon Carbide loadings near 45 vol% are typically achieved. At this loading, moduli near 180 GPa, 4-point bending strengths near 275 MPa, and CTE values near 9 ppm/°C, were reported. These results match those expected from various theories of composite materials. The chief advantages of this approach include the ability to form complex geometries that can not be achieved by the slip casting process usually used to shape the ceramic from slurry, near-net shape capability (less than 1% shrinkage relative to the preform), and minimum machining after infiltration. Some example preforms are shown in Figure 9.15.

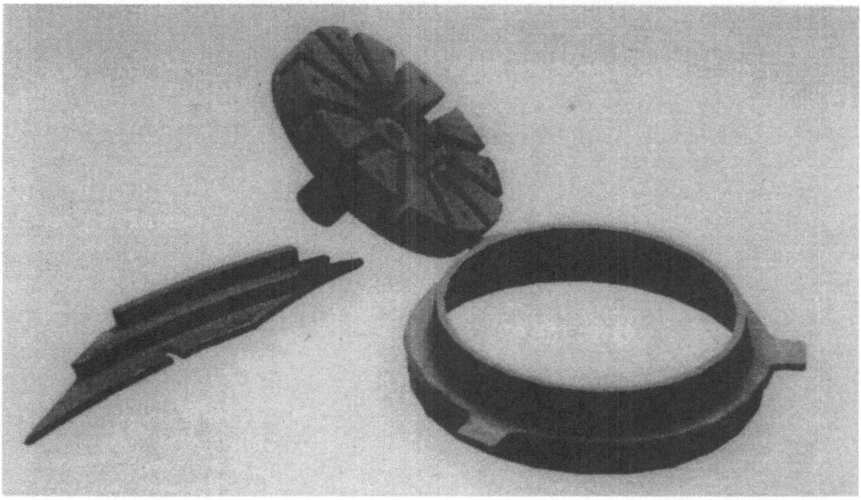

Figure 9.15 Green shapes made of polymer-bound silicon carbide by SLS.

9.4.2 Casting Cores

Figure 9.16 shows a silica (possibly mullite)-cemented alumina casting core that was tested for use in the casting of gas cooled, hollow, turbojet engine turbine blades. This core is shaped from a poly(methyl methacrylate)-coated, 15 μm silica/zircon mixture by SLS (Vail and Barlow, 1992). By a patented process (Barlow and Vail, 1994), colloidal silica in water is infiltrated into the green part before the polymer binder is removed. The infiltrated part is dried and fired at 400°C to remove the polymer and set the colloid. The infiltration and firing steps may be repeated to adjust part density. Finally, the part is fired at 1000°C for 17 hours to achieve 2100 psi (14.5 MPa) bend strengths at a silica content of 22% and a relative density of 61%, with only 1% linear shrinkage relative to the green part.

9.4.3 Bone Implants

Infiltration post-processing is also being used to prepare porous calcium phosphate artificial bones for implantation (Lee and Barlow, 1994; Lee *et al*, 1996). Bend strengths approaching 3000 psi (21 MPa) at 50% relative density have been prepared from modified hydroxyapatite powder by extensive pre- and post- processing. Figure 9.17 shows a replica of a human temporal bone that was prepared by SLS from calcium phosphate powder using slice geometry information that was obtained from CT scans of an actual temporal bone (Levy, 1992). The coarse layers in the replica are a direct result of the coarse CT scans used. Interpolation programs are being developed to eliminate this artifact of the data set.

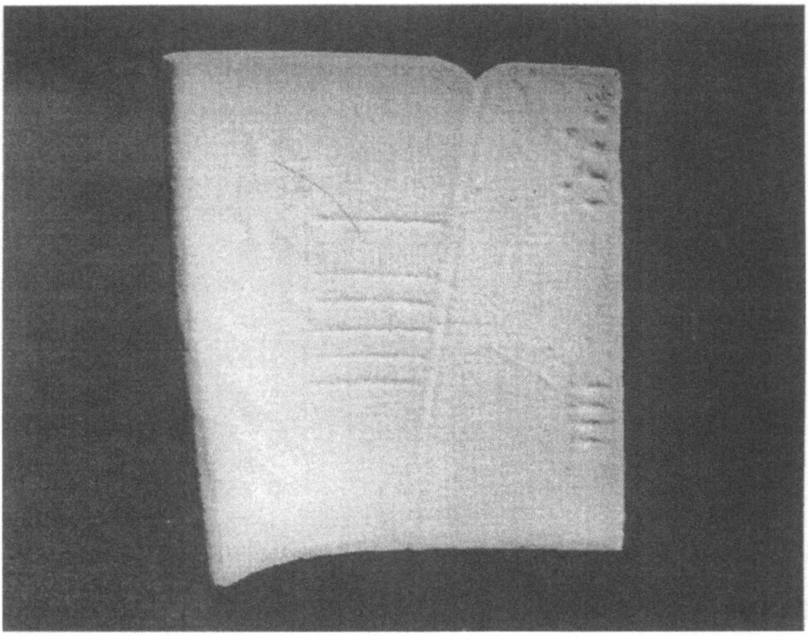

Figure 9.16 Silica cemented alumina core used in casting a hollow turbine blade.

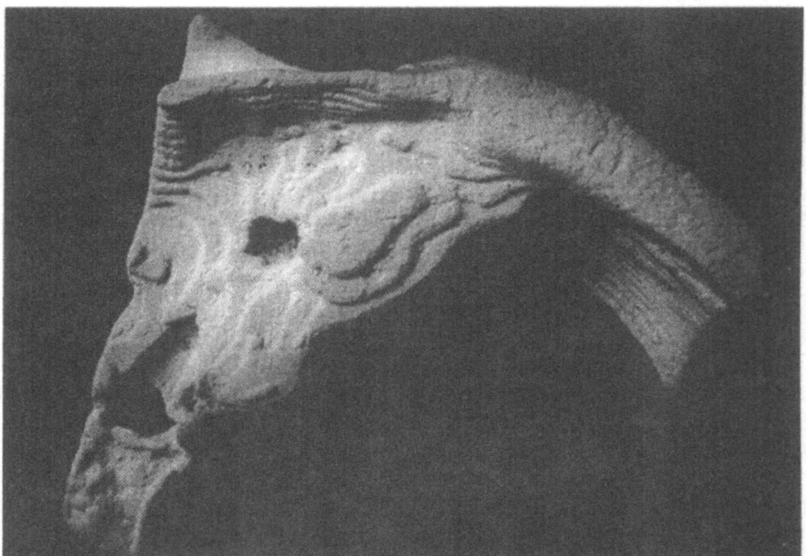

Figure 9.17 Human temporal bone formed by SLS from Computed Tomography data files (*from Lee and Barlow, 1994*)

BioMedical Enterprises, BME, San Antonio, Texas, is presently working with the University of Texas to further develop this technology. Figure 9.18 shows a small test implant, designed by BME, with 2.4 mm hexagonal macropores that are interconnected throughout the structure. The implants are being tested for biocompatibility in dogs, and rabbits.

Figure 9.18 Oral implant test piece (*courtesy of BioMedical Enterprises San Antonio, Texas*)

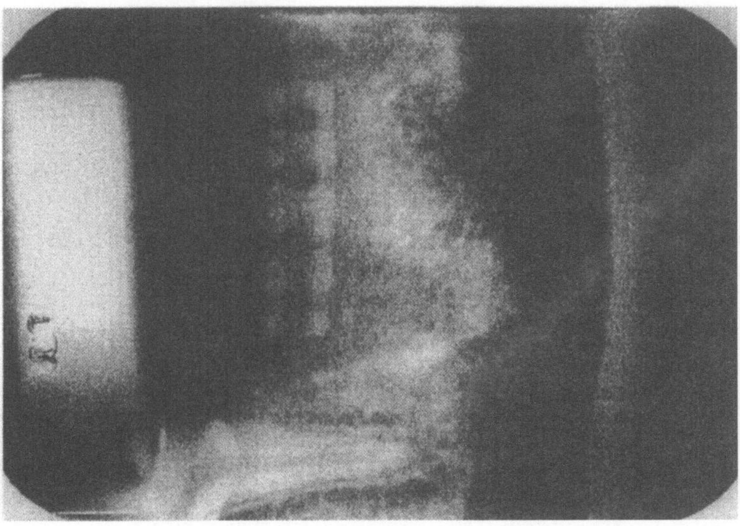

Figure 9.19 Radiograph of an SLS-formed alveolar implant, taken four weeks after placement in a dog (*courtesy of BioMedical Enterprises, Inc., San Antonio, Texas*)

Figure 9.19 shows a radiograph of the test implant taken four weeks after placement in the alveolar ridge of a dog. Crestal resorption and bone infiltration into the apical macropores and into the apical portion of the crestal macropores are evident. Completely lacking is any evidence of chronic or acute inflammation or of foreign body giant cell or macrophage response in both the dog and rabbit studies. These results suggest complete biocompatibility. They also suggest that, over time, the test specimen will completely fill with dense, mineralized bone (Lee, *et al*, 1996). Potential applications of this technology include facial and cranial reconstruction.

9.4.4 3D Fax

In August, 1991, at the Solid Freeform Fabrication Symposium, a three-dimensional fax was accomplished for the first time (Ladendorf, 1991). An automobile piston was scanned at Scientific Measurement Systems Inc, an Austin-based company that manufactures high-power X-ray tomography equipment for industrial applications. This scanned data, which represented sliced geometric information about the piston, were then sent over telephone lines some 8 miles away to the Laboratory for Freeform Fabrication on The University of Texas campus. At this location, an experimental Selective Laser Sintering system was used to recreate the piston slice-by-slice in real time (Figure 9.20). The total part scanning, transfer of data and recreation of the part took approximately two hours. Although the copied part did not show all of the detail of the actual piston since only 60 layers were made in order to reduce demonstration time, the experiment demonstrated the feasibility of such a process.

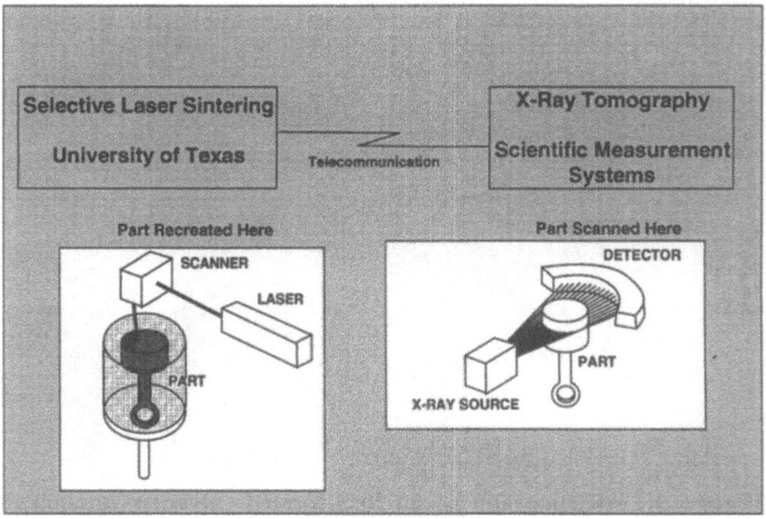

Figure 9.20 Schematic of three-dimensional fax experiment first achieved in Austin in August, 1991.

9.5 SUMMARY

The number of applications for SLS has increased significantly with advances in materials and processing technologies. In particular, the recent introduction of metal-based powders has enabled rapid production of prototype tooling, as was demonstrated in the previous sections. While most current attention has focused on producing models and prototypes of molded plastic parts, future SLS material and process advances will greatly increase the capability to produce metal parts and more durable metal tooling for metals prototype manufacturing.

REFERENCES

Barlow, J. W., and Vail, N. K., "A Method for Preparing High Temperature Parts by Way of Low Temperature Sintering," U. S. Patent No. 5,284,695, Feb. 8, 1994.

Deckard, L. and Claar, T. D., "Fabrication of Ceramic and Metal Matrix Composites from Selective Laser Sintered Ceramic Preforms," *Solid Freeform Fabrication Proceedings*,The University of Texas, Austin, Texas, 1993, pp. 215-222.

Denton, K. and P. Jacobs, "QuickCast & rapid tooling: a case history at Ford Motor Company", *Solid Freeform Fabrication Proceedings*,The University of Texas, Austin, Texas, 1994, pp. 154 - 173.

Forderhase, P. , K. McAlea, M. Michalewicz, M. Gannninger, and K. Firestone, "SLS prototypes from nylon", *Solid Freeform Fabrication Proceedings*,The University of Texas, Austin, Texas, 1994, pp. 102-109.

Forderhase, P. , M. Ganninger, and K. McAlea, "Nylon and Nylon Composite SLS Prototypes", *Proceedings of the 4th European Conference on Rapid Prototyping and Manufacturing*, The University of Nottingham, Nottingham, UK, 1995a.

Forderhase, P. , K. McAlea, and R. Booth, "The development of a SLS plastic composite material", *Solid Freeform Fabrication Proceedings*,The University of Texas, Austin, Texas, 1995b, pp. 287-297.

Fussell, P. , H. Kirchner, F. Prinz, "Sprayed metal shells for tooling: improving the mechanical properties", *Solid Freeform Fabrication Proceedings*,The University of Texas, Austin, Texas, 1994, pp. 74 - 81.

Griffin, E. and S. McMillin, "Selective Laser Sintering and Fused Deposition Modeling processes for functional ceramic parts", *Solid Freeform Fabrication Proceedings*,The University of Texas, Austin, Texas, 1995, pp. 25 - 30.

Ladendorf, K., "World's First '3-D Fax'", *Austin-American-Statesman*, August, 15, 1991.

Lee, G. H., and Barlow, J. W., Fox, W. C., and Aufdermorte, T. B., "Biocompatibility of SLS-Formed Calcium Phosphate Implants," *Solid Freeform Fabrication Proceedings*,The University of Texas, Austin, Texas, 1996, pp. 14-20.

Lee, G. H., and Barlow, J. W., "Selective Laser Sintering of Calcium Phosphate Powders," *Solid Freeform Fabrication Proceedings*,The University of Texas, Austin, Texas, 1994, pp 191-197.

Levy, R. A., Guduri, S., and Crawford, R. H., "Preliminary Experience with Selective Laser Sintigraphic Models of Human Temporal Bone," *Solid Freeform Fabrication Proceedings*,The University of Texas, Austin, Texas, 1992, pp 161-173.

McAlea, K. , U. Lakshminarayan, P. Maruk, "Selective Laser Sintering of Metal Molds: The RapidTool™ Process", *Molding '96*, ECM, Inc., Plymouth, MI, 1996.

Serbin, J. , C. Wilkening, M. Shellabear, "New Developments for Rapid Metal Tooling and Prototyping", *International Conference on Rapid Product Development*, Messe Stuttgart International, Stuttgart, 1995, pp. 117 - 123.

Sindel, M. , T. Pintat, M. Greul, O. Nyrhila, C. Wilkening,"Direct Selective Laser Sintering of metals and metal melt infiltration for near net shape fabrication of components", *Solid Freeform Fabrication Proceedings*,The University of Texas, Austin, Texas, 1994, pp. 94

- 101.

Vail, N. K., Barlow, J. W. , and Marcus, H. L., "Silicon Carbide Preforms for Metal Infiltration by Selective Laser Sintering of Polymer Encapsulated Powders," *Solid Freeform Fabrication Proceedings*,The University of Texas, Austin, Texas, 1993, pp 204-214.

Vail, N. K., and Barlow, J. W.,"Ceramic Structures by Selective Laser Sintering of Microencapsulated, Finely Divided Ceramic Materials," *Solid Freeform Fabrication Proceedings*,The University of Texas, Austin, Texas, 1992, pp 124-130.

INDEX

Glass transition temperature 113-114, 116, 186, 189, 214

Green shape 111, 133, 157

SALD 42, 75, 1, 279-285, 287-289

MIX
Papier aus verantwortungsvollen Quellen
Paper from responsible sources
FSC® C105338

If you have any concerns about our products,
you can contact us on
ProductSafety@springernature.com

In case Publisher is established outside the EU,
the EU authorized representative is:
Springer Nature Customer Service Center GmbH
Europaplatz 3, 69115 Heidelberg, Germany

Printed by Libri Plureos GmbH
in Hamburg, Germany